Helmet-Mounted Displays and Sights

For a complete listing of the *Artech House Optoelectronics Library*,
turn to the back of this book.

Helmet-Mounted Displays and Sights

Mordekhai Velger

Artech House
Boston • London

Library of Congress Cataloging-in-Publication Data
Velger, Mordekhai.
Helmet-mounted displays and sights / Mordekhai Velger.
p. cm. — (Artech House optoelectronics library)
Includes bibliographical references and index.
ISBN 0-89006-822-4 (alk. paper)
1. Helmet-mounted displays. 2. Human-machine systems. I. Title.
II. Series.
TK7882.I6V443 1998
621.381'045—dc21 98-4711
CIP

British Library Cataloguing in Publication Data
Velger, Mordekhai
Helmet-mounted displays and sights. – (Artech House optoelectronics library)
1.Helmet-mounted displays 2. Helmet-mounted displays – Design and construction 3. Helmet-mounted displays – Industrial applications
I. Title
621.3'678
ISBN 0890068224

Cover design by Elaine K. Donnelly

685 Canton Street
Norwood, MA 02062

International Standard Book Number: 0-89006-822-4
Library of Congress Catalog Card Number: 98-4711

10 9 8 7 6 5 4 3 2 1

To Miry, Idan, and Ailon
and to the memory of my late father
Dr. Haim Velger

Contents

Preface

The concept and the potential applications of helmet-mounted displays and sights (HMD/HMS) have fascinated the aviation community for over three decades. The idea of placing a virtual image in the visual path of the viewer and using the head/eyes as a pointing or control device has been proven to be efficient and beneficial for many uses. As time has passed, helmet-mounted displays began emerging outside the military environment and are now being established in medical, training, and virtual reality applications as well as in consumer products.

Many science and technology disciplines are involved in the construction of helmet-mounted displays and sights, and they all find expression in this book. These include the human factors aspects, optics, electronics, display and image source technologies, composite materials, pose measurement methods algorithms and sensors, and others.

The book addresses aspects of the design and use of helmet-mounted displays and sights. Specifically, the book covers the basic human factors associated with the use of helmet-mounted displays, as well as the qualities the helmets should have in order to make them useful to the user. The book underlines the design basics of helmet-mounted displays and sights in terms of optics, mechanics, and electronics. Additionally, the book reviews the various image sources, display technologies, fiber-optic image guides, and holographic optical elements. Also, the various methods of head position measurement are described in detail and analyzed, and the issue of biodynamic interference and HMD image stabilization is introduced and treated in detail.

The book is inclined towards the high-end, sophisticated, and complex pilot helmet-mounted displays and sights, although aspects more common to the less demanding displays are addressed where applicable. It is the author's belief that the topic of the book is better presented through these sophisticated systems while more simpler systems can be readily derived from the more advanced ones.

The book is intended mainly for engineers who are involved in the design of HMD/HMS or in their integration with other systems. Researchers and scientists investigating phenomena related to HMD/HMS will find the book useful as a reference text and as a guide to previous studies. However, the book is addressed not only for HMD/HMS designers and researchers, but is also intended for the general audience of technologists who are interested in becoming familiar with the principles of helmet-mounted displays and sights. The book should also be useful for the community of HMD users such as pilots and aviators.

An effort has been made to create a text that is as self-contained as possible. With this approach in mind, some basic background is included in the text, in particular, concepts of visual system physiology, principles of visual perception, basics of optics, and the physiology of the vestibular system.

Chapter 1 describes and discusses the general head coupled device concept. It summarizes the advantages and qualities of helmet-mounted displays and surveys the evolution of the current systems.

A knowledge and understanding of the human factors associated with the use of HMDs is perhaps the key factor in the successful design of helmet-mounted displays. Chapter 2 is devoted to the human factors issues of HMDs, in particular those related to the visual perception of the display image. It describes the perceptual conflicts in helmet-mounted displays such as binocular rivalry and brightness disparity, and compiles data that have been accumulated over the years by human factors researchers. In this sense, the book is a reference for the human factors issues associated with helmet-mounted displays and sights. More human factors issues are addressed in other parts of the book according to their relevance.

The foundations of the design concepts of HMD/HMS are laid down in Chapter 3. It formulates the "ideal" HMD in terms of the requirements necessary for such a system, and describes and explains the design parameters that should be considered, such as image resolution, brightness, field of view, and type of presentation (i.e., monocular versus binocular, etc.).

Chapter 4 deals with the display characteristics and defines such parameters as modulation transfer function (MTF), resolution, luminance, and contrast ratio. We are now witnessing an era in which promising new technologies of flat panel, high-resolution, miniature displays such as liquid crystal, electroluminescence, and digital micro-mirrors display (DMD) are emerging and improving constantly. The text describes both the conventional CRT display and the flat panel display technologies and evaluates the strengths and weaknesses of each.

Chapter 5 reviews the design parameters and considerations of optical design, reviews the MTF of the optics, and compares refractive versus reflective optics considerations. Also, fiber-optic image relay issues are presented, and methods for increasing image resolution are explained. Additionally, the concept and theory of holographic optical elements (HOE) are introduced.

Chapter 6 concentrates on methods for measuring helmet position. It starts by stating the accuracy, resolution, and update rate requirements considerations and their relation to the specific task definition. A detailed description of commonly used methods follows (i.e., mechanical link systems, principle of operation and design considerations of magnetic sensors, various electro-optical methods, acoustic sensors, and gyroscopic based systems). The chapter concludes with an introduction to eye tracking systems, their principles of operation, and their applications.

The symbology and information presented on a helmet-mounted display is dealt with in Chapter 7. This chapter describes the various graphical symbologies used for different mission types, addresses the issue symbology frames of coordinates, and lists the display control laws. Also, the concepts of *virtual cockpit* and *virtual head-up displays* are introduced and discussed.

Biodynamic interference is the least regarded aspect by designers and users of helmet-mounted displays and sights. This is corrected in Chapter 8, which is devoted to biodynamic effects and image stabilization. The problem of biodynamic interference is explained and supported by relevant research data. The physiological phenomena of biodynamics are surveyed by detailed description of the vestibular system, the vestibulo-ocular reflex, and the pursuit or fixation reflex. The head/eye coordination during vehicle motion is explained by analyzing the mathematical model of the eye movement control system during vibrations. Data on the body's transmission to the head during vibrations and the effect of vibrations during head tracking are presented. The perceptual effect in viewing a display under vibrations and visual acuity of vibration display or head is addressed. Finally, methods of image stabilization on HMD using either conventional methods or by adaptive filtering is introduced and supported by the relevant data.

The integration of HMD/HMS with the helmet is covered in Chapter 9. The mechanical aspects of mounting the display and sight on the helmet is addressed. The helmet design parameters such as helmet shape, size, weight, and balance requirements are presented.

A book about helmet-mounted displays cannot be complete without mentioning the existing or potential applications. In Chapter 10, applications of HMDs other than airborne are described, including wide field of view

HMDs used for simulations and training, vehicular applications, and medical applications for either clinical treatment, aiding low-vision patient, or phobia exposure treatments. A special section introduces low-cost consumer products used for either entertainment, remotely operated systems, or as a portable PC display, fax machine, and pager. The chapter concludes with a review of virtual reality concepts and applications.

Acknowledgments

The writing of a book is a difficult and trying experience and demands spending long weekdays, nights, and weekends. Consequently, its completion is largely dependent on the support and sacrifice of one's family members. For this crucial support, patience, and understanding I am grateful to my wife Miry, to my children Idan and Ailon, and to my mother Halina.

I am indebted to the people who helped me acquire some of the technical reports and references that were used in the preparation of this book. Special thanks are due to Mr. Bruno-Dov Lerer, Mr. Bjorn F. Andersen from ELOP, and Dr. Stephen R. Ellis from NASA-Ames.

I would like to express my appreciation to the anonymous reviewer for the careful review of the manuscript and the helpful suggestions which undoubtedly contributed to improving this text.

I also would like to extend my appreciation to the staff at Artech House for their assistance and patience during the extended period of preparation of this book.

Finally I would like to express my deep appreciation and thank my mentor Prof. Shmuel Merhav from the department of aerospace engineering, Technion Israel Institute of Technology, who acquainted me with the exciting world of helmet-mounted displays and sights in 1980.

1

Introduction

The concept of head-coupled devices is to provide symbolic or pictorial information by introducing into the user's visual pathway a virtual image the user can observe regardless of the direction of gaze. Basically, that is achieved by using a display mounted on the head, usually on a helmet, together with continuous measurement of the head position. The basic helmet-mounted display (HMD) comprises one or a pair of miniature displays and optics that relay a collimated image to infinity or to a sufficiently large distance to be perceived as at infinity to the wearer's eye via partially reflecting combiners (Figure 1.1). As such, the image is superimposed on the outside world scene.

The information displayed on the HMD may vary from simple non-changing symbology, such as a reticle, through more complex dynamically changing information like numerically indicated or symbolically presented speed notation, up to complex graphic imagery superimposed on a video image obtained from a sensor.

Like many other technological innovations and breakthroughs, the evolution and intention of HMDs and helmet-mounted sights (HMSs) were aimed at the aviation community as a potential relief for the high workload and stress imposed on the modern aircraft pilot.

In the early days of aviation, virtually all the functions and flight information were provided by the pilots, who used their senses to assess the aircraft and external world situations. Therefore, the superiority of one pilot over an opponent mainly relied on personal capabilities and skills.

As aircraft grew in complexity and as more sophisticated instruments and weapon systems were introduced into the aircraft, pilots were swamped with data and information those systems collected and presented via a diversity of dedicated displays and gauges. Modern aircraft are equipped with a variety of missiles of dif-

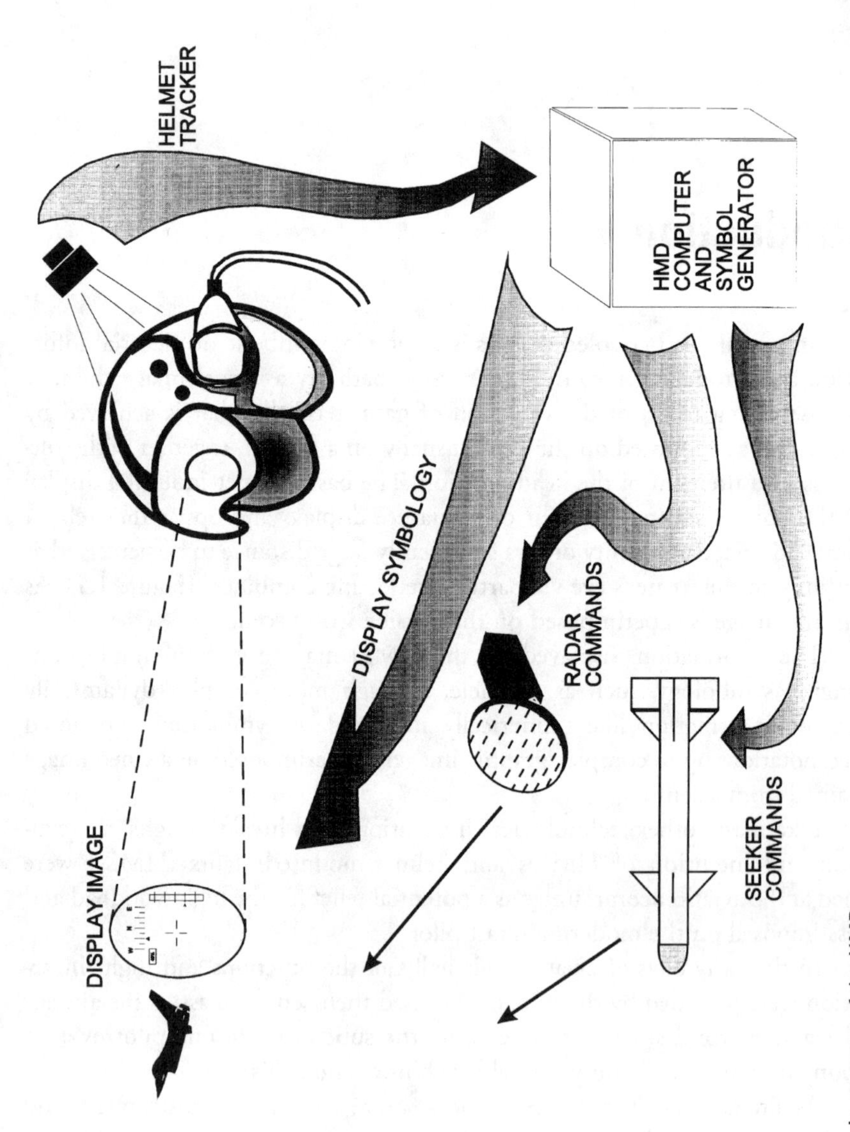

Figure 1.1 Basic concept of the HMD.

ferent types and capabilities, laser- and TV-guided bombs, radar, navigation and targeting systems, threat and countermeasure systems, and battlefield intelligence, and the pilot has to monitor all the information incoming from those devices. The task is demanding, especially in stressful conditions like combat, in which a large number of continuous and discrete events take place simultaneously.

The potential benefits of HMD/HMS have fascinated the aerospace community for more than three decades. With the increase in the complexity of modern aircraft and the subsequent increase in the demand on the pilots' attention, the idea of providing pilots essential information at all time regardless of their direction of looking was very appealing. To reduce pilots' workloads and to quicken their response in stressful conditions, the head-up display (HUD) was the first attempt to put information in a pilot's line of sight. That was followed by consolidation of all major buttons and activating switches on the stick and the throttle so pilots no longer were forced to take their hands off the stick to operate subsystems in the aircraft, a concept known as "hands-on throttle and stick" (HOTAS).

The HUD provided pilots collimated virtual images superimposed on the outside world with bright, sharp, and clear symbology. Soon it became the primary flight instrument, beyond its main role as a sophisticated, high-accuracy, boresighted weapon's sight. A drawback of the HUD, however, was that it required the aircraft to be turned to bring the point of interest into the narrow HUD's field of view. The perception among the aviation community was that that deficiency could be removed by using the "display on the head" concept. As a result, pilots could be free to execute important missions without the need to maneuver the aircraft or to approach or view the instrumentation panel.

The concept of head-coupled devices was extended and transferred to other disciplines that require the full attention of the user to perform a task. Among other military operations, it may be useful to provide the tank commander with the ability to view forward-looking infrared (FLIR) imagery or the gun sight while out of the turret or while completely buttoned up in closed-hatch position [1]. The HMD would allow tank commanders to stay in touch with the "real world" and still retain the freedom to move from one position to another. Such an ability would be even more significantly beneficial for infantry soldiers, who are bound to carry heavy equipment and who must have their hands free to carry guns. HMDs could be used to display gun-sight day/night images, assist soldiers in aiming their guns on the move, and show global positioning system (GPS) information, thus increasing soldiers' survivability and providing them with valuable information that otherwise would be inaccessible [2]. Another useful military application could be for antiaircraft battery gunners, who would be able to point and track low-altitude penetrating fast aircraft.

A whole area of military applications of HMDs is flight simulations. The current large and expensive display system can be replaced by a single large field-of-view high-resolution HMD, giving a capability that cannot be met using the conventional array of cabin-installed large-screen displays. Such a simulator, when combined with eye-position measurement, enables a smaller central area of high-resolution image to be presented, while the peripheral visual field is displayed with lower resolution. In that way, both high resolution and a wide field of view essentially can be obtained using a sensible display and optics without sacrificing image quality.

Many nonmilitary applications also can be envisioned. HMD may prove useful for surgeons by presenting x-ray images in front of their eyes, or for astronauts involved in maintenance tasks in space by presenting repair instructions comfortably and continuously. An HMD could increase the safety of motorcyclists and race car drivers by showing vehicle speeds, shifted gears, oil pressure, and other hazard indications. Even electronic technicians could take advantage of HMDs by being able to probe printed circuits without needing to turn their heads to the oscilloscope screen and risking accidentally short circuiting an electrical circuit during the eye shift.

The range of other possible applications is vast: from simple hand-held personal computer monitors attached to the head, through reading aids for visually impaired people, to virtual reality devices that immerse users in an interactive three-dimensional world [3,4].

Virtual reality devices use HMDs to give users the illusion of displacement to another location. In that sense, they are an extension of the idea that was developed for flight simulations. Both applications share the characteristic that distinguishes them from airborne HMDs, namely, the complete occlusion of the external surrounding world.

Virtual reality is perhaps the most promising area of HMDs, with applications yet to be identified. The great potential use and market of virtual reality, especially for entertainment, undoubtedly will bring advances in many technologies common to HMDs, particularly in the technologies related to high-speed computer graphics and high-resolution, multicolor, low-weight, low-cost flat-panel displays and optics. Such advancements probably will equal the advancements made in personal computers and video games and indubitably will be followed by a similar progress in all other applications of HMDs.

Although the concept of HMDs/HMSs looks simple and intuitive—just hook up a miniature display to the wearer's head in front of his eyes and measure his point of regard—the human physiological and cognitive factors associated with the use of head-attached displays are complex. Both the designer and the user should carefully consider those factors; otherwise, the contribution of such a device may be questionable. Furthermore, the proper design of such devices requires familiarity with and a deep understanding of the most advanced technologies in terms of displays, optics, computer graphics, advanced materials, and electronics. This may be evident from Figure 1.2, which describes the technologies and disciplines embedded in a typical HMD/HMS system.

1.1 Terminology

1.1.1 Helmet-Mounted Sight

A helmet-mounted sight (HMS) is a rather simple optical device that displays to a pilot an aiming mark, or reticle. The sight incorporates a helmet-orientation sensor that measures the pointing direction of the line of sight (LOS) to the target and thus enables designation of the target's position. The HMS is mainly used for launching air-to-air missiles. An HMS usually requires measurement only of the elevation and the azimuth of the LOS.

1.1.2 Helmet-Mounted Display

A helmet-mounted display (HMD) is a more complex optical device that presents dynamic symbology or imagery. There are two main types of HMDs. In the helmet-mounted HUD type, the symbology and imagery orientation is independent (constant) with relation to the head position and orientation. It functions essentially as a HUD with the distinction of being mounted on the helmet. The other type is the true HMD, in which the symbology is compensated as to conform with the outside world regardless of the pilot's head motions.

1.1.3 Visually Coupled System

The visually coupled system (VCS) measures the pilot's head motion and uses the helmet position-tracking sensor to drive steerable imaging sensors, such as a television camera and FLIR mounted on movable and usually inertially stabilized gimbals. The image from the imaging sensors is displayed on the HMD along with graphic symbology. Although the sensors might have a narrow field of view so as to have high resolution, the field of regard of the system is large.

1.2 Benefits of HMDs and HMSs in Aviation

The benefits of HMDs/HMSs have the largest impact on aircraft pilots, who are involved in missions that almost always require aviators to stretch their per-

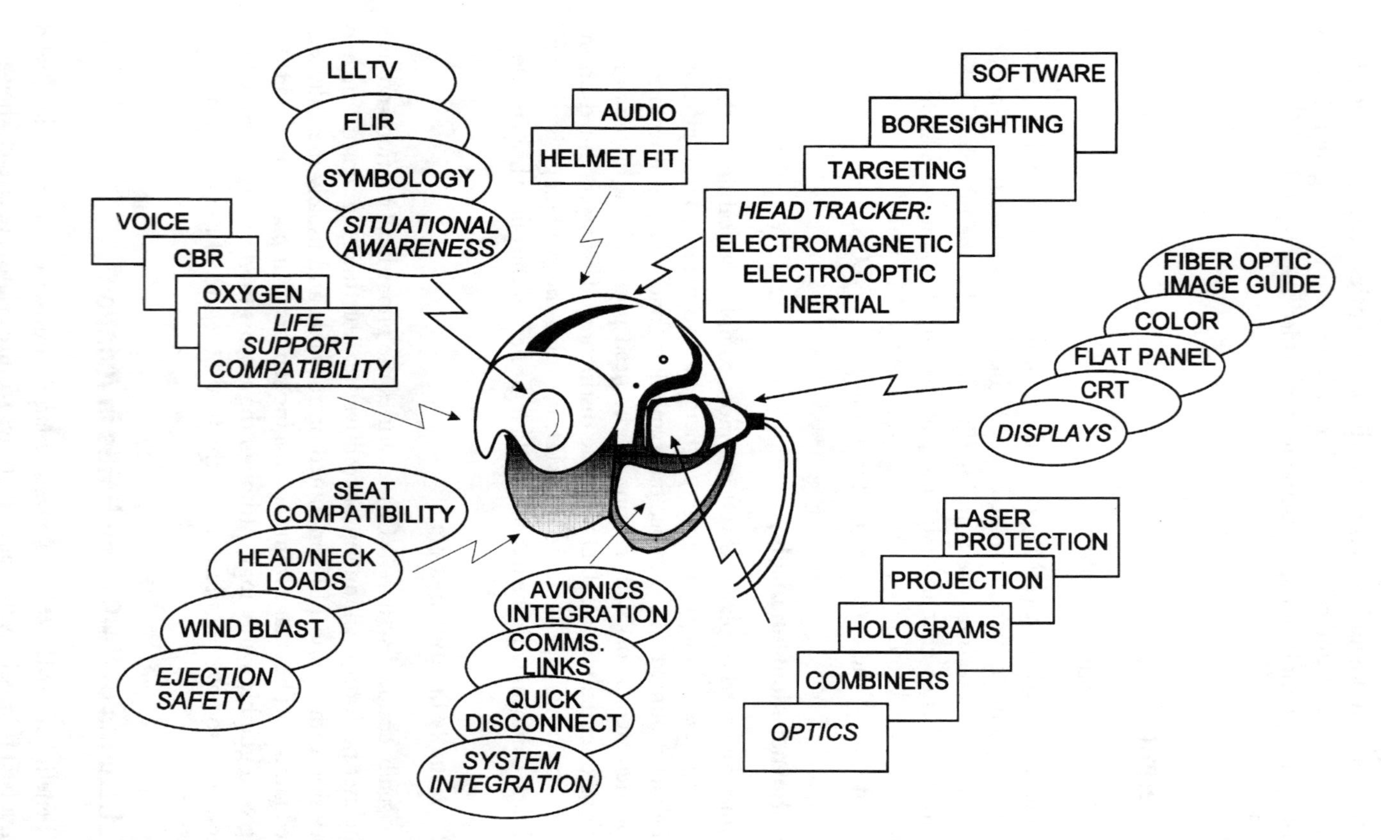

Figure 1.2 The technologies involved in HMDs/HMSs.

formance and capabilities to ultimate limits. To attain superiority, pilots strive to exploit any advantage they can get from every useful device available. Many of those advantages equally apply for other disciplines and can be derived from those that are relevant to aviators. Accordingly, this book focuses on the aerospace applications of HMDs/HMSs.

The basic virtue of HMDs applicable to fighter pilots was recognized long ago, namely, the ability to look and shoot at a target rather than having to point the aircraft toward the target before an air-to-air missile could be launched. Keeping in mind that during a dogfight the hostile contact generally averages only 30 sec to 2 min, any time saved by not needing to perform aircraft-pointing maneuvers gives a pilot a vital advantage. This virtue is crucially significant during engagement of multiple adversaries, especially when the pilot is outnumbered. Using an HMS, a pilot can quickly snapshot the adversary aircraft, launch a missile or lock the radar, and immediately turn to the next target and repeat the process. In that manner, a pilot can designate sequentially several targets to the fire control computer within seconds without being restrained by the maneuverability constraints of the aircraft. This combination of the soon-to-be-introduced high off-axis boresight (HOBS) missile seeker, such as ASRAAM, AIM-9X, and Python-4 [5] (which all have up to a 90-degree off-axis boresight angle capability), with HMS potentially gives pilots a great advantage over opponents.

Figure 1.3 shows the dramatic potential improvement of pilot capability using an HMD during air engagement. The narrow field of view (FOV) of conventional HUDs, or even the wide-FOV HUD, covers only a small fraction of possible threat locations. Modern radar and weapons, although having a wider FOV, still are restricted to the front section of the aircraft. On the other hand, the HMD provides a global view, in the sense that the display-limited FOV can be presented throughout the whole range of the pilot's head positions and is limited only by obstructions of the aircraft's body and wings.

HMSs are useful for other applications as well. One important benefit of the HMS capability to measure head orientation, relative to the aircraft's frame of reference, is its ability to designate targets and hand off their location to other sensors, such as radar or a missile seeker, or to a friendly aircraft. Similarly, the opposite direction is equally useful: a radar-locked target or an infrared search and track (IRST) detected threat can be used to cue the pilot's line of sight to the target by showing on the display directional cues to the threat location [6]. In that way, once the threat warning has been received, the pilot can quickly detect the threat without having to scan the area and spend vital time that otherwise could be used to initiate evasive maneuvers.

In ground attack missions, the pilot can use the HMS to designate the ground target and enter its position to the navigation computer. The computer,

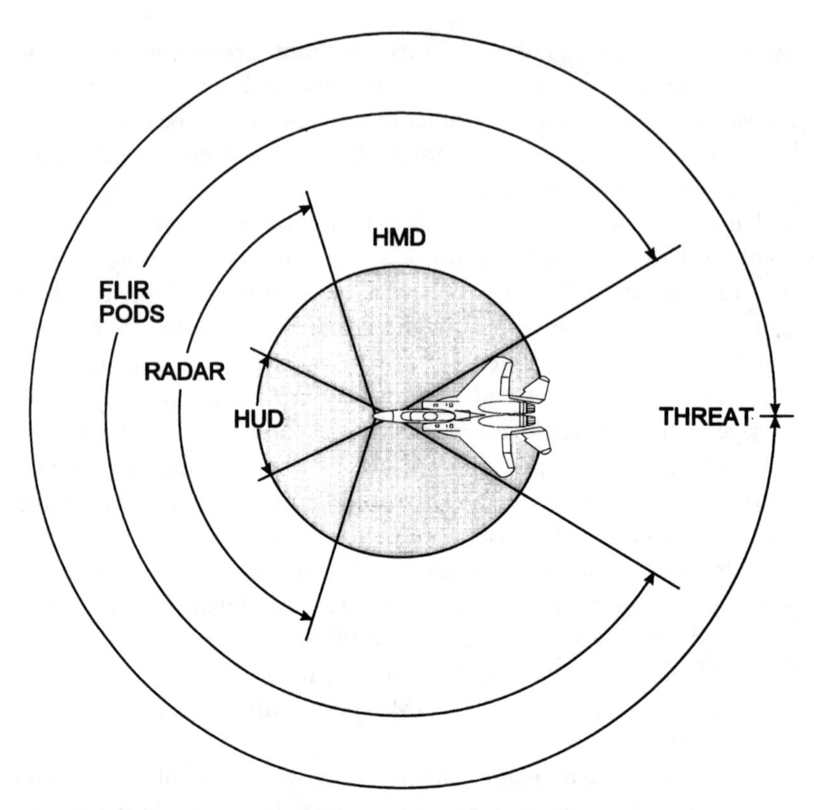

Figure 1.3 HMD field of regard compared to other aircraft instruments.

in turn, generates cueing commands on the HMD that direct the pilot to fly the aircraft to the location of the target, thus eliminating the need to reacquire the target that was passed by and lost in the course of the turning maneuver (Figure 1.4). This feature is important when the target is not prominent, for example, in rescue missions. In such cases, a searching pilot spots the target but then loses it during aircraft-turning maneuvers.

In an air-to-ground missile attack, the HMS enables hitting a target located at an appreciable offset angle to the aircraft flight path. Very often the pilot spots a target to destroy, but that requires either deviating from the intended flight course or approaching the target. In some cases, for example, when the target is an anti-aircraft battery, that may impose a high risk to the aircraft and the pilot. In such cases, the HMS capability to launch off-track missiles is advantageous. Once the pilot designates the target, the LOS to the target is determined, transmitted to the missile guidance system, which in turn commands the missile to the location of the desired target. The concept of an offset missile launch is depicted in Figure 1.5.

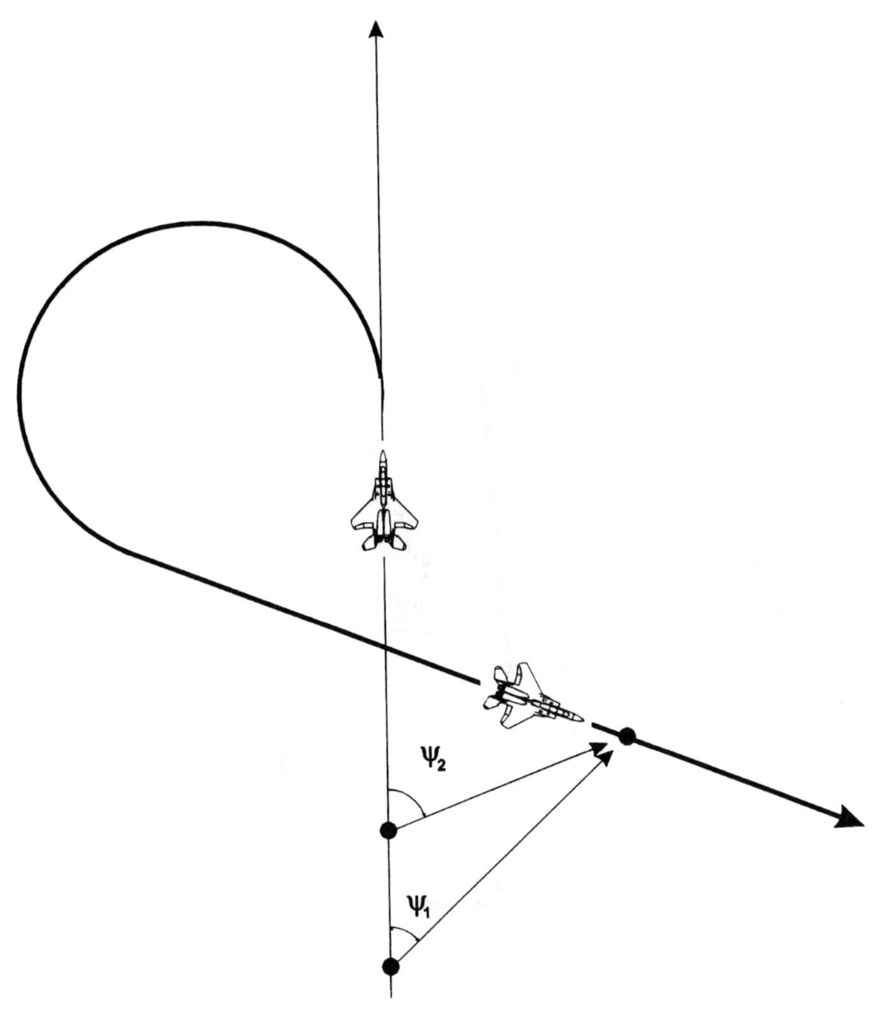

Figure 1.4 Target reacquisition using an HMS.

Another useful application of the HMS is position updates of an inertial navigation system (INS). This can be done by just looking at a remote object whose coordinates are known. The principle of operation of position update using an HMS is illustrated in Figure 1.6. By merely pointing at the object from two distinct positions along the flight path and inserting the head angular orientation to the system's computer, the navigation system can compute by simple triangulation the range to the object and thus the current position of the aircraft. Beyond the position correction, the errors of the inertial sensors are computed and used to calibrate the INS until the next position update. The

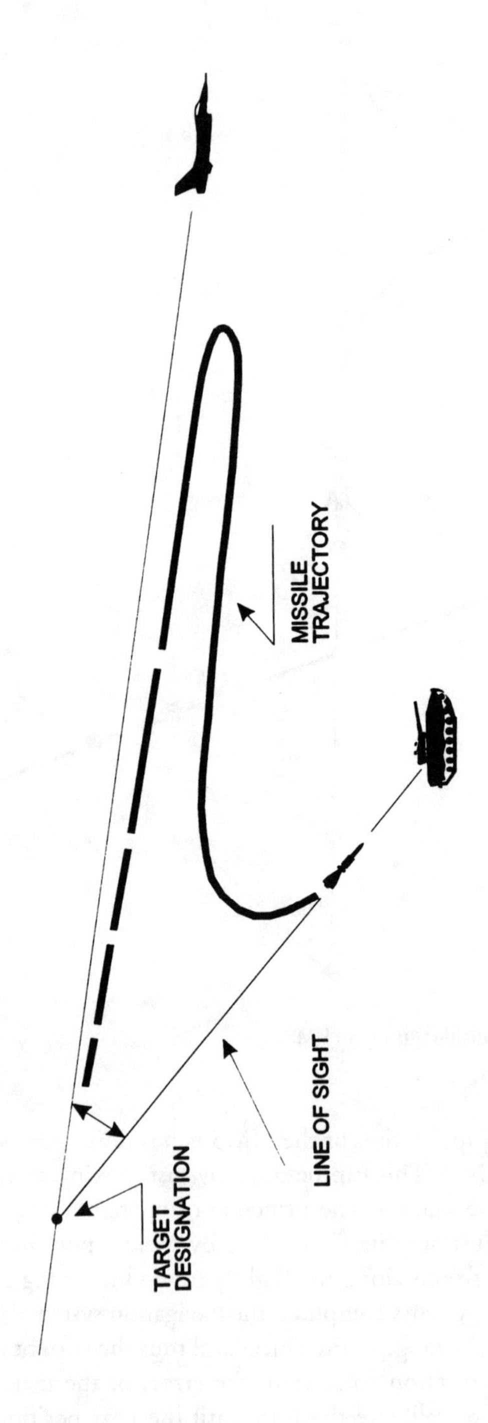

Figure 1.5 Offset missile launch in an air-to-ground attack mission.

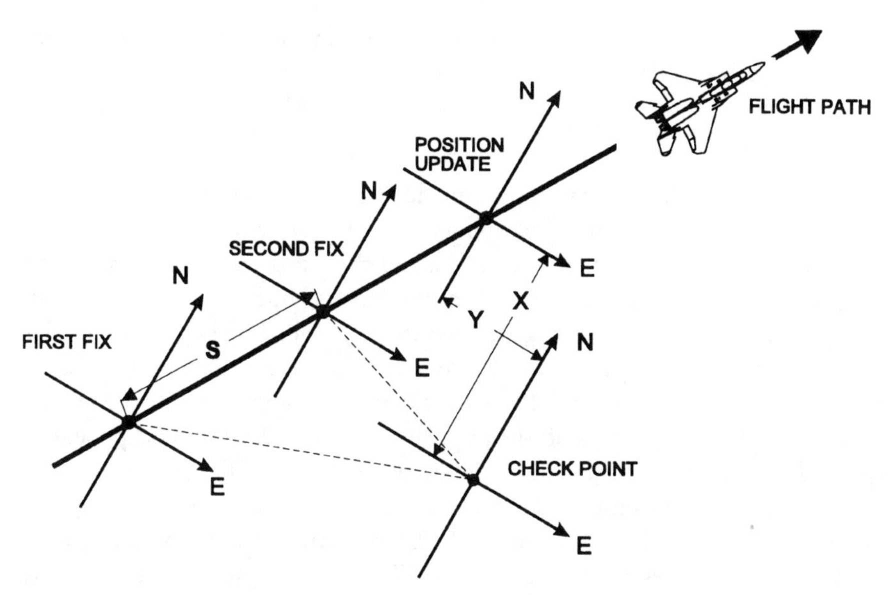

Figure 1.6 The principle of updating an INS using an HMS and offset checkpoint.

process is implemented in a Kalman filter, which compares the computed position of the aircraft with the position calculated by the INS to compute the inertial sensor error that led to the position error. Using this position-fix procedure, the pilot can avoid the currently used procedure of flying over the checkpoint, a situation that often is inconvenient or even tactically unwise [7].

Beyond the fundamental capability of aiming a missile at a target, as part of the HMS, the HMD enhances both fixed-wing aircraft and helicopter operational capabilities in many ways. The HMD enables pilots to fly the aircraft and simultaneously view important navigation, flight, and fire control data without turning their heads or shifting their eyes to view flight instrumentation or cockpit displays.

Perhaps, the most important contribution of the HMD is in the increase of situational awareness (SA) it provides pilots. SA is the knowledge pilots have on the combat arena in terms of the surrounding threats relative to their own aircraft, the spatial position and orientation of the aircraft, the energy status, weapon availability, shooting opportunities, hazards, and other aircraft relevant information. SA can be classified into two subcategories:

- *Global SA* includes the close arena and the far, beyond-visual-range arena and covers ranges up to 200 miles. As such, it should be presented on a large, graphics-intensive, high-resolution panel display [8].

- *Tactical SA* covers the close-in visual combat and navigation arena, which dictates the pilot's out-of-cockpit attention and therefore is the best method to exploit the advantages of HMDs.

This capability is very important during night operations, especially at very low altitude flights in which pilots devote most of their attention to the task of avoiding close-proximity obstacles and thus must view the outside scene constantly. In helicopters, during nap of the Earth (NOE) flights, in which the pilot takes advantage of the terrain features (trees, bushes, buildings), this problem is aggravated. Because the obstacles surround the pilots, they cannot focus attention solely ahead of the aircraft [9] without compromising safety. Before HMDs, pilots either had to quickly shift their view between outside of the cockpit and the panel display or give up the valuable information shown on the display.

A major enhancement to the aircraft flight capability during night is the addition of night vision devices (NVDs) in form of night vision goggles (NVG), low-light-level TV (LLLTV) or FLIR. NVDs are beneficial because they extend the pilot's SA during flight at night.

In night combat missions, such as target bombing, the pilots enter the bomb release maneuver, keeping their heads out of the cockpits, while avoiding looking at the tracers from the antiaircraft fire and attempting to keep their bearing with no horizon available. This situation often leads to disorientation effects such as vertigo [10]. The use of an HMD-shown image of the outside scene easily would prevent such circumstances.

Obviously, the benefits of which pilots can take advantage depend on the combination-type of HMD/HMS. While a simple type of display can be used for target designation only, and a HMD without head tracking capability can be used for aircraft data type display only, the full advantages of HMD/HMS can be exploited only if the system has day/night dynamic symbology, wide FOV, and high-resolution display combined with a high-accuracy head-tracking sensor. Table 1.1 categorizes the system requirements for each type of HMD/HMS application.

1.3 Evolution of HMDs/HMSs

The concept of HMDs/HMSs goes back to the beginning of the century with Albert Bacon Pratt's 1916 patent for a helmet-mounted gun and sight. In Pratt's early invention, the wearer would trigger the gun by blowing air into a tiny tube held in the mouth and connected to a rubber balloon that when inflated activated the trigger [11,12]. There is no evidence that Pratt's HMS, shown in

Table 1.1
System Requirements of HMD/HMS Application Categories

	Application Categories				
System Requirement	**I**	**II**	**III**	**IV**	**V**
	Flight data display	Missile aiming, cueing, INS updating, offset attack	Night-vision enhancement	Visually coupled sensor	Simulations
Aircraft type	Fighter jet	Fighter jet	Helicopter	Helicopter/ fighter	All aircraft
Helmet tracker	Needed	Needed	Not needed	Needed	Needed
Raster image	Not needed	Not needed	Needed	Needed	Needed
Stroke symbology	Needed	Needed	Optional	Needed	Not needed
FOV	Narrow (25°)	Narrow (2° to 3°)	Wide (40°)	Wide (40°)	Very wide (60°)
Night-vision device	Not needed	Not needed	LLLTV/IIT*	FLIR or LLLTV/IIT	Not applicable

*IIT = image intensifier tube.

Figure 1.7, was ever implemented as an operational weapon or even tested to prove the idea. Thus, it should be regarded as a gimmick more than a truly useful weapon. Nevertheless, it was the first HMS that recognized and embodied the same philosophy that has motivated today's sophisticated systems, namely, the look-and-shoot concept.

Serious development began in the early 1960s. Several companies were involved in the effort to come up with operational HMD/HMS systems for both fixed-wing fighters and helicopters [13]. Bell Helicopter developed a dual CRT biocular HMD with a mechanical head tracker and installed it in the Bell AH-1G Huey Cobra attack helicopter. The Bell HMD was licensed to Electro Optical Systems, which adapted it to fixed-wing aircraft and later tested it in the Ling-Temco-Vought A-7 attack aircraft.

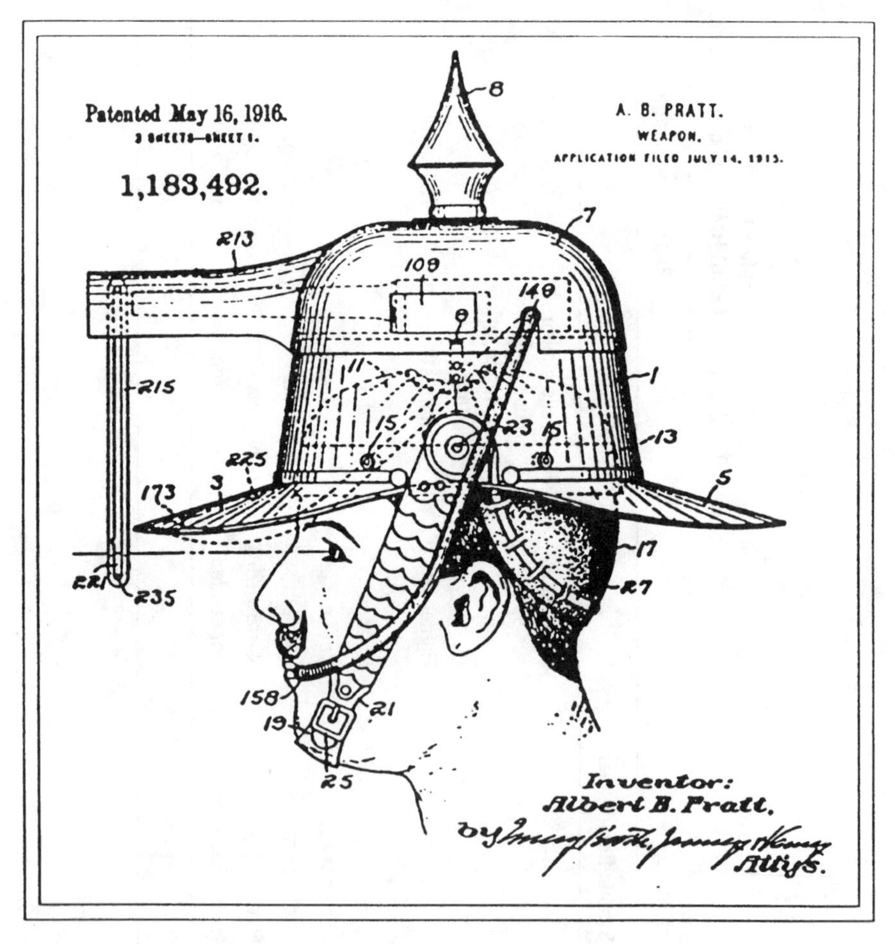

Figure 1.7 Pratt's invention of a helmet-mounted gun and sight.

In the early days of HMDs, the brightness of CRTs was very low and insufficient for practical airborne daylight use. To overcome that deficiency, Hughes Aircraft developed a monocular occluded display that provided a much brighter image and had a better display visibility against the bright sky. However, the viewing of separate images by the two eyes was tiring and confusing. Other HMDs were developed by Texas Instruments, Honeywell, Autonetics, and Perkin Elmer.

Among the pioneering achievements was the HMD built by Ivan Sutherland: a binocular HMD that used two miniature CRTs to present a 40-degree FOV. Sutherland's HMD was the first known computer-generated three-dimensional stroke image coupled with a head-position tracker (the tracker was initially mechanical then ultrasonic) [14]. The first flyable prototypes were tested in 1969, and helmet-coupled systems have been in operational use since the mid-1970s in the Navy F-4 [15], using the visual target acquisition system (VTAS), and the Army Apache, using Honeywell's integrated helmet display and sighting system (IHADSS).

As the need for displays with a wider FOV grew and following the recognition that an FOV wider than 30 to 35 degrees could not be obtained even with the most advanced holographic HUDs, the endeavor was diverted toward the development of HMDs. Moreover, wide-FOV HUDs are not achieved without high penalties of cost, weight, and size, and ultimately they are in a fixed position relative to the aircraft axis.

Much effort, time, and money were invested during the next 20 years to develop HMD/HMS with the ultimate goal of integrating operational systems into fighter aircraft. That objective, however, was elusive, and most designs languished in the laboratory.

The initial lack of acceptance of HMD and HMS systems in fighters resulted from users being reluctant to adopt systems that were heavy and bulky, had insufficient resolution and brightness, and were often accompanied by perceptual conflicts. The general feeling among both users and designers was that the suitable technology was not yet mature and the knowledge of the basic human factors that affect the systems' use had yet to be studied [16]. On the other hand, support in pushing the technology to come up with a satisfactory system was modest, which resulted in slow progress in the development of HMDs. Weight and comfort were persistent problems with early HMDs, and the serious problem of wind blast impact caused concern about the early eyepiece designs. There also was a tendency to try for the 24-hr HMD for the fighters from the start, the most complicated solution. Also, the high cost of the more sophisticated systems has always been a stumbling block in the acceptance of HMDs.

Above all, however, there was a lack of agreement among users as to the characteristics that would guarantee the operational acceptance and use of

HMDs/HMSs systems. To resolve the requirements issue, the main air forces of the Western powers have been occupied in endless specification, evaluation, and testing programs of head-coupled systems. As a consequence, the introduction of more sophisticated HMDs/HMSs has been very slow.

Somewhat more success has been achieved in attack helicopters, primarily because helicopters' HMDs generally are not required to withstand high-g maneuvers, ejection, and air blast. Also contributing to the acceptance of HMDs in helicopters was the fact that those aircraft lack HUDs and must provide some sort of NVDs. Moreover, the relatively small and fixed HUD FOV has significant disadvantages for night flying with synthetic video, particularly in the helicopter NOE role. As a result, the only HMD/HMS currently operational in a production aircraft in the United States is the IHADSS, incorporated in the AH-64 Apache helicopter. The IHADSS is a part of the Apache pilot night vision system (PNVS) and relays a head-steered infrared (IR) image from a sensor mounted on gimbals in the nose of the helicopter overlaid with flight symbology. The symbology is a HUD type, that is, fixed to the display coordinates.

A significant deficiency of the IHADSS is that it is a monocular HMD. For helicopters, a wide-FOV binocular HMD is considered a basic requirement for NOE flight. Figure 1.8 shows an advanced binocular wide-FOV HMD that has been proposed for the future LHX helicopter (LHX stands for light helicopter, experimental).

Figure 1.8 An advanced wide-FOV binocular HMD proposed for the LHX helicopter (courtesy Kaiser Electronics).

A stimulus to the development program came in the late 1980s, with the advent of the Russian helmet-mounted sight that is part of the MIG-29 equipment and used to launch the Vympel R-73 high off-boresight seeker air-to-air missiles [17]. Although the Russian HMS has only a simple light-emitting diode (LED) reticle for aiming, it can be combined with high off-boresight seeker missiles, a combination that is widely acknowledged as significant combat advantage [18]. In parallel, Western high off-boresight missiles have been developed and have raised the need for wide-field-of-regard aiming capabilities.

The early-generation HMDs displayed only a simple aiming reticle with an FOV of 2 to 3 degrees, often generated by using LEDs, allowing the pilot to track and designate a target. Using this type of display, the pilot is able to aim missile seekers without the need to maneuver the aircraft, as is required when a conventional HUD having a narrow FOV is used. Such HMDs typically had an FOV on the order of 10 to 12 degrees, similar to the HUDs of that generation. A typical early-generation HMD is shown in Figure 1.9. Although sim-

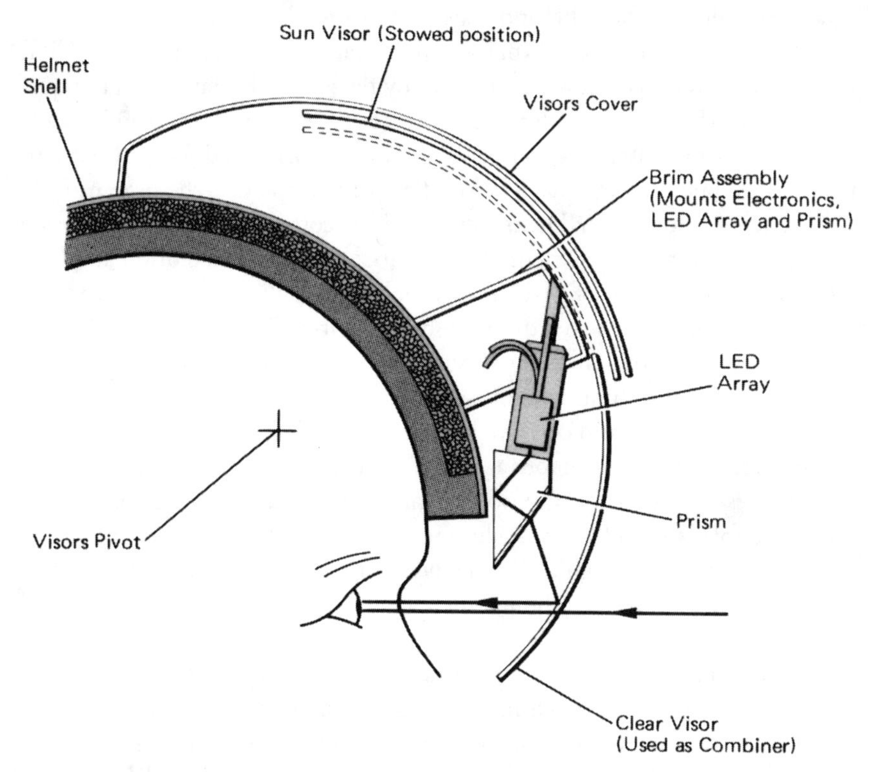

Figure 1.9 Cross-section of an early-generation LED array–type HMD (courtesy GEC-Marconi Avionics).

ple, first-generation HMDs have proven effective and beneficial during air-to-air combat.

The next generation of HMDs had more sophisticated displays, combining raster images from imaging sensors with a stroke overlaid symbology. That symbology, as well as the raster image, was fix-oriented to the display. Being fixed, it did not compensate for pilot head orientation and therefore did not preserve a true correspondence between the horizon line projected on the display and the true horizon.

The early fixed-wing aircraft HMDs had only day capability. Aircraft were not equipped with FLIR pods, and NVG did not qualify for high-g maneuvering aircraft because of their high weight and the risk they imposed to the pilot's neck in the case of pilot ejection. However, the need for NVDs still existed, so an approach to integrate NVDs into the helmets was taken, although they added considerably to the weight of the HMD. Several HMDs were developed that incorporate an integrated night vision sensor such as an LLLTV or an image intensifier tube (IIT) into the helmet, coupled with the HMD [19–21]. Among those sensors are GEC's Knighthelm and Crusader, Sextant Avionique's Topsight, and Kaiser Electronics' Agile Eye.

If the first HMDs relied mainly on miniature cathode ray tubes (CRTs) as a display source, the latest designs employ flat-panel displays, such as liquid-crystal diode (LCD), electroluminescence (EL), light emitting diode (LED) junctions, plasma discharge, and even images transmitted via a fiber-optic image guide. The driving force to adopt flat-panel displays is the recognition of the benefits of color in HMDs, especially for helicopter use. When other favorable factors, such as lower weight, the potential for higher resolution and lower cost, and the avoidance of placing high-voltage supplies on pilots' heads, are taken into account, flat-panel displays show merits [17].

Figure 1.10 shows an early version of an advanced HMD developed by ELOP Electro Optics Industries of Israel. The HMD image source is generated on an off-helmet CRT and conveyed to the HMD via a fiber-optic image guide and projected on the visor on a holographic combiner. The ELOP HMD demonstrates several of the advanced technologies needed for the construction of high-performance HMDs, including the use of holographic optical elements, fiber-optic image guides, and advanced materials and coatings.

Table 1.2 summarizes the main characteristics of some the most notable HMD/HMS systems.

In spite of the tremendous progress that has been made recently, significant development and research into the human factors associated with the use of HMDs and into new technologies of image sources, optical-element materials and design methods, and life-support equipment are needed before fully operational systems become common pilot equipment.

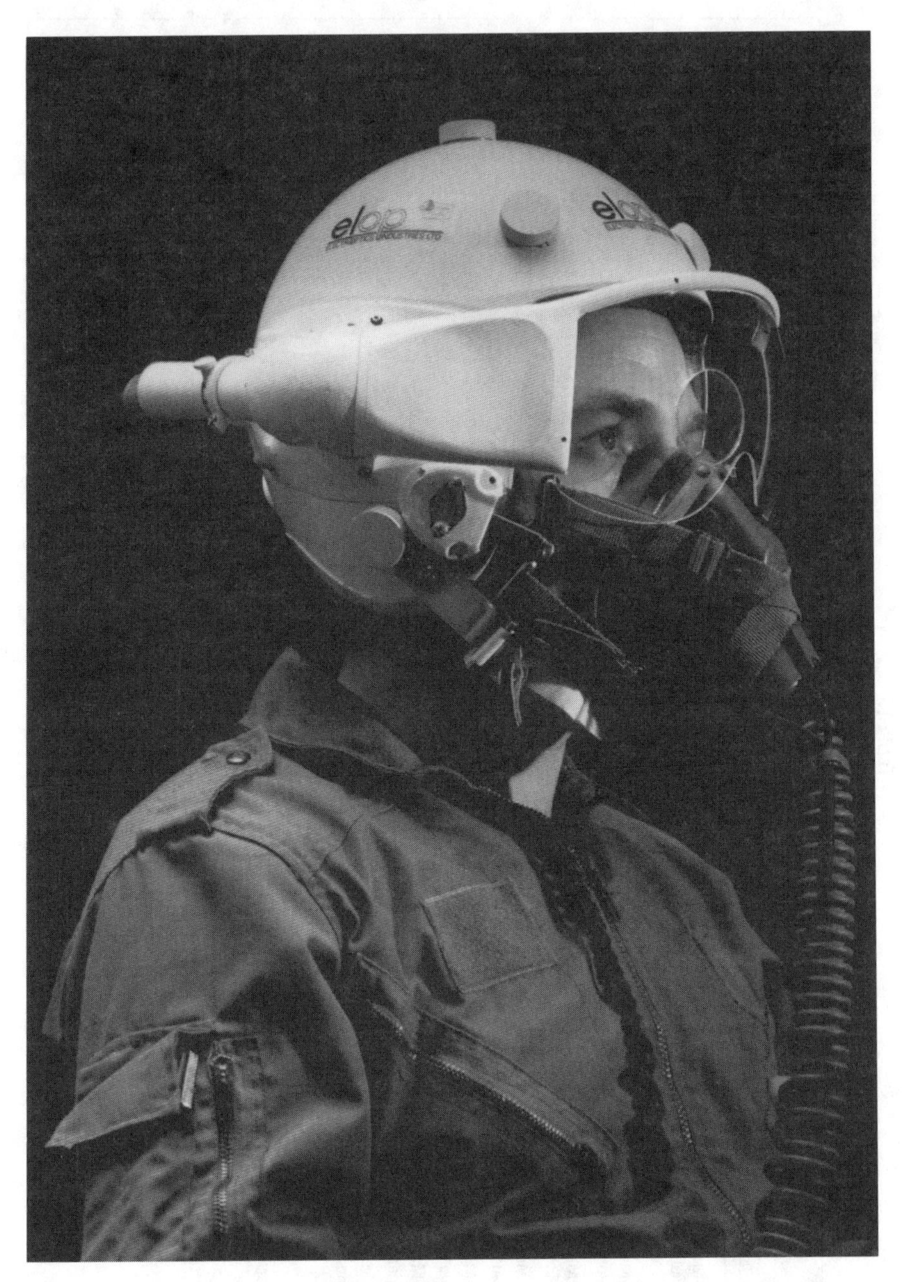

Figure 1.10 Early model of an advanced HMD for fighters (courtesy ELOP Electro-Optics Ind.).

Table 1.2
Characteristics of Some Notable HMD/HMS Systems

Device	Manufacturer	Application[a]	FOV (°)	Ocularity	Exit Pupil (mm)	Eye Relief (mm)	Luminance (ft-L)	See-through (%)	Display Type	Helmet Weight (kg)[b]
HMD MK IV	GEC-Marconi Avionics	II	10	Monocular	>10	Visor projection	400	78	32×32 LED matrix	0.5 D
Knighthelm I	GEC-Marconi Avionics	I, III, IV	40	Binocular	15	35	180 raster; 3,000 stroke	40	CRT	2.3 D+H+M
Alpha HMS	GEC-Marconi Avionics	II	3.5	Monocular	16	47.5	2,000	80	LED array	1.4 D+H+M
Falcon Eye	GEC-Marconi Avionics	I, IV	30	Biocular	10	25	250 raster; 1,000 stroke	30	CRT	1.9 D+H+M
Viper I	GEC-Marconi Avionics	I, II	22	Monocular	20	Visor projection	800 raster; 1,500 stroke	70	CRT	1.96 D+H+M
Viper II	GEC-Marconi Avionics	I, II	40	Binocular	>15	Visor projection	800 raster; 1,500 stroke	70	CRT	1.6 D+H
Crusader	GEC-Marconi Avionics	I, II, III, IV	30, 40	Monocular, binocular	12	Visor projection	800 raster; 1,500 stroke	70%	CRT	2.0 D+H+M
Agile Eye	Kaiser Electronics	I, II	20	Monocular	15H, 12V	53	800 raster; 1,500 stroke	8; 65	CRT	1.3 D+H
Strike Eye	Kaiser Electronics	I, II, III, IV	30	Binocular	12	25	230 raster; 2,000 stroke	10; 40	CRT	1.86 D+H
Wide Eye	Kaiser Electronics	I, II, IV	40, 40x60	Monocular, binocular	10–15	31	100 raster; 2,000 stroke	6.5; 30	CRT	2.0 D+H
IHADSS	Honeywell	I, IV	30x40	Monocular	10	50	400	75	CRT	1.8 D+H
HMCS	Honeywell	I, II	20	Monocular	>25	Visor projection	1,300; 700	9 clear visor; 70 with 13% tinted visor	CRT	1.54 D+H+M

Table 1.2 (Continued)

Device	Manufacturer	Application[a]	FOV (°)	Ocularity	Exit Pupil (mm)	Eye Relief (mm)	Luminance (ft-L)	See-through (%)	Display Type	Helmet Weight (kg)[b]
						Visor projection	1,500	Yes	CRT	1.45 D+H+M
Topnight	Sextant Avionique	I, III, IV	40x30	Binocular		Visor projection	800; 1,500	Yes	CRT	1.8 D+H+M
FOHMD	CAE	V	65x125; 60x120	Binocular	15	38	50 light valve; 10 CRT	10	CRT or light valve	2.4 D+H
DASH	Elbit Computers	I, II	22	Monocular	12V, 18H	50	3,000	Yes	CRT	1.2 D+H
HADAS	ELOP	I, II, III	30x22	Monocular	>11	25	3,000	15; 80	CRT in cabin + fiber-optic image guide	1.8 D+H
HMDD	ELOP	I, II	28x20	Monocular	10	25	2,800	40	CRT	0.27 D

[a]Application categories refer to the definitions in Table 1.1.
[b]D = display; H = helmet; M = oxygen mask.

1.4 Summary

HMDs/HMSs have long been recognized as potentially powerful and effective devices for increasing the users' functionality and safety in conditions that require full attention to the outside world together with vital system information. Because of their apparent benefits, HMD/HMS systems can be used in both military and civil applications.

The history of HMDs shows admirable inventiveness in the design and progress in the technologies embedded in these devices. They evolved from clumsy, heavy, and bulky add-ons to lightweight, integrated, and technically well-balanced systems. However, in spite of the impressive progress evident in recent designs, many issues have yet to be resolved before HMD/HMS systems become operationally reliable and effective.

In recent years, new technologies that are potentially capable of bridging the gap between users' requirements and technology barriers have emerged, including low-cost flat-panel color displays, lightweight optics, fast and accurate head trackers, and powerful graphic-symbol generators. Many of those technologies have since been adopted for consumer-oriented products.

References

[1] Brooks, R. L., "Helmet Mounted Display for Tank Applications," *Imaging Sensors and Displays*, SPIE, Vol. 765, 1987, pp. 19–21.

[2] Kennedy, A. J., "Helmet Mounted Display for Infantry Applications," *Imaging Sensors and Displays*, SPIE, Vol. 765, 1987, pp. 26–28.

[3] Adam, J. A., "Virtual Reality Is for Real," *IEEE Spectrum*, October 1993, pp. 22–29.

[4] Ellis, S. R., "What Are Virtual Environments?" *IEEE Computer Graphics and Applications*, January 1994, pp. 17–22.

[5] Beal, C., and B. Sweetman, "Helmet Mounted Displays: Are We Jumping the Gun?" *International Defense Review*, September 1994, pp. 69–75.

[6] Chapman, F. W., and G. J. N. Clarkson, "The Advent of Helmet-Mounted Devices in the Combat Aircraft Cockpit—An Operator Viewpoint," in T. M. Lippert, ed., *Helmet-Mounted Displays III,* SPIE, Vol. 1695, April 1992, pp. 26–37.

[7] Honeywell, "Introduction to Honeywell Helmet Sight Systems Including Applications Data and Experience," 1977.

[8] Adam, E. C., "Tactical Cockpits—The Coming Revolution," *IEEE Sys. Mag.*, Vol. 9, No. 3, March 1994, pp. 20–26.

[9] Hart, S. G., "Helicopter Human Factors," in E. L. Wiener and D. C. Nagel, eds., *Human Factors in Aviation,* San Diego: Academic Press, 1988, pp. 591–638.

[10] Antonio, C., "USAF/USN Fixed Wing Night Vision: The Mission," in T. M. Lippert, ed., *Helmet-Mounted Displays III,* SPIE, Vol. 1695, April 1992, pp. 21–25.

[11] Pratt, A. B., "Weapon," U.S. Patent No. 1,183,492, May 16, 1916.

[12] Marshall, G. F., "Back From the Past: The Helmet Integrated System of Albert Bacon Pratt (1916)," *Optical Engineering*, Vol. 28, No. 11, November 1989, pp. 1247–1253.

[13] Miller, B., "Helmet-Mounted Display Interest Revives," *Aviation Week & Space Technology*, February 24, 1969, pp. 71–83.

[14] Sutherland, I. V., "A Head-Mounted Three Dimensional Display," 1968 Fall Joint Computer Conference, *AFIPS Conf. Proc.*, 33, 1968, pp. 757–764,.

[15] Klass, P. J., "Navy Pilots to Use Helmet Sight," *Aviation Week & Space Technology*, January 31, 1972, pp. 37–40.

[16] Lucas, T., "Advances in Helmet Mounted Displays," *Avionics*, June 1994, pp. 20–27.

[17] Beal, C., and B. Sweetman, "A New Vision: Helmet-Mounted Displays Take Multi-Track Approach," *Janes International Defence Review*, August 1997, pp. 41–47.

[18] Merryman, R. F. K., "Vista Sabre II: Integration of Helmet-Mounted Tracker/Display and High-Boresight Missile Seeker Into F-15 Aircraft," in R. J. Lewandowski, W. Stephens, and L. A. Haworth, eds., *Helmet and Head Mounted Displays and Symbology Design Requirements*, SPIE, Vol. 2218, 1994, pp. 173–184.

[19] Wanstall, B., "HUD on the Head for Combat Pilots," *Interavia*, Vol. 4, 1989, pp. 334–338.

[20] Jarrett, D. N., and A. Karavis, "Integrated Flying Helmets," *Proc. Institute of Mechanical Engineers*, Vol. 206, 1992, pp. 47–61.

[21] Cameron, A. A., "The 24 Hour Helmet Mounted Display," in C. T. Bartlett and M. D. Cowan, eds., *Display Systems*, SPIE, Vol. 1988, 1993, pp. 181–192.

2

Human Factors Associated With HMDs

HMDs interact with the most vital human sensory channel: vision. Vision is our primary sense for derivation of real-world data. It plays a dominant role in sensing spatial orientation and combines with the other senses to provide the kinesthetic and gravitic interaction that enables a full range of human activities.

The fundamental concept of the HMD is to introduce additional visual information to the visual field by inserting a virtual image without interfering with the normal function of perceiving the outside world. That puts HMD wearers in a unique situation that is unnatural and uncommon with regard to many aspects that they have never experienced or encountered in other normal situations or activities. For example, when using a monocular display, the HMD alters the illumination of one eye relative to the other or even causes one eye to see a different image than the other eye sees. In some cases, it may be beyond the capacity of the wearer to fuse two images that differ both in brightness and in information. Similarly, the HMD image generally is displayed on a CRT or a calligraphic display. Both display images are refreshed periodically, thus stimulating the eye with temporally changing illumination, which may be perceived to flicker. All those aspects of HMDs affect both the sensory and the cognitive processes of visual perception of the display. Figure 2.1 illustrates the basic concept of the HMD and insinuates some of the potential perceptual problems associated with viewing a monocular HMD. In the figure, a virtual image to the right eye is introduced through a semitransparent visor, while the left eye is exposed to the outside-world scenery. The visual system is required to process simultaneously two independent channels of visual information. The channels may differ in many aspects, including brightness, resolution, and type of information (e.g., structured versus homogenous, accommodation requirements differences, etc.).

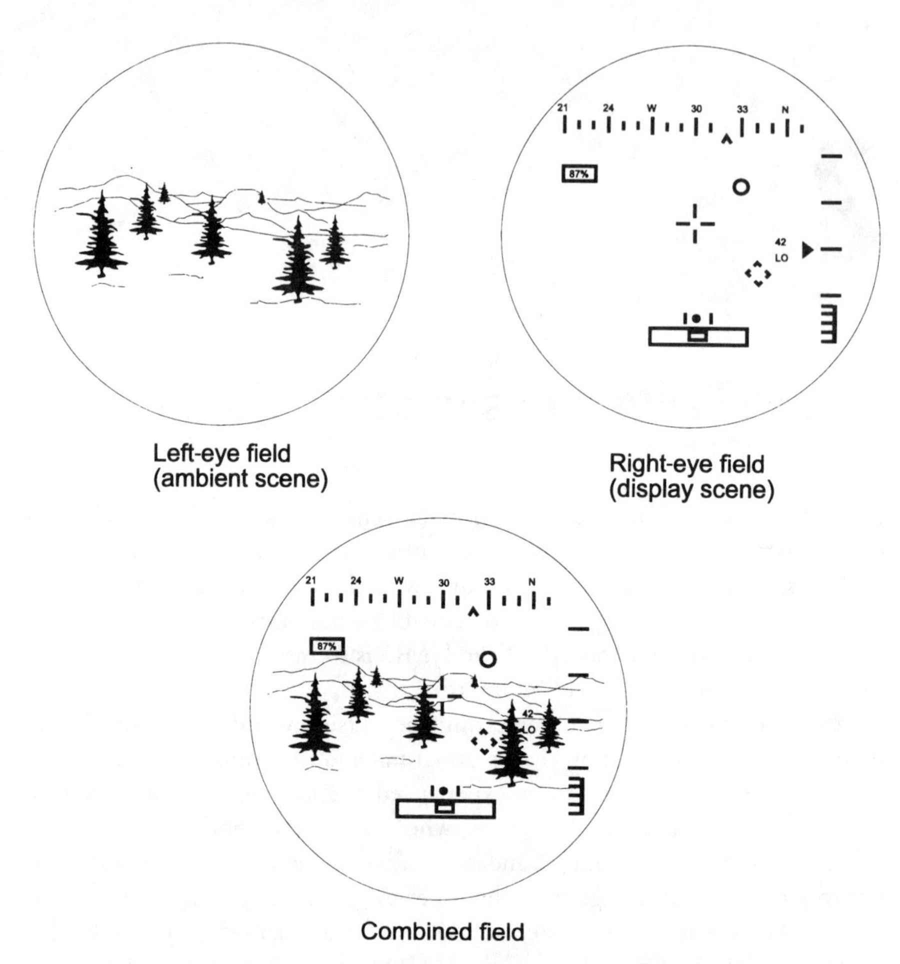

Figure 2.1 Right and left visual fields when viewing a typical right-eye-occluded monocular HMD against the outside-world scenery. The bottom figure shows the display imagery overlaid on the scenery, with complete fusion of both visual fields.

Unlike photography or electro-optical sensing, vision is more than just a direct translation of incoming sensory data. It involves recognition and interpretation of incoming data, a process that requires transformation and integration of retinal images, and is based on information processing, expectations, previous experience, and knowledge of the current situation.

The human observer is very sensitive and thus very critical to the quality of the displayed image. Any degradation in quality of the displayed image results in the viewer's dissatisfaction and even annoyance [1].

Many factors affect the quality of the viewed display: FOV, resolution, contrast, collimation, brightness, image stability, exit pupil, registration, color, distortion, and more. Each parameter may influence the perception of the display in obscure ways. Designers and users of HMDs must be fully aware of the limitations and the capabilities of visual systems.

This chapter describes the principal human factors associated with the perception in HMDs. The human factors that are related to the interaction of the HMD with other systems, such as the helmet, voice and sound system, comfort, and safety, are described in depth in Chapter 9, while Chapter 8 is devoted to the effects of image and observer vibration and motion.

2.1 Visual Perception

Perception refers to the way in which messages from the senses are interpreted. It is a process that integrates the processes of receiving, selecting, transforming, organizing, and processing information through the senses [2].

The human visual system can be regarded as having optical, photosensitive, transduction, conduction, and central physiological processing mechanisms. Stimulation of the visual system by light rays results in brain activity. Essentially, perception is the brain's interpretation of sensation.

Visual perception consists of sensory and cognitive processes. In the sensory process, light is collected by the optical system of the eye, sensed by visual receptors, and transformed to electrical signals, which are transmitted to the brain. In the cognitive process, the signals are processed so as to extract the information, contained in the light rays, that is useful to the human.

If some of the sensory processes can be identified and studied from the anatomy and physiology of the eye, the mechanism of perception can be studied almost only by psychophysical methods (with the exception that in some cases electrophysiological tests can be used). The psychophysical method looks at some responses of the observer as an indication of light stimuli. For example, in visual acuity tests, subjects are required to detect a gap between lines or identify a shape out of a sequence of randomly presented shapes.

Many theories have been suggested over the years to explain the relation between sensation and perception. Two main theories have received the most attention: the image and cue theory and the direct perception theory [3]. The image and cue theory suggests that perceptions are learned by acquiring an understanding of the cues contained within the image sensed by the eye. The direct perception theory suggests that perceptions are a function of biological organization and innate perceptual mechanisms.

Because researchers still are arguing over which theory is correct, equal attention should be paid to both of them. The design of HMDs should be consistent in terms of providing or retaining the necessary visual cues as well as with the physiology of the visual system.

Visual perception involves space (or depth) perception, form perception, and movement (or motion) perception. Depth perception makes use of several cues, some monocular and some binocular. The monocular cues are:

- *Relative size:* Comparison of objects of known size reveals which is closer (closer objects appear bigger).
- *Interposition:* Overlapped objects are said to be farther away than overlapping objects.
- *Linear perspective:* Parallel lines, such as roads and railway tracks, appear to converge in the distance and diverge nearby.
- *Aerial perspective:* Closer objects appear bright and sharp, while distant objects appear pastel and hazy.
- *Monocular movement parallax:* During relative motion between the head and the environment, a differential angular velocity exists between the LOS to a fixated object and the LOS to any other object in the visual field (i.e., when you move your head, nearby objects seem to move a great distance, while distant objects seem to move hardly at all).
- *Height on a plane:* Objects that from the viewer's point of view are higher on a plane are perceived as farther away than objects lower on the plane.
- *Accommodation:* Closer objects are seen with more details than more distant objects.

Table 2.1 summarizes monocular cues and their learned meanings.

Along with the monocular depth cues, two binocular cues exist. Binocular cues, which require the coordinated activity of the two eyes, are:

- *Convergence:* The lines of fixation of both eyes to an object converge on the object. For closer objects, the convergence is large; for very distant objects, both eyes' LOSs are parallel. Convergence may serve as a minor cue for depth responses [4], and convergence cues cannot be differentially effective for objects at distances great than several meters.
- *Stereoscopic vision:* Because the eyes are set apart, the retinal image of objects closer than about 7m to 8m are sensed on slightly different locations on the left and right retinas. More than any other cue, stereoscopic vision, or retinal disparity, gives a strong perception of depth at close distances.

Table 2.1
Monocular Depth Cues and Their Learned Meaning

Cue	Learned Meaning
Relative size	Closer objects appear bigger
Interposition	Overlapping object is closer than overlapped one
Linear perspective	Parallel lines converge in distance and diverge nearby
Aerial perspective	Sharp and bright objects are closer, far objects are pastel and hazy
Monocular movement parallax	During head movements, objects that move more are nearer than objects that move less
Height on a plane	Objects higher on the plane are farther away
Accommodation	Objects with more details are closer

Apart from the visual cues described above, depth perception also relies on learned perception based on three constancies:

1. *Size constancy:* When the retinal image of a known object changes its size, the interpretation is that the distance to the object has changed while the size of the object remained constant.
2. *Shape constancy:* An object remains the same shape, despite the fact that the image it casts on the retina may vary in shape, depending on the viewing angle.
3. *Brightness constancy:* Objects appear to have the same brightness independent of the lighting conditions in which they are seen. White objects in low-light conditions, for example, appear brighter than gray objects in higher-light conditions, regardless of the amount of reflected light actually coming from the object.

Form perception, which enables the viewer to distinguish objects or symbols from the background, consists of the abilities to detect contours and to group image elements to structured or meaningful forms. It is a perceptual rather than an objective quality, because different forms may be perceived for the same stimuli. Humans can detect and recognize objects by identifying contours in the image. The contour does not have to be continuous, and the observer is able to group several image features and attribute them to the object shape. Assemblage of orderly arranged dots, for example, is not perceived singly or as a chaotic group but rather is easily perceived as a broken line or curve.

Motion perception is attributed mainly to the peripheral vision. It enables humans to detect an object's motion, its velocity, and direction of motion. When the viewer is stationary, motion cues usually are reliable and only a few problems are encountered [5]. However, when the viewer is in motion, there are conflicts between the perception of self-motion and object motion. That is due to the interaction with the other sensory systems, namely, the vestibular system, which senses the body accelerations. Such conflicts may lead to illusions or symptoms of motion sickness.

2.2 Sensory Processes

2.2.1 The Structure of the Human Eye

Figure 2.2 illustrates the structure of the human eye, showing the horizontal cross-section. The human eye is nearly spherical with a diameter of about 20 mm. It is enclosed by three membranes. The outer membrane consists of the transparent cornea over the anterior surface of the eye and the sclera, which is an opaque membrane continuous with the cornea and encloses the remainder of the optic globe [6].

Directly below the sclera lies the choroid coat, which contains a network of blood vessels that are the main source of nutrition to the eye. The choroid coat is heavily pigmented and serves to reduce the amount of light entering the eye as well as back scatter within the eye.

At its anterior extreme, the choroid coat differentiates into the iris diaphragm and the ciliary body. The lens is suspended by zonular fibers that attach to the ciliary body. The amount of light entering the crystalline lens is controlled by the iris diaphragm.

The innermost membrane of the eye is the retina. Part of the retina, the retina propria, which lines the entire posterior portion of the eye and extends forward to the ora serrata, contains the visual receptors and their neural connections. The retina consists of 10 layers, as shown in Figure 2.3. The innermost layer, which encounters the light rays entering through the lens, is the inner limiting membrane, and the next layer includes the optic nerve fibers. The following five layers, in sequence, are the ganglion cell layer, the inner plexiform layer, the inner nuclear layer, the outer plexiform layer, and the outer nuclear layer. The last three layers, which include the receptors, are the outer limiting membrane, the bacillary layer (or the rod and cone layer), and the pigment epithelium layer. The light that stimulates the receptors traverses the optic media (i.e., the cornea, the lens, and the vitreous humor) and passes through the blood vessels and the optic nerve fibers, embedded in the retina, before it reaches the photoreceptors.

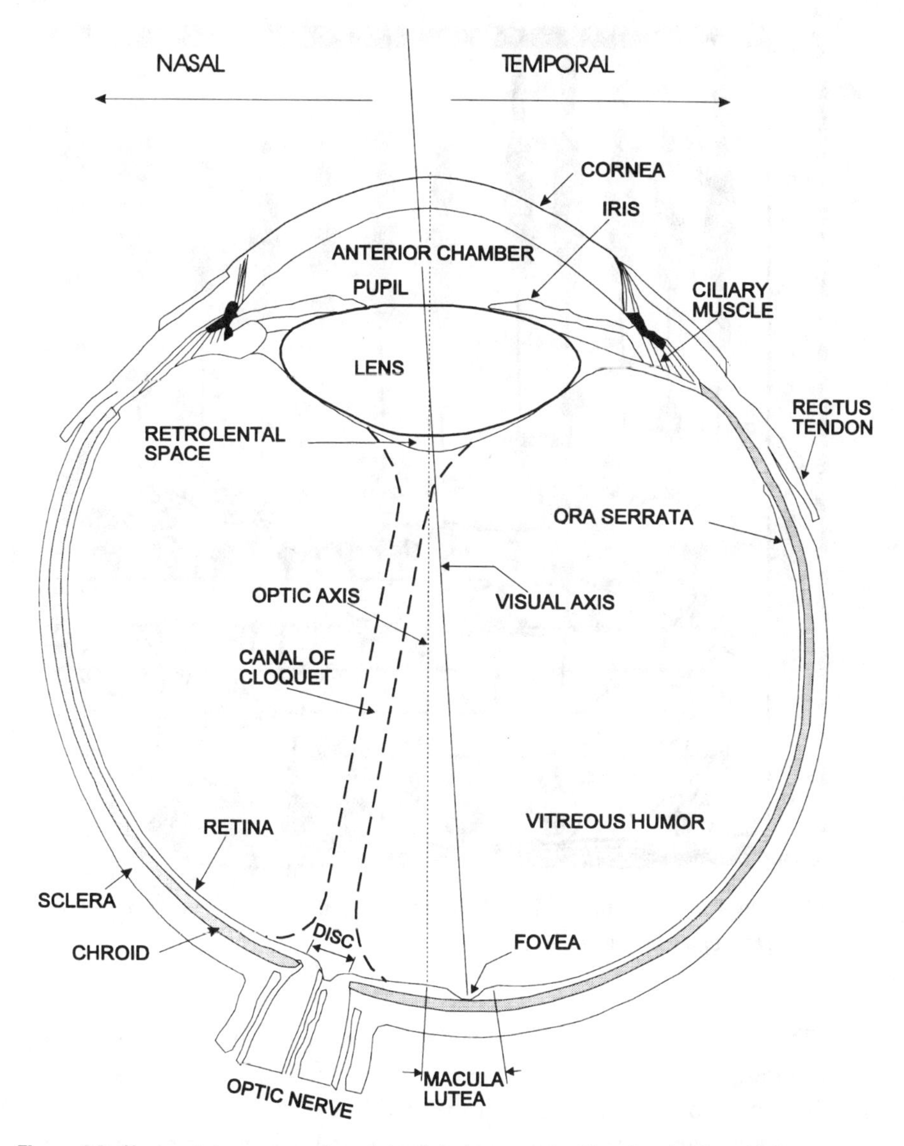

Figure 2.2 Horizontal cross-section of the right human eye. The visual axis, which connects the center of the fovea through the center of the lens to the fixation point, divides the visual field into the nasal and temporal fields (*From:* [6]).

The central 2- or 3-mm-diameter area of the retina, which is equivalent to 6 to 10 degrees, is the macula lutea and is marked by a yellow pigment. The depression in the center of the macula lutea is the fovea. The center region of the fovea, the fovea centralis, extends 500 to 600 μm in diameter and contains most of the 6 or 7 million cones found in the human eye.

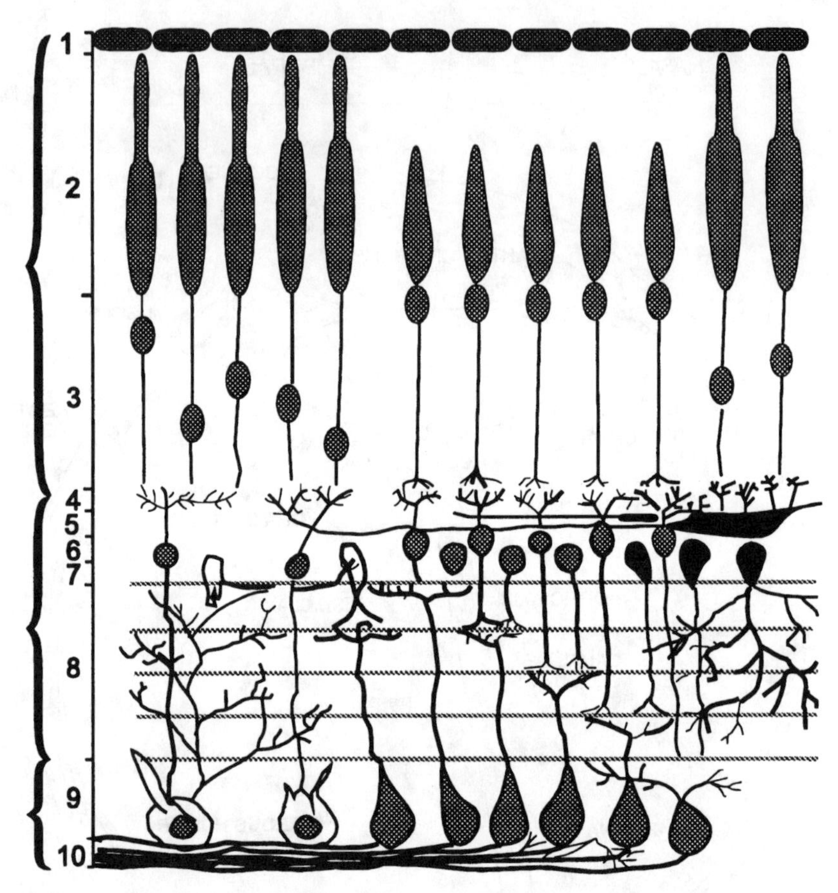

Figure 2.3 Cross-section of the retina with the cellular relations. The numbers on the left refer to the layers of the retina.

The second kind of photoreceptors, the rods, number between 75 and 150 million and are distributed over the retina surface. Figure 2.4 shows the distribution of rods and cones along the meridian of the retina.

The fovea, which nominally extends the central 2 degrees of vision, contains very few rods. The populations of rods and cones are approximately equal at 10 degrees around the visual center. Visual acuity is the highest in the densely cone-populated area and sharply falls off to a very low level in the periphery. The significance of that to the design of HMDs is that the important information that must be continuously seen sharp should be confined to the central 2- to 10-degree area of the display. That extends a field of view similarly to the conventional HUD.

The optic nerves emerge from the eye at a point that is about 15 degrees from the fovea and converge at the disk of the optic nerve. Because there are no

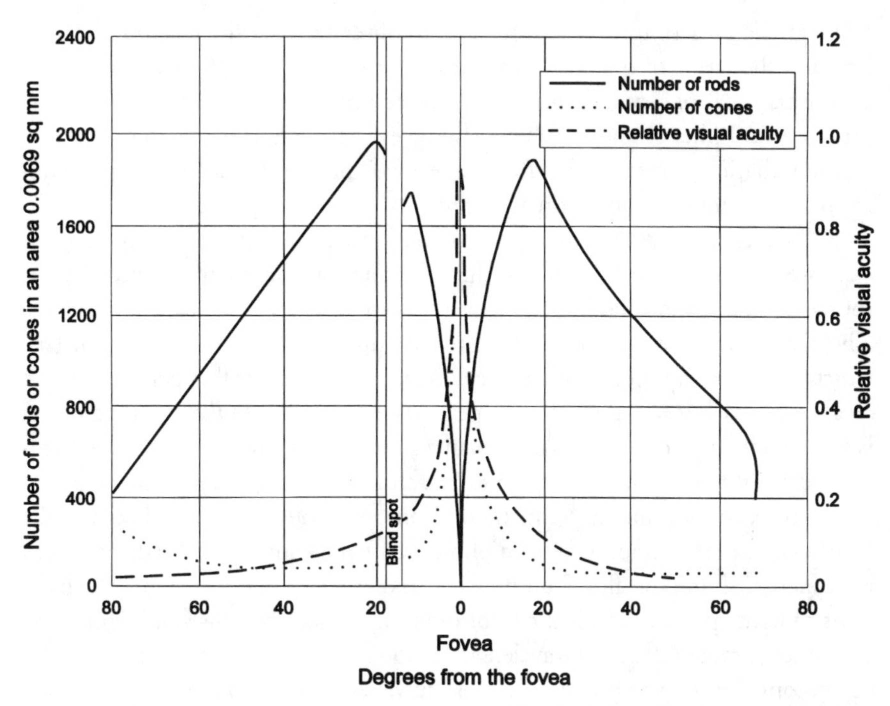

Figure 2.4 Distribution of rods and cones along the horizontal meridian of the retina. Also shown is the relative visual acuity for high luminance (*From:* [7]; *After:* [6]).

photoreceptors of any kind in the optic disk, this point (the two parallel vertical lines in Figure 2.4) is termed the blind spot. Obviously, no information that the viewer should see while fixating at the center of the display (e.g., at the sight's aiming reticle) should be displayed in the vicinity of the blind spot on a monocular display.

The imaginary line extending from the fovea and passing through the center of the lens to the fixation point on the viewed object is called the visual axis. It divides the visual field from the nasal field, which is left of the visual axis for the right eye and right of the visual axis for the left eye, and to the temporal field, which is right of the visual axis for the right eye and left for the left eye. In a similar manner, the retina is divided into nasal and temporal retinas. Due to the reversing effect of the eye's optics, images of the temporal fields fall on the nasal retina, and images of the nasal fields fall on the temporal retina.

2.2.2 Light as a Visual Stimulus and Photometry

Photometry relates to the measurement and rating of light as it affects vision according to the sensitivity of the eye to the light. Light is the aspect of the electromagnetic radiation to which the human observer is sensitive through stimulation of the retina of the eye. In that sense, light is a psychophysical rather than a pure physical or psychological property, as opposed to radiation, which is a pure physical phenomenon. Obviously the eye retina is sensitive to only a part of the radiation spectrum, commonly termed the visible spectrum. It extends from a wavelength of approximately 390 to 760 nm (Figure 2.5).

The energy of electromagnetic radiation of a point source typically is expressed in joules (J). The radiant flux, the total radiant energy emitted by the source in all directions per unit time, is given in joules per second, or watts (W). The radiant intensity is the radiant flux per unit solid angle, or steradian (sr), which is $1/4\pi$ of a sphere. Similarly, irradiance is the radiant flux per surface unit area, and as such it has units of watts per square meter. When the radiating object is an extended source, that is, larger than a source point, then the radiance that is the radiant flux per unit area of the source and per solid angle is considered.

For each radiometric term, there is an analogous term that describes the equivalent aspect of photometry. In photometry, the equivalent of radiant flux in radiometry is luminous flux. Luminous flux is expressed in lumens (lm), the basic units of light "power," and is used, for example, to specify household lightbulbs. One watt of radiant flux at a wavelength of 555 nm (a wavelength that is the peak of photopic luminosity function) is the equivalent of 685 lm of luminous flux.

The intensity of light sources that are focused or emit light in a particular direction is specified by the luminous intensity, which is expressed in lumens per steradian, or candelas (cd). Illuminance is a measure of the visible light falling on a surface, such as the light level in a cockpit or work environment. Illuminance usually is expressed in units of foot-candles, or lux (lx). The measure of the visible light emitted or reflected from a surface is the luminance and is given in units of foot-lamberts (ft-L) or candelas per square meter. Luminance is used to specify how bright a display is.

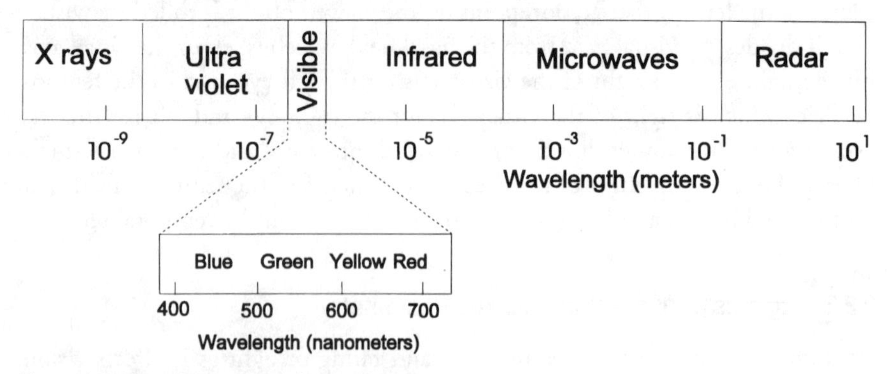

Figure 2.5 The electromagnetic spectrum.

Luminance is not identical to brightness. Luminance is a physical property, while brightness is a perceptual quantity. Because the eye responds in a logarithmic function, if the luminance increases by a given factor, the brightness is not perceived to have increased by the same amount.

Typical military HMD, which employs a miniature CRT, has a luminance of about 1,500 ft-L at the eye for symbology and about 800 ft-L for imagery. Color LCDs, which are used mainly for virtual reality displays, have luminance on the order of 3 to 30 ft-L.

Table 2.2 summarizes the radiometric, their analogous photometric terms, and the main units used to express those terms.

2.2.3 Spectral Sensitivity of the Eye

Visual sensation involves electrical and photochemical activities. When light strikes the eye, a photochemical event is initiated. Two photosensitive substances, rhodopsin and iodopsin, undergo chemical reaction when struck by light. Both substances are pigments. Rhodopsin is found in the cones, while iodopsin is found in the rods.

The visual pigment is a protein molecule with one or more chromophores attached to it. The chromophore, retinene, is photosensitive, while the protein

Table 2.2
Radiometric and Photometric Terms and Units

	Symbol	Units
Radiometric terms		
Radiant energy	Q_e	Joules (J)
Radiant flux	Φ_e	Watts (W)
Radiant intensity	I_e	Watts per steradian (W/Ω)
Irradiance	E_e	Watts per square meter (W/m^2)
Radiance	L_e	Watts per steradian per square meter (W/Ω/m^2)
Photometric terms		
Luminous energy	Q_v	Lumen-seconds
Luminous flux	Φ_v	Lumen
Luminous intensity	I_v	Lumens/steradian (lm/Ω); 1 lm/Ω = 1 candela (cd)
Illuminance	E_v	Lumens per square meter (lm/m^2); 1 lm/m^2 = 1 lux = 1 cd-m
Luminance	L_v	Lumens/steradian/square meter (lm/Ω/m^2); 1 lm/Ω/m^2 = 1 cd/m^2 = 0.3142 millilambert (mL) = 0.2919 foot-lambert (ft-L)

core is not. The protein core is called opsin. The opsin in the rods is called scotopsin, and the opsin in the cones is called photopsin [8]. The retinene is an isomer of an aldehyde of Vitamin A, which is a complex alcohol.

There are three kinds of cones in the parafovea (the area that is enclosed within about half the diameter of the macular region), each of which has predominantly a single pigment [9–11]. The pigments of the cones have concentrations and photosensitivities similar to the pigments of the rods [12]. The cone pigments absorb light maximally in three regions of the visible spectrum, as shown in Figure 2.6.

2.2.3.1 The Duplicity Theory

Daylight and color vision are functions of the cones, whereas low-light vision is a function of the rods. The day or intense-light vision, which involves mainly cone vision, is called photopic vision (*photos* is Greek for "light"). Photopic

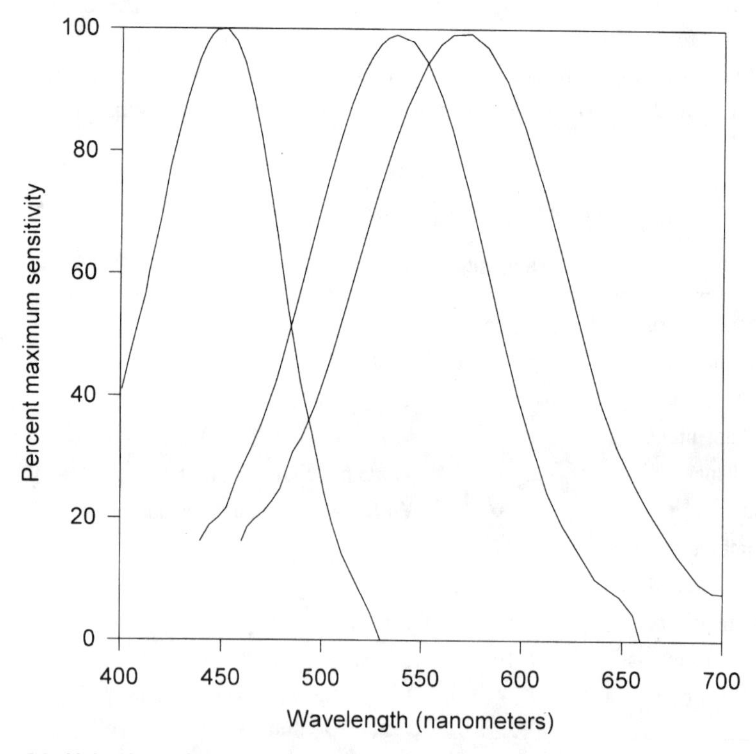

Figure 2.6 Light absorption by the cone pigments in the visible spectrum. The blue-sensitive, green-sensitive, and red-sensitive cones are shown. All three types of cones are sensitive to light of wavelength between 460 and 530 nm (*From:* [13]).

vision is for luminance levels above 10 cd/m^2. Low-light vision is called scotopic vision (*skotos* is Greek for "darkness"). It involves rods vision only and is for luminance levels below 10^{-3} cd/m^2. Intermediate vision, that is, for luminance levels between 10^{-3} and 10 cd/m^2, involves both rods and cones and is called mesopic vision (after the Greek word *mesos,* for "middle"). The boundary regions of vision are shown in Figure 2.7.

The spectral sensitivity of the eye is different for scotopic and photopic vision [14]. Peak sensitivity in the scotopic vision occurs at the wavelength of 510 nm; for photopic vision, the peak is at 505 nm (Figure 2.8). This change in spectral response from scotopic to photopic vision is called Purkinje shift.

Table 2.3 summarizes the main characteristics of photopic and scotopic vision responses.

2.2.4 Light and Dark Adaptation

When the eye is exposed to light, it is being light-adapted. If the light is turned off, the eye becomes dark-adapted. During light adaptation, the sensitivity of the eye decreases. During dark adaptation, the sensitivity of the eye increases, and it can see objects of very low intensity [8]. The ability to see in the dark is

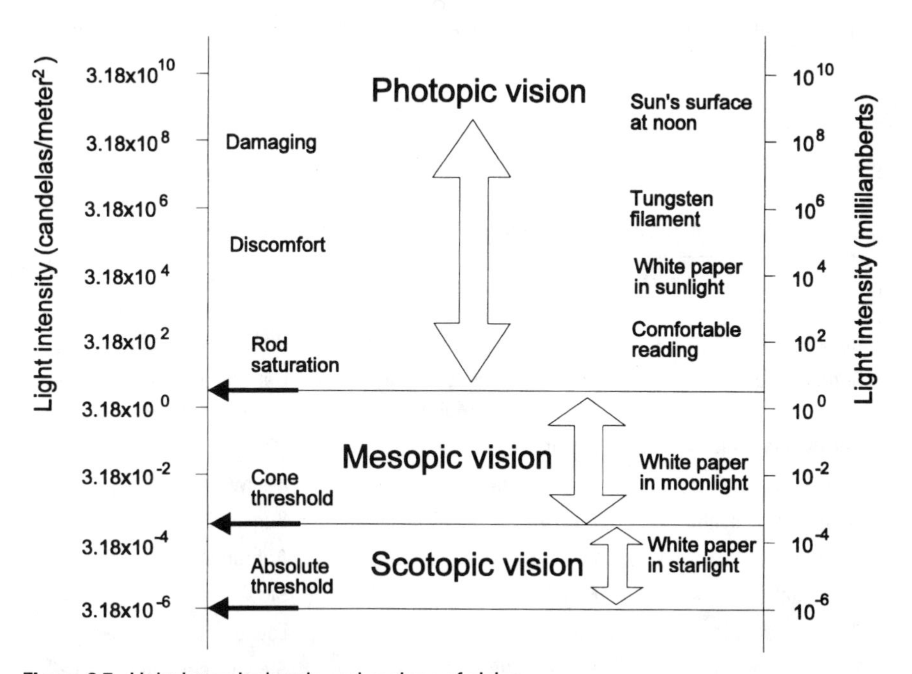

Figure 2.7 Light-intensity bands and regions of vision.

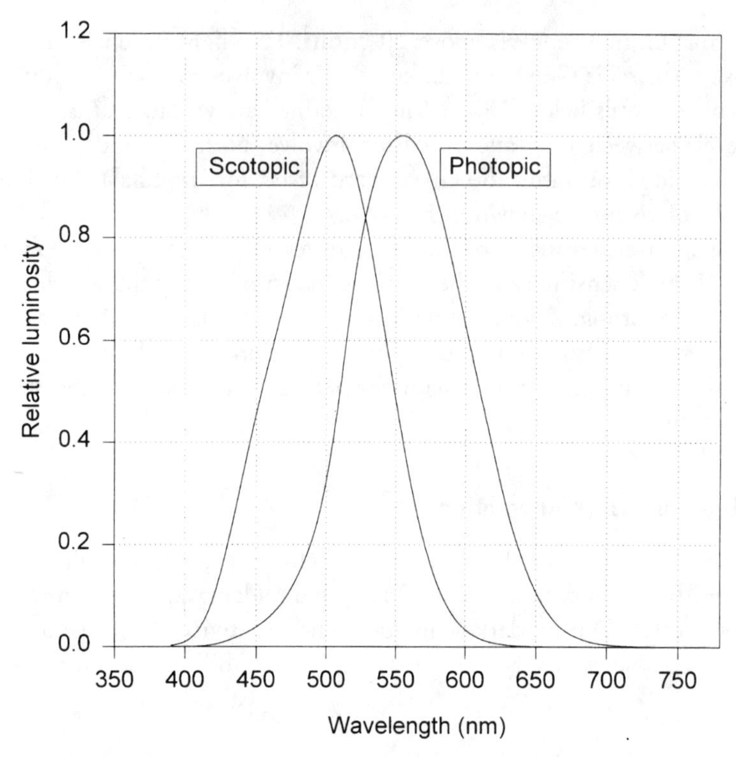

Figure 2.8 Sensitivity of the eye for scotopic and photopic vision.

Table 2.3
Main Characteristics of Photopic and Scotopic Vision

	Photopic Vision	**Scotopic Vision**
Receptor	Cones (6–7 million)	Rods (75–150 million)
Retinal location	Concentrated at center; fewer at periphery	General at periphery; none in fovea
Peak wavelength	555 nm	505 nm
Luminance level	Above 10 cd/m^2	Below 10^{-3} cd/m^2
Neural processing	Discriminative	Summative
Color vision	Normally trichromatic	Achromatic
Dark adaptation	Rapid	Slow
Spatial resolution	High acuity	Low acuity
Temporal response	Fast reacting	Slower reacting

due mainly to the rods, which are very sensitive to small amounts of illumination. The rods are not sensitive to red light, so the eyes can be exposed to red light and still retain dark adaptation. The change in the state of adaptation, either dark or light, is a slow process and takes several minutes to change the state of adaptation significantly. A brief exposure to bright light may disrupt night vision but for only a short time, and the glare recovery is quick.

2.2.5 Retinal Illuminance

Retinal illuminance is the amount of the light entering the eye in terms of the luminance of the viewed scene. As shown in (2.1), the retinal illuminance (E) is the scene luminance (L) multiplied by the effective pupil area (A). E is expressed in trolands (Td), L in candelas per square meter, and A in square millimeters.

$$E = L \cdot A \tag{2.1}$$

Light passage to the eye is controlled by the iris, which adjusts the diameter of the pupil [15]. Stiles and Crawford found in 1933 that in cones the visual sensation of equal amounts of light passing through different parts of the pupil is not equal, even though they hit the same patch on the retina [16]. Light rays incoming from the edge of the pupil are less effective by a factor of up to 8 than rays passing through the pupil center [15,17,18]. Figure 2.9 describes the Stiles-Crawford effect, showing the relative luminous efficiency of rays entering the pupil away from the center to rays entering the pupil at the center.

The practical implication of the Stiles-Crawford effect is that the retinal illumination per se is not a sufficient indicator of the effectiveness of visual stimulation. Another implication, for displays with small exit pupil, is the need to align the display with the eye to ensure optimum visibility. A way to compensate for the Stiles-Crawford effect is to correct the retinal illuminance equation by using an effective pupil area instead of the true pupil area. The best known expression for the ratio between the true pupil and the effective pupil is the effectivity ratio [15]:

$$E_R = 1 - 0.0106 \cdot d^2 + 0.0000419 \cdot d^4 \tag{2.2}$$

where d is the true pupil diameter in millimeters. Farrell and Booth [15] applied (2.2) to data from several sources and computed the eye pupil size as a

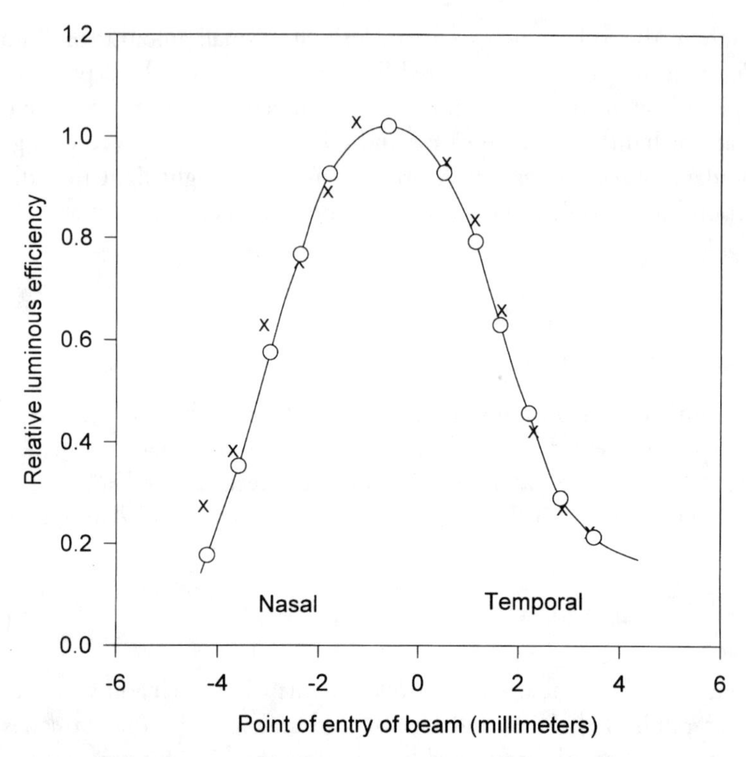

Figure 2.9 The Stiles-Crawford effect reflecting the relative luminous efficiency of light entering the pupil at different points on the eye pupil. The X symbols indicate measurements for the left eye, the circles are the measurements for the right eye for a single subject (*After*: [13]).

function of the scene luminance. The correction was used for the cones vision only and is shown in Figure 2.10.

2.2.6 Sensitivity to Brightness Contrast

The term brightness contrast (as opposed to color contrast) indicates the relationship between neighboring areas within the visual field that are differently illuminated [19]. Brightness contrast is expressed as the ratio of the difference between the higher (L_H) and the lower (L_L) luminance levels to the lower luminance level and usually is expressed as a percentage:

$$C = \frac{L_H - L_L}{L_L} \cdot 100 \tag{2.3}$$

The most common way to quantify contrast is to look at the modulation contrast, known also as the Michelson contrast, which is defined as

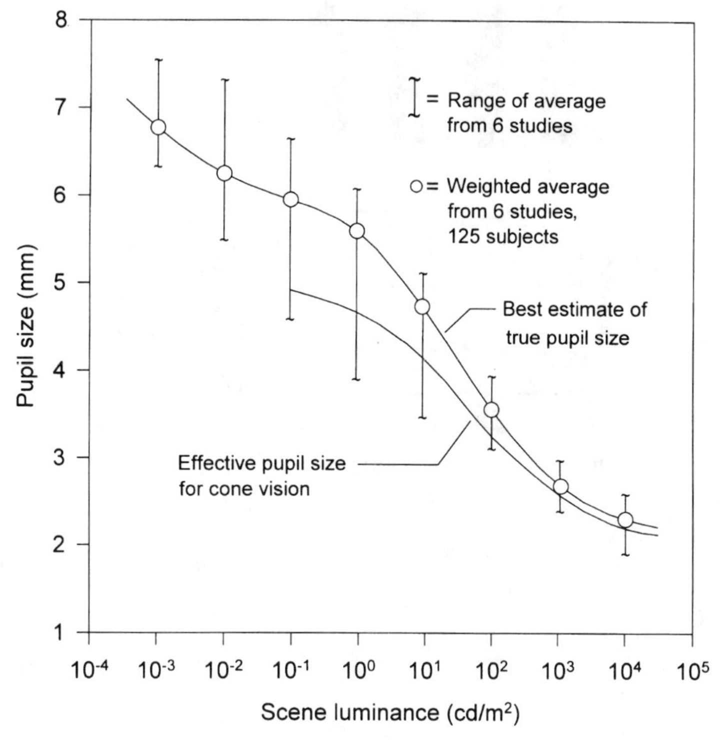

Figure 2.10. Eye pupil size as function of scene illuminance (*After:* [15]).

$$C_m = \frac{L_1 - L_2}{L_1 + L_2} \tag{2.4}$$

where L_1 is the target maximum luminance and L_2 is the target minimum luminance. The modulation contrast is tested using either a square-wave or a sinusoidal-wave grating, similar to that shown in Figure 2.11.

The ability of the eye to resolve cyclical targets in terms of modulation contrast and spatial frequency is shown in Figure 2.12. The curve, referred to as the J curve [15], describes the contrast sensitivity function (CSF) and is the most widely accepted model of how the visual system processes real-world display images [20]. The CSF relates the contrast modulation required to detect an object to the spatial frequency or equivalently to the size of the object.

Contrast discrimination is another property of the visual system. In HMDs, the displayed information is seen against backgrounds that may vary from total darkness to the brightest daylight. Also, out-world scenes must be seen against a bright display. In daylight conditions, the luminance of the sky may reach levels of 10,000 ft-L. To discriminate an image of the HMD against

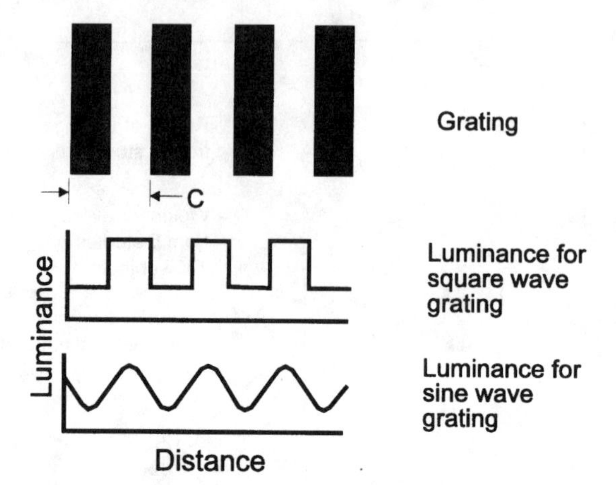

Figure 2.11 Modulation contrast of a grating target.

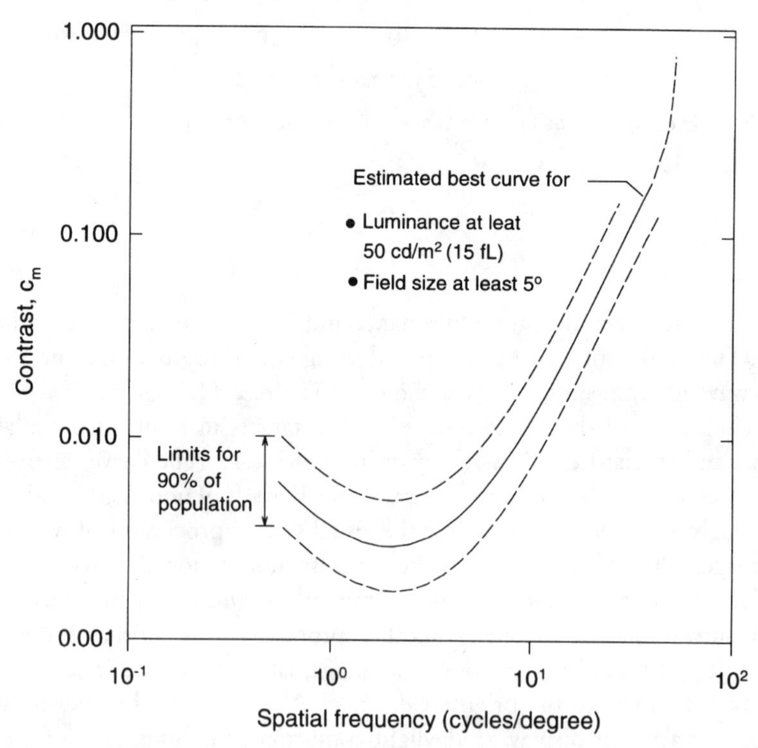

Figure 2.12 The contrast sensitivity function (CSF) of the visual system.

the ambient light, the luminance of the display should be at least 500 ft-L to be at the threshold of detectability, as indicated in Figure 2.13. Because the threshold levels are statistical and vary among people [21], a rule of thumb suggests provision of a luminance of 50% or higher to ensure image detectability. Typical values for the luminance to the eye for symbology are 1,500 ft-L and for imagery (video) 800 ft-L.

In monocular viewing, the contrast sensitivity threshold is about 40% higher than in binocular viewing [15], implying that for low-luminance-level conditions the use of binocular HMD can increase the visibility of objects.

2.2.7 Sensitivity to Temporal Stimulation

In most displays, the luminance of areas of the images varies with time. If the time variations are at sufficiently high frequencies, the image appears constant. However, if the frequency is too low, the image appears to flicker. Due to the

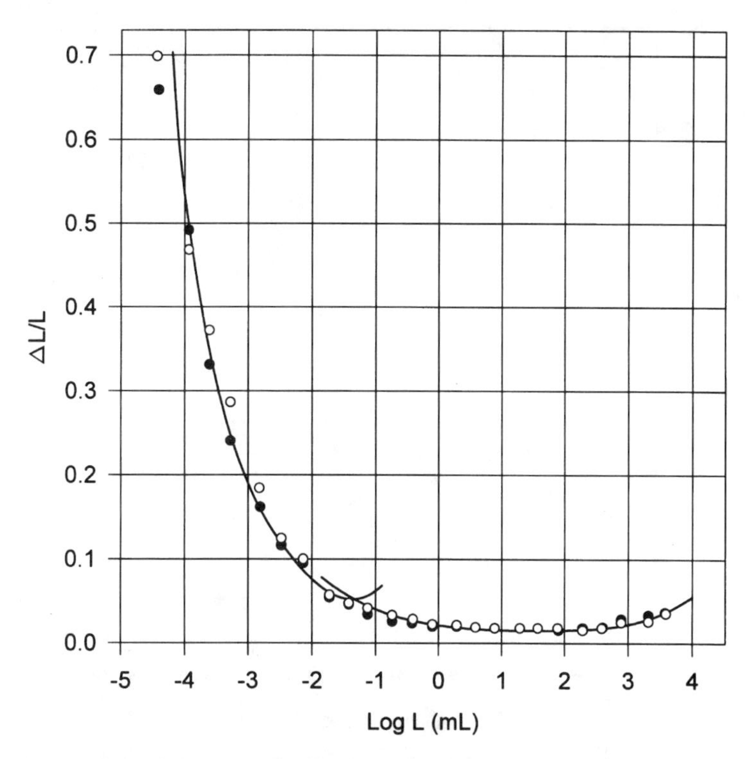

Figure 2.13 Ratio of the luminance threshold increment, ΔL, to the adapting light, L, as a function of the adapting light (*After:* [22]).

intermittent nature of image generation in HMDs based on CRTs, the eye sensitivity to flicker plays an important role. Image flicker is annoying, particularly in HMDs, because the image is constantly present in the viewer's visual path and thus should be avoided if possible. Prolonged viewing of a flickering display increases tiredness and discomfort.

Flicker discrimination refers to the perception of a change in the brightness as the visual stimulation alternates. The point of transition between the appearance of constant and flickering image is known as the critical flicker frequency (CFF). The CFF increases as the luminance or the viewed area increases. Also, CFF for illuminance higher than 0.1 Td may decrease significantly with the radial distance from the eye fixation point [6].

Figure 2.14 shows the CFF for three retinal image locations: 0, 5, and 15 degrees. It is apparent that at high luminances the CFF decreases from the peak of about 45 Hz at the fovea to less than 20 Hz for target locations at 15 degrees.

CFF is also highly dependent on the modulation amplitude of the alternating image. For high modulation amplitudes (10% to 100%), it is related to

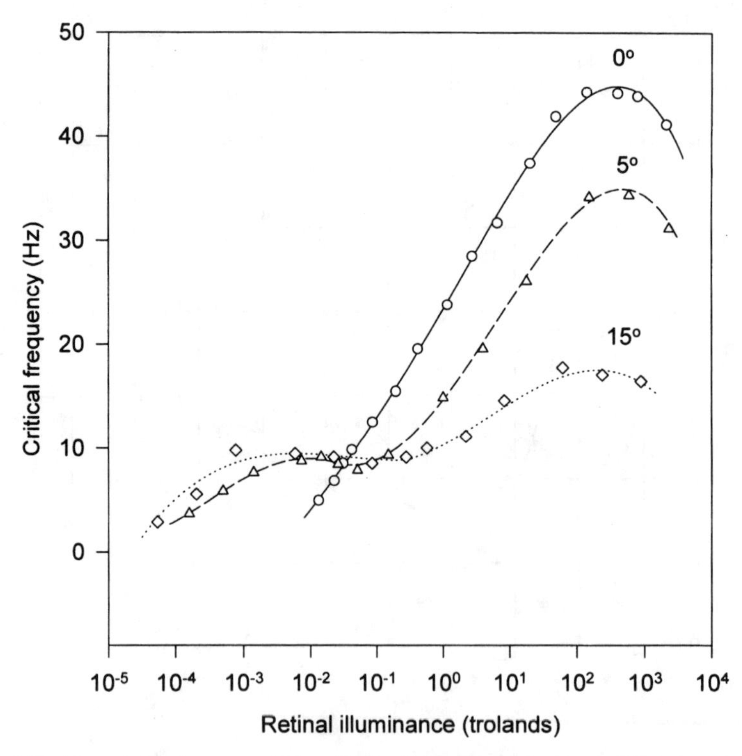

Figure 2.14 The relation between CFF and retinal illuminance for three different retinal target locations [23].

luminance by the Ferry-Porter law, which states a linear relation between the logarithm of the retinal illuminance and CFF. Figure 2.15 shows the relation between flicker fusion frequency to the retinal illuminance for different modulation amplitudes. It can be seen that at low modulation amplitudes there may be two CFF values for a given retinal illuminance.

For most display viewing cases, a frequency of 60 Hz is not perceived flicker. However, if the display extends more than 20 degrees, and the luminance exceeds 100 fL, a situation that is frequently common to HMDs, then the display appears to flicker for frequencies up to about 80 Hz [24].

2.2.8 Visual Acuity

Visual acuity is the capability to discriminate the fine details of objects in the field of view [25]. Good visual acuity signifies that fine details can be discriminated, whereas poor visual acuity means that only gross features in the image can be seen.

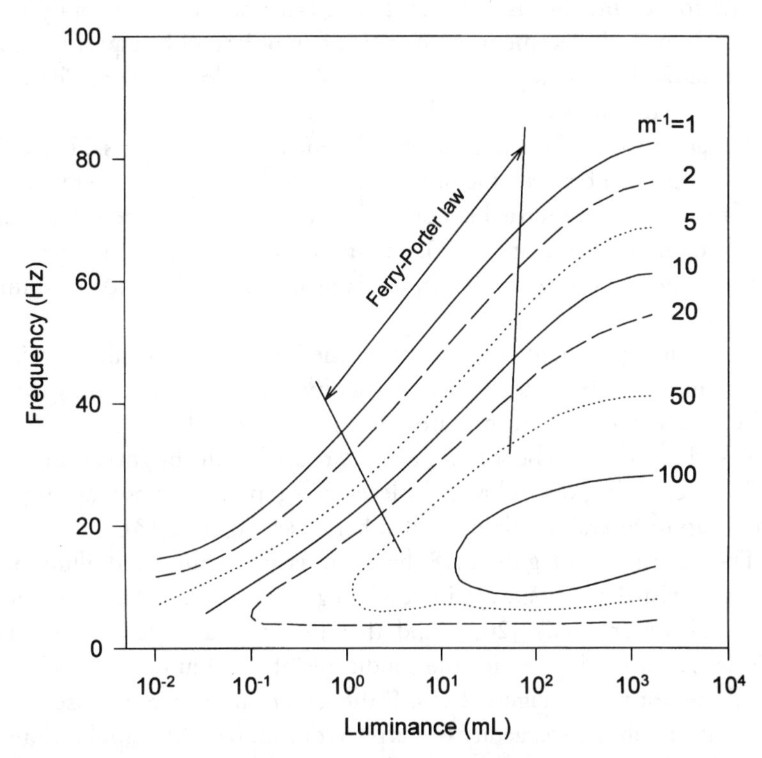

Figure 2.15 Flicker fusion frequency as a function of retinal illuminance for various modulation amplitudes (*After:* [24]).

Visual acuity usually is defined as the smallest size of an object that can be seen in terms of the visual angle it subtends at the viewer's eye. It is commonly expressed in units of arc minutes. In clinical tests, which ordinarily are used by optometrists to test vision, subjects who can correctly respond to objects that extend 1 arc-min are regarded as having "normal" visual acuity.

Visual acuity commonly is measured by one of four fundamental methods, which are categorized by the visual acuity task. The first method is the minimum visible or detection task in which the subject is required to determine whether an object is present in the visual field. In this test, a dark line or disc is presented against a bright background or a bright line or disc is presented against a dark background. The acuity in this test is determined by the smallest width of the line or disc that can be detected.

The second and perhaps the most familiar method is the recognition task. The Snellen letters and the Landolt ring, or C, tests belong to this category.

The third method is the minimum separable or resolution method, in which the observer is asked to detect a gap between two or more lines or discs or between two parts of a figure.

The fourth method is the localization task that tests the ability to discriminate a small displacement of one part of an object with respect to another part. Vernier and stereoscopic tasks are typical tests of localization. The various types of the acuity tests are shown in Figure 2.16.

The ability to resolve a target highly depends on the acuity task. Figure 2.17 shows a comparison between the minimum resolvable target for the main acuity tasks. If for a minimum separable test, such as the Landolt ring test, the smallest resolvable target is on the order of 0.5 arc-min, then in a minimum perceptible test a black line or a disc with a width or diameter smaller than 0.01 arc-min can be detected.

Visual acuity is highly dependent on the intensity of the illumination on the eye retina. The dependence of visual acuity, which is expressed by the reciprocal of the visual angle in minutes, on luminance is shown in Figure 2.18. The finest detail that can be seen is inversely related to the brightness of the display [22]. Acuity is gross in low-light levels but rapidly increases as brightness increases, up to several hundred millilamberts (see Figure 2.13).

The relation in Figure 2.18 between visual acuity and illumination implies that visual acuity always increases regardless to the adaptation level of the eyes. However, Craik [26] found that for luminance levels from 10 to 10,000 ft-L acuity is highest for the conditions of equal luminances of adaptation and the test target (Figure 2.19). If the target luminance increases beyond the adaptation luminance, acuity is sharply reduced. In HMD applications, the ambient luminance changes rapidly due to glare, which may cause a pronounced decline in visual acuity.

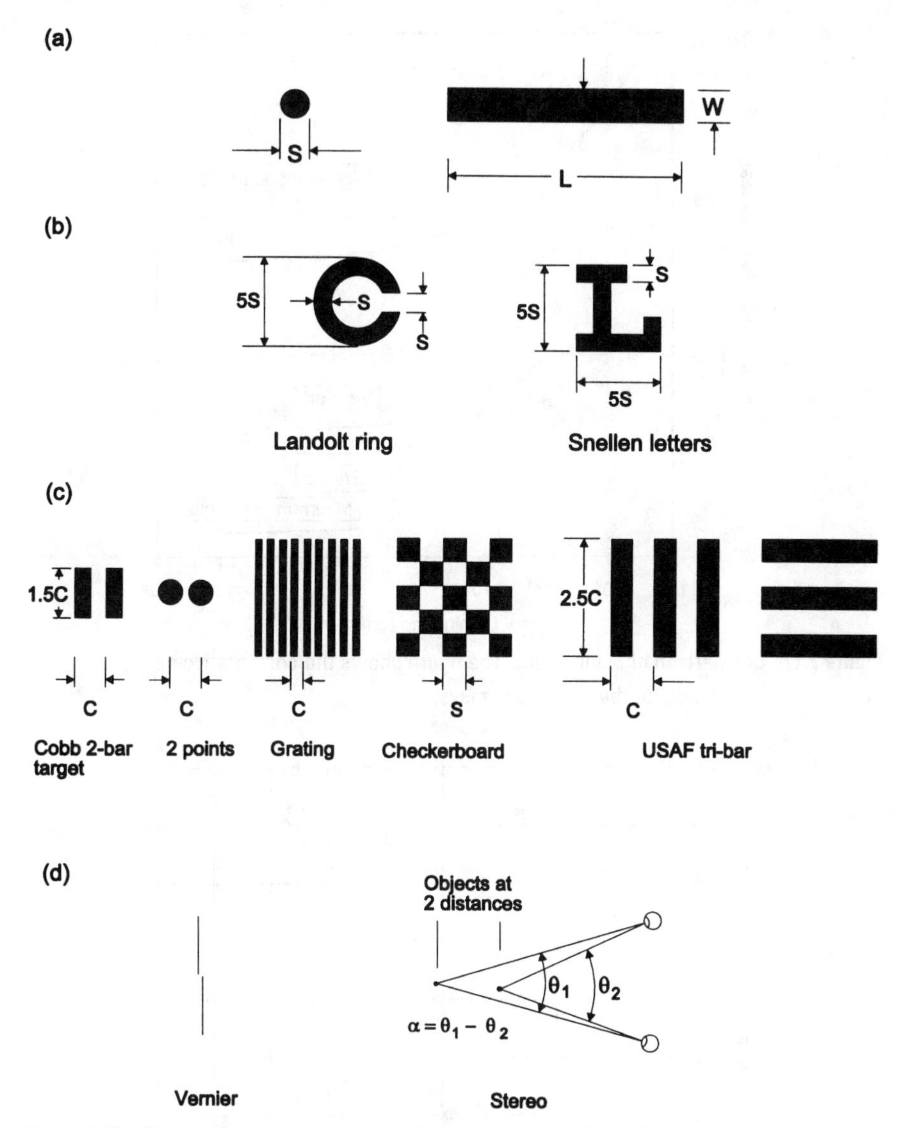

Figure 2.16 Common types of targets used to test visual acuity: (a) targets of minimum visible or detection test, (b) the common targets used in the recognition test, (c) several targets for the resolution test, and (d) techniques for the localization test.

2.2.9 Accommodation

Accommodation is the change in the shape of the lens of the eye. Because the distance from the cornea to the retina essentially is fixed, to focus images of different distances on the retina, the lens must change its power, which is done by

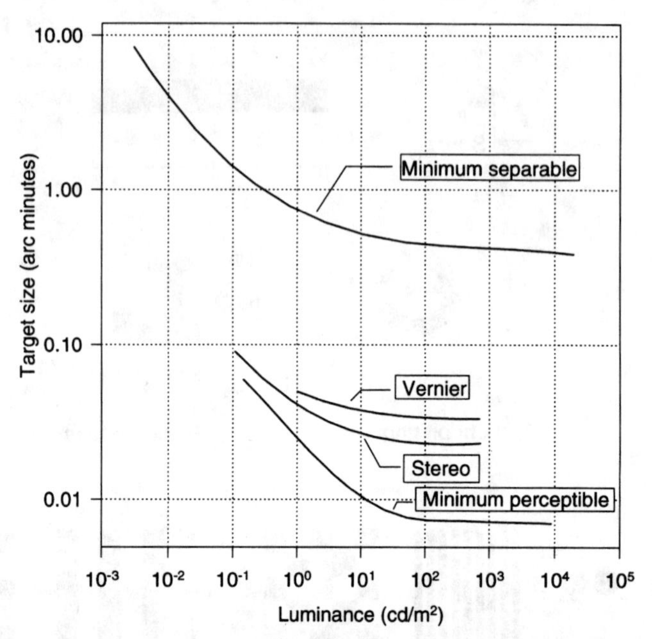

Figure 2.17 Comparison of acuity tasks. The figure shows the comparative acuity thresholds for the four acuity tasks.

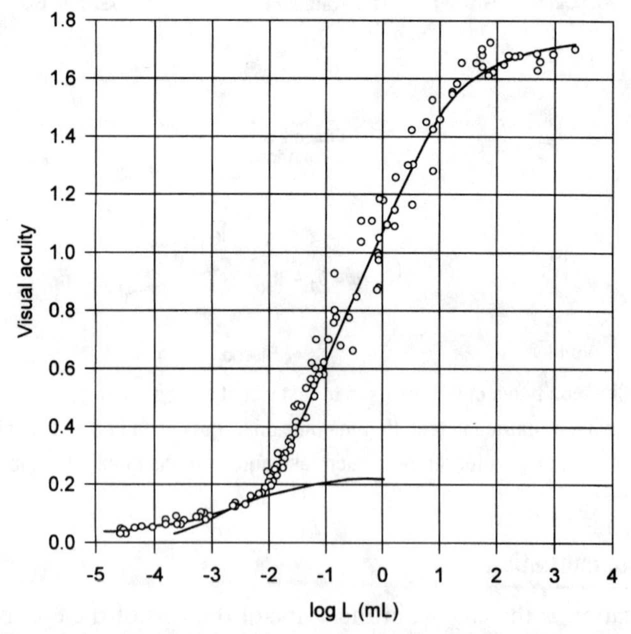

Figure 2.18 The relationship between visual acuity and illumination. The lower curve is a function for rod vision. The upper curve is a function for cone vision (*From* [18]).

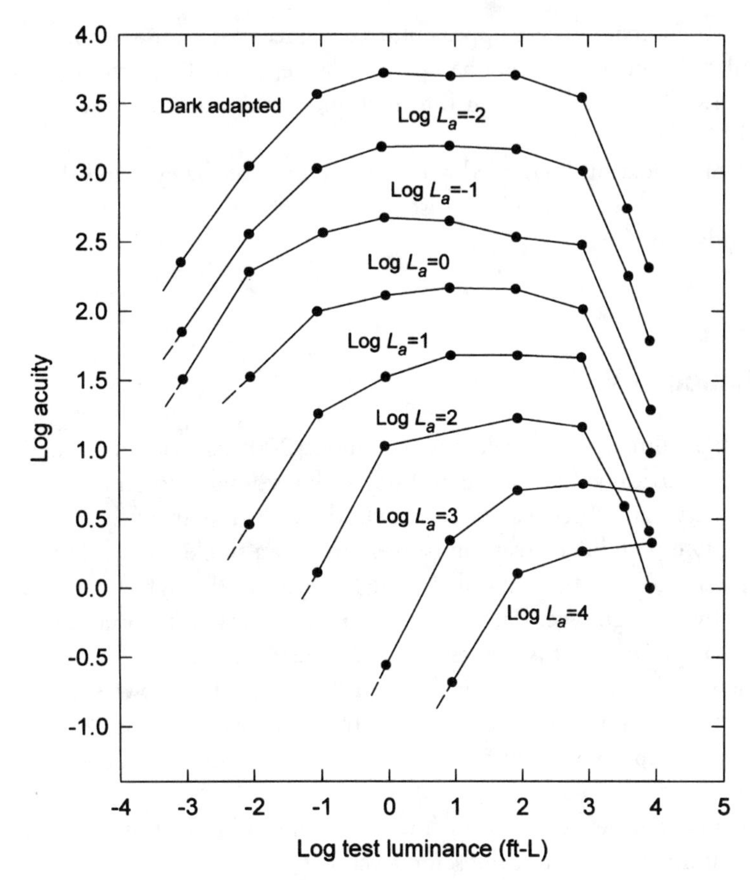

Figure 2.19 Visual acuity as functions of target luminances for various adapting luminances.

changing its curvature. The accommodation usually is expressed in diopters, which is the inverse of the distance of the object from the eyes in meters (1/m). Thus, an eye focused on an object at a distance of 0.2m from the eyes is accommodated 5 diopters. Zero-diopter accommodation is for an eye focused on an object at infinity.

In darkness or the viewing of an empty field, such as clear skies, the accommodation lapses to a resting position. The typical resting position is at distances of 0.5m to 2m.

Accommodation is both a voluntary and a reflexive process. It changes reflexively by the image blur but also is affected by a knowledge that there is an object close to the eyes. The implication of accommodation in the use of HMDs is that the wearer who is aware of the existence of the visor and the display may sometimes accommodate at the visor or display distance [27]. In such instances, the outside scenes appear blurred; hence, with a lower contrast and as a result, distant objects may be not seen or detected. Thus, the whole purpose of HMD to overlay useful information on the outside scenery may be missed [28–30].

Accommodation is coupled with the vergence of the eyes and thus tends to change with vergence [31]. Also, mental effect may cause a lapse in accommodation.

2.3 Binocular Vision

The normal human visual field extends about 200 degrees horizontally and 130 degrees vertically. Out of the 200-degree horizontal field, only the central approximately 120 degrees can be seen by both eyes simultaneously. Each of the remaining two peripheral areas can be seen only by a single eye: the left far area of the visual FOV can be seen only by the left eye and the right far area can be seen only by the right eye. Each eye transmits an image to the brain, but only one image is perceived: it is the resulting binocular vision.

The visual field of humans is shown in Figure 2.20. The two small elliptical patches are the blind spots, each of which are seen each by only one eye. The blind spots are approximately 5 degrees in diameter and occur between 13 to 18 degrees in the periphery.

In binocular vision, the eyes interact at high illumination levels. If the eyes are illuminated at about the same level, the scene is perceived as brighter; thus, there is a brightness summation in binocular fusion [32]. The opposite also may happen: when one eye is illuminated at a different level than the other, the sensation is of reduction of binocular fusion. That effect is known as the Fechner paradox. Its effect in the use of HMDs may be apparent when an occluded display is used during dark conditions: it may be seen dimmer than it really is.

Binocular vision provides stereoscopic depth perception [33]. Such perception, termed stereopsis, results from retinal disparity due to the horizontally separated eyes [34]. The locus of all images that stimulate the corresponding retinal points defines a curved line called the longitudinal horopter. The region surrounding the horopter, called Panum's fusional area, relates to the area for which the eyes have the ability to fuse disparate images. Images outside the Panum's area may be seen as diplopic or double images.

Binocular vision improves the visibility of low-contrast objects by a factor of $\sqrt{2}$ over single-eye viewing owing to a summation effect that doubles the visual signal but increases only by a factor of $\sqrt{2}$ the retinal random noise [35].

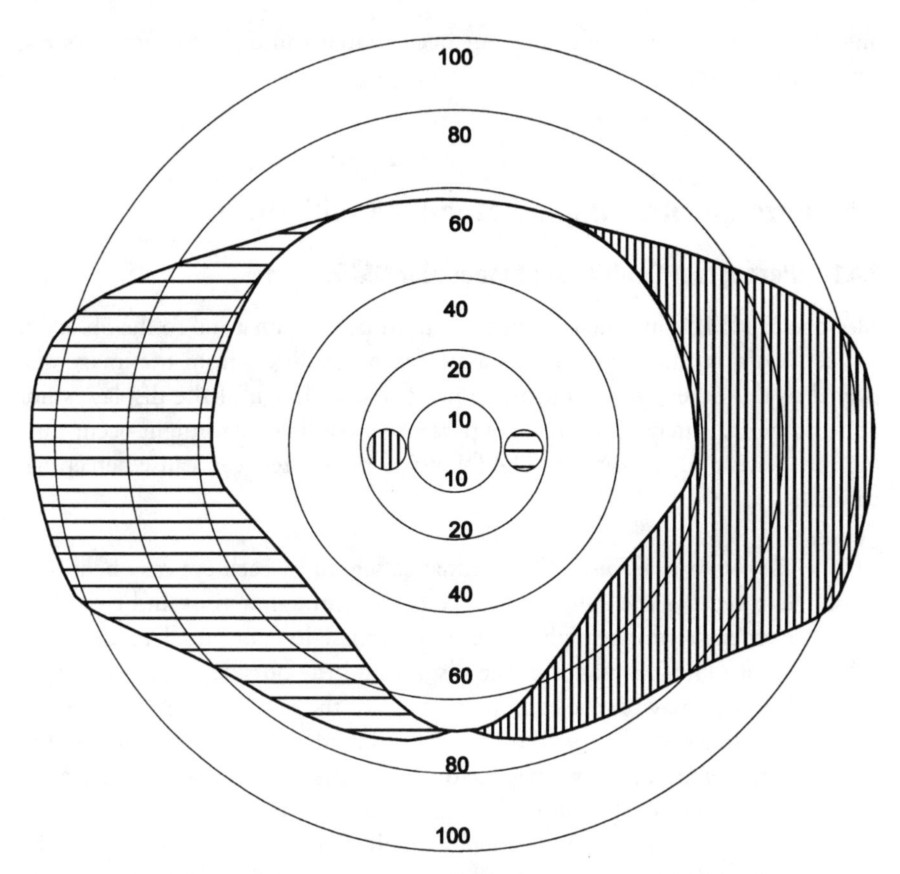

Figure 2.20 The visual field of the normal human. The vertically shaded area is the right-eye monocular visual field, while the horizontally shaded area is the left-eye monocular visual field. The white central area is the binocular visual field.

Furthermore, binocular vision increases acuity (see Figure 2.17), improves form recognition, provides stereopsis, and improves visual detection.

For good binocular vision, the eyes must operate together and both converge on the same point in the scene. Then the two images appear fused into one image and stereoscopic vision is realized. If for some reason, improper convergence is adopted or imposed, the viewer sees two separate images, a condition known as diplopia. In that case, one image often predominates and the other fades away with slow alternation between the two, eventually leading to discomfort and stress. Similar phenomena occur if the two eyes do not maintain the same vertical angular position. The difference between the vertical angles of the two eyes is called dipvergence and may cause discomfort and stress similar to diplopia.

2.4 Perceptual Conflicts Associated With HMDs

2.4.1 Perceptual Conflicts in Monocular HMDs

Monocular HMDs impose a unique situation that is unnatural to the observer. In such a display, the image presented to each eye differs from the image displayed to the other eye with some respects. If the design of the display is not done properly, it may result in some perceptual conflicts that might occur during the viewing of the display. The following parameters can cause perceptual conflicts.

- *Brightness differences.* The luminance sensed by each eye may be different because the light to one eye passes through a visor and combiner elements, which might have a transmissivity not equal to the visor alone. The luminance of the display and the outside scene is additive. One eye may see the combined image of the outside scene and the display image, while the second eye sees only the outside scene. In another arrangement, one eye may see only the outside scene, while the second eye is exposed to the display image only. The luminance of the eye exposed to the display depends on the luminance of the display, the transitivity of the optics, and the reflectivity of the combiner, whereas the outside scene luminance depends only on the visor and combiner transmissivities. In some conditions, the luminance of the outside scene is dominant, such in daylight conditions; in other conditions, such as at twilight or night conditions, the luminance of the display may be dominant.
- *Image characteristic differences.* The image of the outside scene is pictorial and usually contains large parts of homogenous background. The display image, on the other hand, contains mainly graphic symbolics and alphanumeric information.
- *Scale differences.* Due to the magnification or minification of the optical elements of the HMD, the image of the display may differ from the image of the outside world.
- *Viewing-distance differences.* The HMD image usually is fixed at a distance that varies from 0.5m to infinity, while the outside image is located at a distance that may vary from several meters to infinity.

- *Contrast differences.* The contrast of the display image that is determined by the characteristics of the display may differ from the outside world image.
- *Resolution differences.* The resolution of the displayed image is determined by the resolution of the image source and the optics of the HMD and may differ from the resolution of the image of the outside scene.
- *Color differences.* The color of the image of the display is determined by the spectral characteristics of the display and might be either monochrome or limited color, by any spectral filtering in the optical path, whereas the outside scene may contain the full color spectrum.
- *Motion differences.* The movements of the image presented on the display may be different from the image coming from the outside world.
- *Orientation differences.* In most displays, the information on the display is fixed within a frame of reference related to the helmet even for a dynamic display. The information moves with the head and changes in orientation with the head. That is in contrast to the outside world, which is fixed in space and changes orientation as the orientation of the head changes.

2.4.2 Binocular Rivalry

In HMDs, different images may be presented to each eye. One image may be the HMD image exclusively, while the other may be the outside scene. In another case, one image may be the HMD imagery superimposed on the outside world scene, while the other image may be the outside scenery only. The eyes are expected to fuse both images, regardless of the contents of each visual field. In some HMD uses, the fusion of the images may not be achieved because of binocular rivalry. In general, all the perceptual conflicts that result from the disparity between the visual fields of the two eyes are categorized under the overall title of binocular rivalry.

When corresponding areas of the retinas of the two eyes are stimulated simultaneously with images that differ in some respects, one of the following three phenomena occurs [36]:

- The two visual fields fuse.
- One visual field is seen to the occlusion of the other.
- The two visual fields alternate.

The alternation of the visual fields is called binocular rivalry or retinal rivalry. The rivalry may be seen in one of two forms:

- *Successive rivalry*, in which the alternation is complete and the two visual fields supersede one another completely;
- *Simultaneous rivalry*, in which both images may be seen at the same time but only one of the two fields is visible at any place.

Binocular rivalry is affected by many physical parameters, including: contrast, brightness, contour, continuity, illumination, interest, movement, and color of the opposing visual fields [37,38].

Binocular rivalry has been studied extensively over the years, and several theories attempted to explain it [39]. Helmholtz in 1886 suggested the attention theory, which considers a competition between two perceptions in the central processes and attention determines which one dominates at each particular time and place. Hering proposed in 1864 an alternative theory that suggests that the binocular impression is a mixture of monocular excitations that are not summative but depend on the nature of the image; thus one excitation always dominates. According to Hering, parts of the image sensed by one retina that contain contours always dominate over the respective parts of the other retina that contain textured image.

Another theory proposed by Levelt [40] suggests that binocular rivalry results from the conflict between two mechanisms: one mechanism, which he termed the monocular brightness averaging mechanism, averages the brightness of the two corresponding points on the retina. The second mechanism, the contour mechanism, ensures that areas in the vicinity of contours are kept intact.

Studies of binocular rivalry pointed out some rules regarding binocular rivalry:

1. During rivalry, the visual field with the higher contrast dominates over the visual field with the lower contrast, so the image with the higher contrast is perceived for longer time periods than the image with the lower contrast.
2. The field with greater contour density dominates over the field with more texture density.
3. The brighter field dominates over the less bright field.
4. The rate of alternation between the two fields tends to increase as the size of the items in the two fields increases.
5. The alternation is not under complete voluntary control.

6. The alternation of visual fields typically takes 1 to 4 sec.
7. Time on task reduces the alternation rate.

2.4.2.1 Troxler's Effect

Another effect attributed to binocular rivalry is observed when a small image brighter or darker than the surrounding scene is presented at the periphery of the visual field. After some time, the image fades and merges into the background [41,42]. The faded image is filled in by the brightness or the color of the surrounding field. After a while, the image may reappear and disappear again, producing the impression of periodic fluctuation of the image.

In the use of HMDs, the Troxler effect can be meaningful because, in certain conditions, if the HMD symbology is too small and cluttered, some of it may disappear and thus is not seen or is ignored by the viewer. That is especially important for symbols that are not continuously presented (such as certain indicators or hazard symbols). Therefore, the Troxler effect should be carefully regarded in the design of the display symbology.

2.4.3 Brightness Disparity

Brightness differences perceived by the two eyes can result from differences in the transmittances of the optical paths, and that disparity may affect vision. Observers viewing an occluded HMD often find it difficult to view the display when the other eye is exposed to high ambient luminance [33]. In such instances, viewers tend to close the nondisplay eye to increase the legibility of the display. Furthermore, when the two eyes are exposed to different levels of brightness, retinal rivalry may be induced. This special form of rivalry is called brightness disparity [40]. Because the monocular HMD is currently the most adopted type of display, control of the brightness differences is warranted so the viewer is capable of viewing the display without having to quit looking at the outside scenery.

2.4.4 Differential Eye Adaptation

The mechanism of light adaptation is independent for each eye [43]. Therefore, exposure to different light levels (e.g., when using a monocular HMD) causes a differential state of adaptation between the two eyes, which in turn causes a differential light sensitivity [44]. When the illumination on each eye differs greatly, or when the state of adaptation of the eyes differs significantly, an illusion called the Pulfrich effect may be experienced [45,46]. The Pulfrich effect accounts for the illusion in which an object that moves in a frontal plane is seen by the differently adapted two eyes. The effect is that the object is perceived to displace in depth.

2.4.5 Luning

Binocular HMDs are designed either with full overlap, in which both eyes view the same image, or with partial overlap, in which only the central area of the display is shown to both eyes, while the regions to either side are seen only by one eye.

Partial-overlap displays can be convergent or divergent, as shown in Figure 2.21. In a convergent display, the left monocular region can be seen only by the right eye, while the right monocular region can be seen only by the left eye. In divergent display, the opposite is the case: the left eye views the left monocular region, and the right eye views the right monocular region.

The main benefit of the partial-overlap display is the ability to achieve a wider FOV with the same resolution and optical design. However, the partial-overlap display imposes some problems. Apart from the obvious limited binocular FOV, a perceptual conflict known as luning may be experienced. Luning is a subjective darkening of a crescent shape in the monocular regions near the binocular overlap borders. The luning phenomenon is illustrated in Figure 2.22. The circular aspect of luning is due to the circular images of the HMD's images.

The dark areas of the images tend to alternate over time, which implies that luning probably is due to binocular rivalry. Luning can be reduced in three ways: (1) by reducing the luminance of the display near the edges of the binocular region, thus softening the transition between the binocular and the monocular fields; (2) by using convergent overlap image configuration; or (3) by adding contour lines separating the binocular from the monocular regions, thus emphasizing the monocular regions [47–49].

2.5 Summary

In view of the fact that vision plays such an important role in all human activities, and because we tend to rely on and have high confidence in our visual system, it is essential that the use of an HMD does not impair that vital source of information. The introduction of an HMD in the visual path, if done incautiously, may produce perceptual illusions or conflicts.

The visual perception of an HMD depends on many parameters, including resolution, FOV, type of displayed information, contrast, illuminance, spectral and temporal characteristics of the displayed image, optical structure, light transmission, and operational conditions. Some of these parameters frequently are dictated by the available technology and by engineering trade-offs. As a result, it often is difficult to predict how the relevant human factors will be

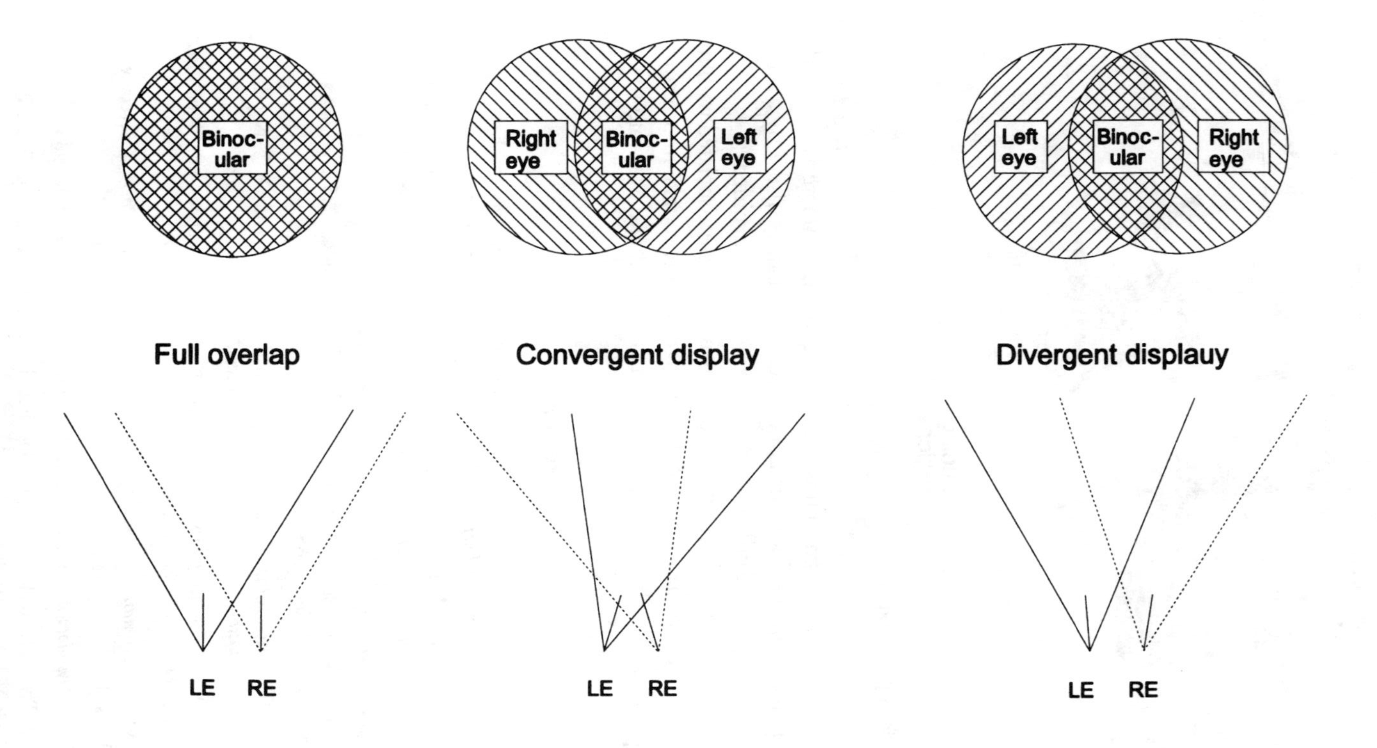

Figure 2.21 Partial-overlap and full-overlap HMD arrangements. The circles show the visual fields of each eye and the extent of the binocular overlap. Below each configuration is the angular visual field of each eye. The solid lines describe the visual field of the left eye; the dashed lines show the visual field of the right eye.

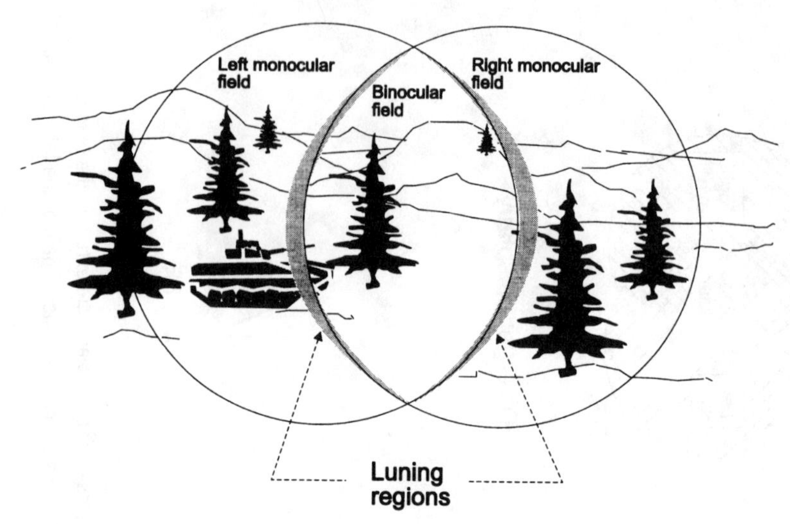

Figure 2.22 The luning phenomenon.

reflected in a particular design. However, if the basic characteristics of the human visual perception mechanisms are obeyed, one can confidently assume that most visual perception conflicts and problems will be avoided.

Understanding the complex human factors issues that are associated with the use of HMDs is the key to the success of their design. Failure to take into account human factors ultimately will lead to a design that is difficult and awkward to use.

References

[1] Clapp, R. E., "Field of View, Resolution and Brightness Parameters for Eye Limited Displays," *Imaging Sensors and Displays*, SPIE, Vol. 765, 1987, pp. 10–17.

[2] Wickens, C. D, and J. M. Flach, "Information Processing," in E. L. Wiener and D. C. Nagel, eds., *Human Factors in Aviation*, San Diego: Academic Press, 1988.

[3] Dworetzky, J. P., *Psychology*, St. Paul, MN: West Publishing, 1982.

[4] Graham, C. H., "Visual Space Perception," in C. H. Graham, ed., *Vision and Visual Perception*, New York: Wiley, 1965.

[5] Leibowitz, H. W., "The Human Senses in Flight," in E. L. Wiener and D. C. Nagel, eds., *Human Factors in Aviation*, San Diego: Academic Press, 1988.

[6] Brown, J. L., "The Structure of the Visual System," in C. H. Graham, ed., *Vision and Visual Perception*, New York: Wiley, 1965.

[7] Woodson, W. E., *Human Engineering Guide for Equipment Designers*, University of California Press, 1954.

[8] Morgan, C. T., *Physiological Psychology*, 3rd ed., New York: McGraw-Hill, 1965.

[9] Marks, W. B., W. H. Dobelle, and E. F. MacNichol, "Visual Pigments of Single Primate Cones," *Science*, Vol. 143, 1964, pp. 1181–1183.

[10] Wald, G., "Human Vision and the Spectrum," *Science*, Vol. 101, 1945, pp. 653–658.

[11] Brown, P. K., and G. Wald, "Visual Pigments in Single Rods and Cones of the Human Retina," *Science*, Vol. 144, 1964, pp. 145–151.

[12] Land, E. H., and J. J. McCann, "Lightness and Retinex Theory," *J. Optical Society of America*, Vol. 61, January 1971, pp. 1–11.

[13] Land, E. H., "The Retinex Theory of Color Vision," *Scientific American*, Vol. 237, No. 6, 1977, pp. 108–128.

[14] Smith Kinney, J. A., "Sensitivity of the Eye to Spectral Radiation at Scotopic and Mesopic Intensity Levels," *J. Optical Society of America*, Vol. 45, July 1955, pp. 507–514.

[15] Farrell, R. J., and J. M. Booth, "Design Handbook for Imagery Interpretation Equipment," Seattle: Boeing Aerospace Co., February 1984.

[16] Stiles, W. S., and B. H. Crawford, "The Luminous Efficiency of Rays Entering the Eye Pupil at Different Points," *Proceedings of the Royal Society (London)*, 112B, 1933, pp. 428–450.

[17] Alpern, M., "The Eyes and Vision," in W. G. Driscoll and W. Vaughan, eds., *Handbook of Optics*, New York: McGraw-Hill, 1978.

[18] Riggs, L. A., "Light as a Stimulus for Vision," in C. H. Graham, ed., *Vision and Visual Perception*, New York: Wiley, 1965.

[19] Brown, J. L., and C. G. Mueller, "Brightness Discrimination and Brightness Contrast," in C. H. Graham, ed., *Vision and Visual Perception*, New York: Wiley, 1965.

[20] Seeman, J., et al., "Advanced Helicopter Pilotage Visual Requirements," presented at the American Helicopter Society 48th Annual Forum, Washington, DC, June 3–5, 1992.

[21] Chisum, G. T., "Head-Coupled Display Visual Design Considerations," *Aviation, Space, and Environmental Medicine*, Vol. 46, No. 11, 1975, pp. 1373–1377.

[22] Hecht, S., "Vision II. The Nature of the Photoreceptor Process," in C. Murchinson, ed., *Handbook of General Experimental Psychology*, Worcester, MA: Clark University Press, 1934, pp. 704–828.

[23] Brown, L. B., "Flicker and Intermittent Stimulation," in C. H. Graham, ed., *Vision and Visual Perception*, New York: Wiley, 1965.

[24] Kelly, D. H. "Visual Responses to Time-Dependent Stimuli: I. Amplitude Sensitivity Measurements," *J. Optical Society of America*, Vol. 51, 1961, pp. 422–429.

[25] Riggs, L. A., "Visual Acuity," in C. H. Graham, ed., *Vision and Visual Perception*, New York: Wiley, 1965.

[26] Craik, K. J. W., "The Effect of Adaptation on Visual Acuity," *Brit. J. Psychology*, Vol. 29, 1939, pp. 252–266.

[27] Edgar, G. K., C. D. Pope, and I. Craig, "Visual Accommodation Problems With Head-Up and Helmet-Mounted Displays," in *Display Systems*, SPIE, Vol. 1988, 1993, pp. 100–107.

[28] Roscoe, S. R., "The Problem With HUDs and HMDs," *Bull. Human Factors Soc.*, Vol. 30, 1987, pp. 1–3.

[29] Iavecchia, J. H., H. P. Iavecchia, and S. R. Roscoe, "Eye Accommodation to Head-Up Virtual Images," *Human Factors*, Vol. 30, 1988, pp. 698–702.

[30] Norman, J., and S. Ehrlich, "Visual Accommodation and Virtual Image Displays: Target Detection and Recognition," *Human Factors*, Vol. 28, April 1986, pp. 135–151.

[31] Edgar, G. K., C. D. Pope, and I. Craig, "Visual Accommodation Problems With Head-Up and Helmet-Mounted Displays," in *Display Systems*, SPIE, Vol. 1988, 1993, pp. 100–107.

[32] Bartlett, N. R., "Thresholds as Dependent on Some Energy Relations and Characteristics of the Subject," in C. H. Graham, ed., *Vision and Visual Perception*, New York: Wiley, 1965.

[33] Patterson, R., et al., "Depth Perception in Stereoscopic Displays," *J. Soc. Info. Display*, Vol. 2, No. 2, 1994, pp. 105–112.

[34] Patterson, R., and W. L. Martin, "Human Stereopsis," *Human Factors*, Vol. 34, December 1992, pp. 669–692.

[35] Rogers, P. J., and M. H. Freeman, "Biocular Display Optics," in M. A. Karim, ed., *Electro Optical Displays*, New York: Marcel Dekker, 1992.

[36] Alexander, L. T., "The Influence of Figure-Ground Relationships in Binocular Rivalry," *J. Experimental Psychology*, Vol. 41, 1951, pp. 376–381.

[37] Hughes, R. L., L. R. Chason, and S. C. H. Schwank, "Psychological Considerations in the Design of Helmet-Mounted Displays and Sights," USAF Aerospace Medical Research Laboratory, Wright-Patterson AFB, Ohio, Tech. Report AMRL-TR-73-16, August 1973.

[38] Julesz, B., "Stereopsis and Binocular Rivalry of Contours," *J. Optical Society of America*, Vol. 53, 1963, pp. 994–999.

[39] Jacobs, R. S., T. J. Triggs, and J. W. Aldrich, "Helmet Mounted Display/Sight System Study," Vol. 1, USAF Flight Dynamics Laboratory, Wright-Patterson AFB, Ohio, Tech. Report No. AFFDL-TR-70-83, March 1971.

[40] Levelt, W. J. M., "The Alternation Process in Binocular Rivalry," *Brit. J. Psychology*, Vol. 57, 1966, pp. 225–238.

[41] Levelt, W. J. M., *On Binocular Rivalry*, The Hague: Mouton, 1968.

[42] Goldstein, A. G., "Retinal Rivalry and Troxler's Effect: A Correlation," *Perception and Psychophysics*, Vol. 4, No. 5, 1968, pp. 261–263.

[43] Mitchell, R. T., and Liaudansky, L. H., "Effect of Differential Adaptation of the Eyes Upon Threshold Sensitivity," *J. Optical Society of America*, Vol. 45, October 1955, pp. 831–834.

[44] Furness, T. A., "The Effects of Whole-Body Vibration on the Perception of the Helmet-Mounted Display," Ph.D. thesis, University of Southampton, 1981.

[45] Katz, M. S., and I. Schwartz, "New Observation of the Pulfrich Effect," *J. Optical Society of America*, Vol. 45, July 1955, pp. 523–524.

[46] Diamond, A. L., "Simultaneous Brightness Contrast and the Pulfrich Phenomenon," *J. Optical Society of America*, Vol. 48, 1958, pp. 887–890.

[47] Melzer, J. E., and K. Moffitt, "An Ecological Approach to Partial Binocular-Overlap," in *Large-Screen-Projection, Avionics, and Helmet-Mounted Displays*, SPIE, Vol. 1456, 1991, pp. 124–131.

[48] Klymenko, V., et al., "Convergent and Divergent Viewing Affect Luning, Visual Thresholds and Field-of-View Fragmentation in Partial Binocular Overlap Helmet Mounted Displays," in R. J. Lewandowski, W. Stephens, and L. A. Haworth, eds., *Helmet- and Head-Mounted Displays and Symbology Design Requirements*, SPIE, Vol. 2218, April 1994, pp. 82–96.

[49] Grigsby, S. S., and B. H. Tsou, "Visual Processing and Partial-Overlap Head-Mounted Displays," *J. Soc. of Info. Display*, Vol. 2, No. 2, 1994, pp. 69–74.

3

Characteristics of Head-Coupled Devices

Head-coupled devices usually are composed of a display device attached to the helmet or constructed as head-worn gear and a sensor for measuring head motions. The display itself comprises several components:

- A miniature display device that is an emissive source on which the image is presented;
- An optics assembly of relay and collimating lenses that relay the image onto a combiner, where the image is reflected as a distant image to the viewer's eye;
- An image generator such as a graphics computer or a video source from some imaging sensor.

Because the image is reflected from a partially reflecting mirror, it is perceived as superimposed on the outside world scene.

The second essential part of the system is the head position measurement system, or the head-motion tracker. It is used either to measure the LOS to a target, to measure the head orientation so as to enable the display symbology to conform with the head orientation, or to generate commands to a VCS.

A VCS employs weapon and sensors mounted on a steerable turret installed in front of the aircraft. The turret houses a variety of sensors, such as FLIR and LLLTV or day TV camera, and is used for targeting and night vision [1,2]. A typical VCS system is depicted in Figure 3.1.

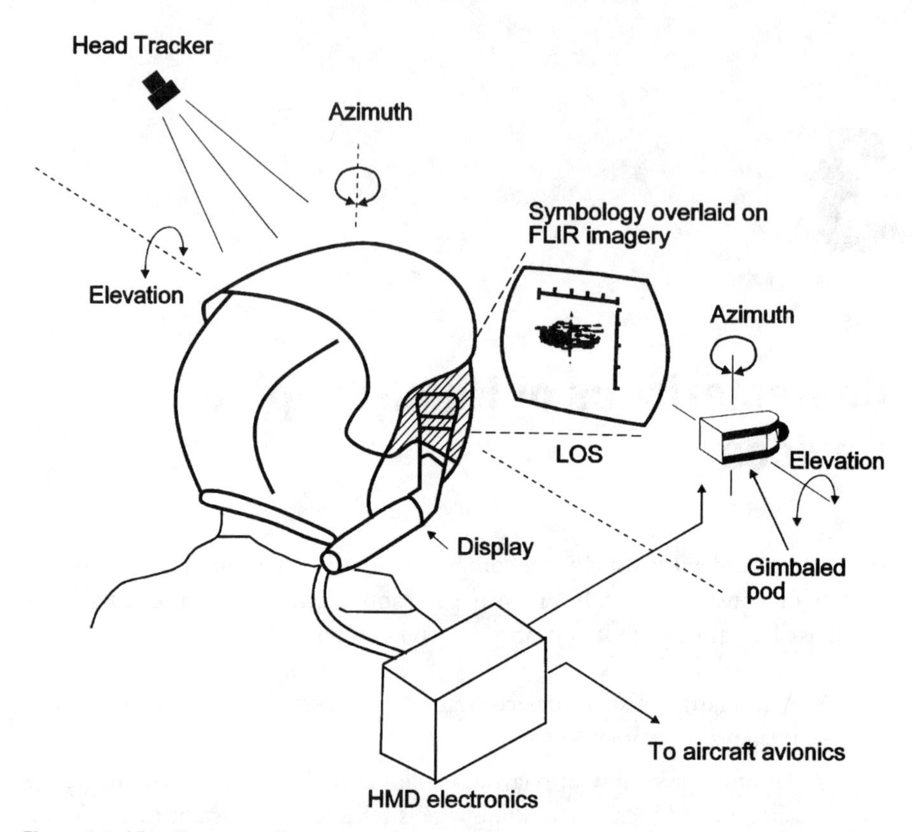

Figure 3.1 Visually coupled system.

The HMD/HMS can be an add-on system that mounts on a standard aviator's helmet or is a completely custom-designed helmet in which the display and perhaps additional night vision sensors are embedded. The HMD is arranged as one of the following configurations: monocular, biocular, and binocular. Monocular configuration is used by one eye only, while in biocular and binocular systems the image is presented to both eyes.

The image typically is projected on a semitransparent combiner, which is used to combine the display image with the outside-world scene. The combiner may be placed closely in front of the eyes or may use the helmet visor as a projection combiner. A typical display is used to present some graphical imagery or symbology overlaid on the external scene.

Some systems, mainly those used by helicopter pilots, integrate NVDs to display on the HMD the image of the night vision sensor.

The design of a head-coupled device usually is tailored for a specific use. Systems designed for fighter pilots usually are lighter in weight, more compact, and monocular and often use the helmet visor as part of the optical arrangement, acting as the image-combining element. Helicopter HMDs tend to be binocular and almost always incorporate some type of NVD [3]. They also are constructed to have a wider FOV than their fixed-wing counterparts. To achieve the wider FOV, the displays use larger combiners and are placed closer to the wearer's eyes.

3.1 Requirements: The Ideal HMD/HMS

The design of head-coupled devices, like that of all technology-intensive devices, is eventually a compromise and a trade-off in many design parameters, dictated by technology limitations, physical properties, and cost. Before a design is executed, it is important to recognize what would be the requirements for the ideal system should the technology limitations be lifted. Understanding the ultimate requirements would help in eventually stating the design goals of head-coupled devices.

The ideal HMD has a large FOV, preferably as large as the human visual FOV. The display is optimized for day and night use in terms of brightness. For night use, it incorporates an NVD capable of providing images of the outside-world scene. The NVD is integrated with the display device on the helmet. The contrast of the display is high enough to be seen against the high ambient luminance scene. The resolution of the display is sufficiently high for comfortable viewing and the symbology clearly and legibly seen against the ambient scenery in all light conditions. The display has a clear view of the outside scene with no attenuation and does not introduce any blind spots or obscurations in the user's visual field.

The ideal display is lightweight and properly balanced. The added weight is compensated for by weight reduction of the helmet, so the total weight does not exceed that of currently used helmets and may be even less. Any element of the HMD in no circumstance intrudes on the normal vision of the wearer. It does not obscure the real world or block the periphery of the visual field. Obviously, the ideal HMD retains compatibility with all other head-worn equipment such as oxygen masks, nuclear biological chemical (NBC) headgear, communications mikes, and so on. It is easy to adapt and does not require intensive training before use.

The ideal HMD system is as accurate as current HUDs, at least on axis, to permit elimination of HUDs [4]. The head-tracker lag and symbol-generation lag are not noticeable. Finally, the ideal system provides laser protection without jeopardizing outside-world transmission at other times.

3.2 Design Considerations

3.2.1 Type of Display Device

The display device is the image source on which the image is generated before it is relayed and projected to the eye. The selection of the display device usually is determined by the type of information that is to be displayed [5]. The information to be displayed may vary from a simple fixed aiming reticle, through more complex graphics and alphanumeric display, up to a full-color imagery display.

However, the type of information to be displayed is not the only factor that prescribes the selection of the display. The task conditions also have a major impact. Other factors are the illumination conditions, the FOV requirements, and the resolution. All those factors eventually reflect on the size and the weight of the optics of the display. The aircraft or vehicle type may also influence the display selection, especially if it requires provisions for quick disconnect of high-voltage supply lines.

In military and aviation applications, miniature CRT displays still dominate because of their high brightness and resolution characteristics. Common display sizes have a faceplate of 1/2 or 1 in. The older LED displays are almost no longer used, although they still have their advantages in terms of size, weight, and cost, and as such may be considered good candidates for automotive and other commercial applications that require only a simple type of symbology. For commercial applications such as virtual reality, flat-panel displays are being used, thanks to their low weight, color-displaying capabilities, and the fact that their relatively lower brightness can be tolerated. In the high-quality and high-resolution HMD-based simulators, LCD light valve–based image sources are used.

Several alternative miniature flat-panel display technologies are emerging. These include innovative technologies such as field emission display (FED), digital micromirror display (DMD), and even virtual retinal display (VRD), in which the image is scanned directly on the viewer's eye. These technologies are still in various developmental stages, but undoubtedly they will play a major role in future HMDs.

3.2.2 Field of View

The HMD's instantaneous FOV is perhaps the single most fundamental parameter. The instantaneous FOV is the angular subtense of the display, for a given head field of regard (FOR). In general, the larger the FOV, the better. However, a larger FOV requires larger optics; therefore, a trade-off must be made to achieve a sufficient FOV while keeping a reasonable size and weight of the optics [6,7].

When using conventional optics, the only way to obtain a wider FOV is either to increase the magnification factor of the image (which requires large optics) or to reduce the eye relief (i.e., bring the optics closer to the eye). The basic optical relation between FOV, optics size, and eye relief is illustrated in Figure 3.2, which shows the basic compound microscope-type viewing optics of an HMD.

The FOV requirement of the HMD is task dependent. For a simple aiming reticle, a limited FOV of about 4 to 5 degrees is sufficient. When flight and weapon-aiming information is to be displayed, a FOV of at least 20 degrees is needed. During night operations and if the display is used to show sensor imagery, peripheral vision is vital, and a much wider FOV—at least 40 degrees—is essential.

3.2.3 Resolution

Second to the FOV, the most important key parameter that characterizes HMDs is the image or display resolution. The two parameters are inversely related: the increase in FOV evidently is accompanied by a decrease in resolution [6].

The image source display has a fixed linear resolution characterized by the number of picture elements, or pixels, or by the number of lines. The geometrical resolution (R) of the display is simply the ratio between the FOV (α) to the total number of pixels or lines (N) and is given by the following relation:

$$R = \frac{\alpha}{N} \tag{3.1}$$

Equation (3.1) elucidates the basic difficulty in constructing high-resolution wide-FOV displays. Figure 3.3 gives plots of the geometrical resolution,

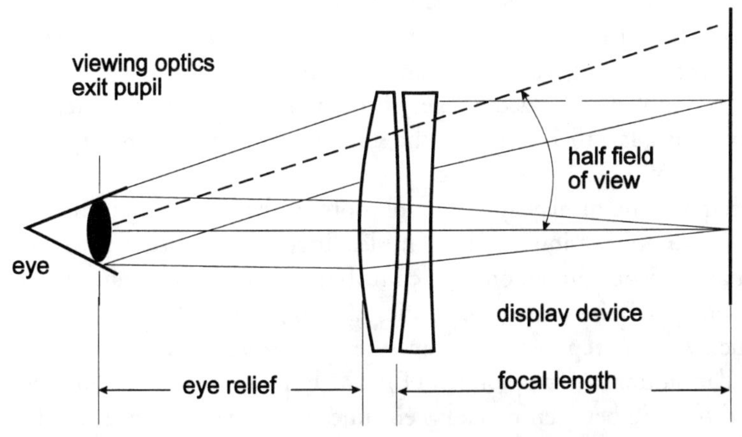

Figure 3.2 The basic compound microscope optics principle of viewing an HMD.

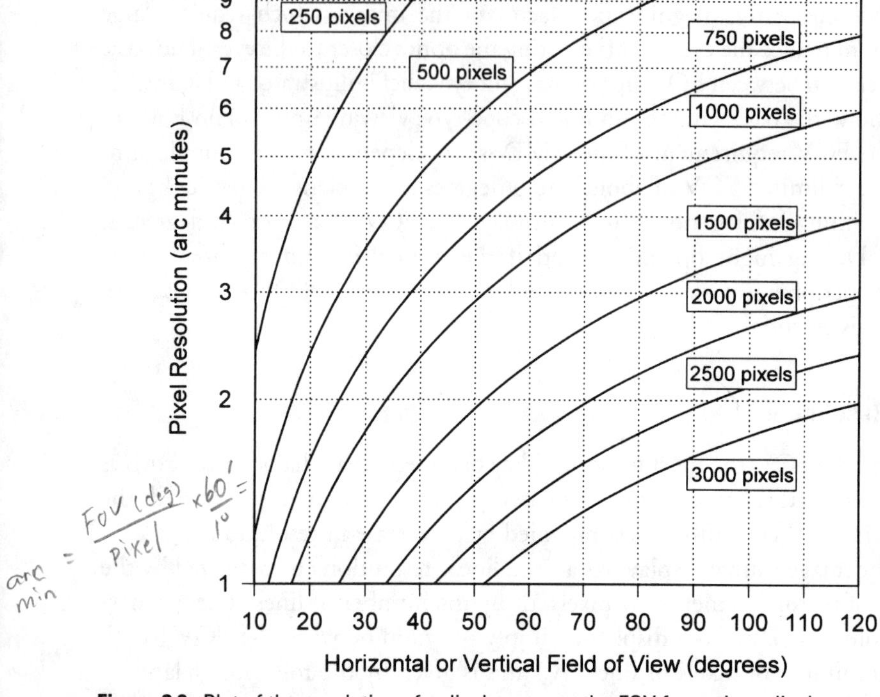

Figure 3.3 Plot of the resolution of a display versus the FOV for various display numbers of pixels.

expressed as arc minutes subtended per pixel or line, as a function of the FOV of the display for displays having a different number of pixels. From those plots, it can be deduced that to obtain graphical image comparable to a conventional HUD displaying a line width of 1 mrad ~ 3.5 arc-min and a similar FOV (about 20 degrees), a display device having more than 350 pixels is required. If the FOV is doubled and video imagery is desired, to meet the requirement for a helicopter's display, then to match the human observer visual acuity of 1 arc-min, almost 2,500 pixels are needed.

Specifying the display system resolution by the geometrical resolution is often an oversimplification and even misleading, since it describes the display source characteristics and ignores the contribution of other elements of the display and image systems such as the optics, the contrast, and even the eye visual capabilities. A more representative measure of the display system resolution is the modulation transfer function (MTF), which indicates how a modulation or contrast difference between two adjacent lines or image features is obtained at a particular spatial frequency. The spatial frequency is expressed in line pairs per millimeter or cycles per unit distance or angle. Spatial resolution is perhaps the most important factor that determines the picture quality of the display [8,9].

Indication of the limit of the visual performance of a display can be described by the modulation transfer function area (MTFA), which is the area enclosed between the display MTF and the contrast sensitivity function (see Figure 3.4). The MTFA is defined by

$$\text{MTFA} = \int_0^{u_0} [S(u) - T(u)]\,du \tag{3.2}$$

where $S(u)$ is the MTF at the spatial frequency, $T(u)$ is the contrast threshold function, and u_0 is the limiting resolution.

The MTFA indicates the range of spatial frequency in which the display modulation is higher than the contrast sensitivity function. The point where the two curves intersect is the limiting resolution, which can be seen by the viewer for a specific display and at a specified luminance (Figure 3.4).

The overall MTF of a display is determined by the MTF of all individual components of the display system and therefore is the product of the MTFs of

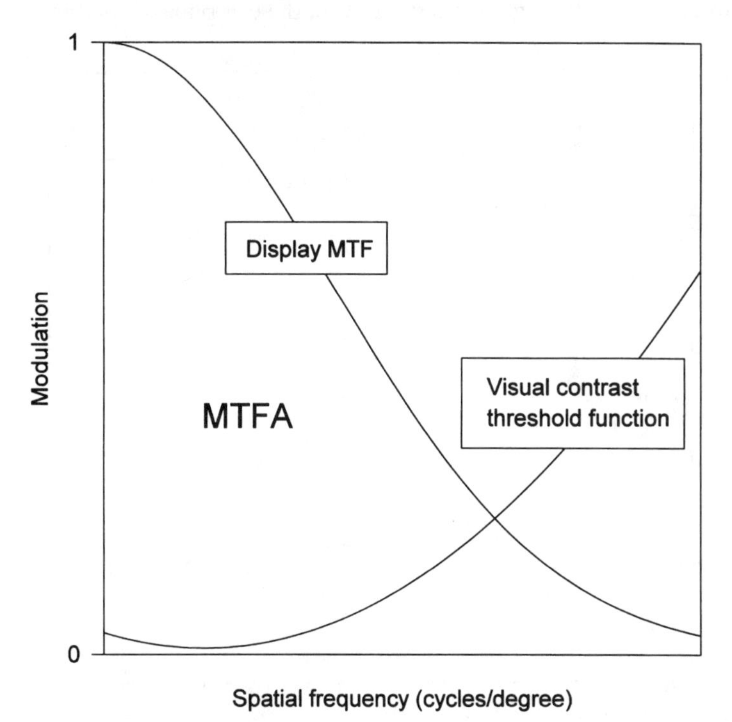

Figure 3.4 The concept of the MFTA (the area enclosed between the two curves). The upper curve indicates the display total MTF and the lower curve the contrast sensitivity curve.

all the components comprising the display, such as the image source, the relay optics, and the combiner, as follows:

$$MTF_{display\ system} = MTF_{image\ source} \cdot MTF_{relay\ optics} \cdot MTF_{combiner} \quad (3.3)$$

Some of the individual MTF components are themselves products of MTFs of subelements. For example, the MTF of the image source is the product of the MTFs of the screen phosphor, the deflection and focusing electronics, the face plate, and so on. That implies that a poor choice of a single component can reduce drastically the quality of the display. Figure 3.5 depicts typical MTF curves of an HMD.

3.2.4 Exit Pupil

The exit pupil of any optical device is the volume in space in which users must place their eyes to utilize the full available FOV. The exit pupil must be larger than the eye pupil, which varies in diameter from about 2 to 5 mm in bright light conditions to 10 mm in dim light conditions. Also, the exit pupil of the HMD must

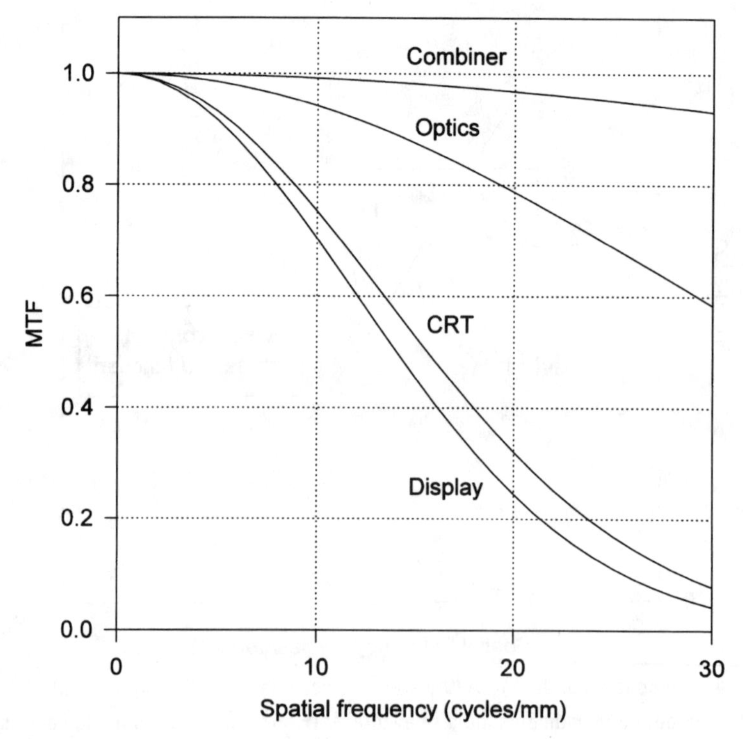

Figure 3.5 Typical MTF curves of an HMD system.

account for eye motions as well as for helmet slippage, which often occurs when helmets are used. From that aspect, the larger the exit pupil the better. However, there is a practical limit as to the extent that the exit pupil may be increased. When a display with too large an exit pupil moves relative to the viewer's instantaneous eye position, principal rays from the relay optics are incoming at different angles, which causes the image perceived as moving in a wavy motion.

In principal, an increase of the exit pupil can be achieved by either decreasing the eye relief or employing a larger optics aperture.

3.2.5 Eye Relief

Eye relief is the distance between the nearest optical element to the eye. Eye relief is an important design parameter, because the smaller the eye relief is, the larger the FOV that is attainable. Larger eye relief requires, for given angular FOV, larger viewing optics. Figure 3.6 shows the relation between the viewing optics diameter (using the optical configuration in Figure 3.2) and the FOV for various values of

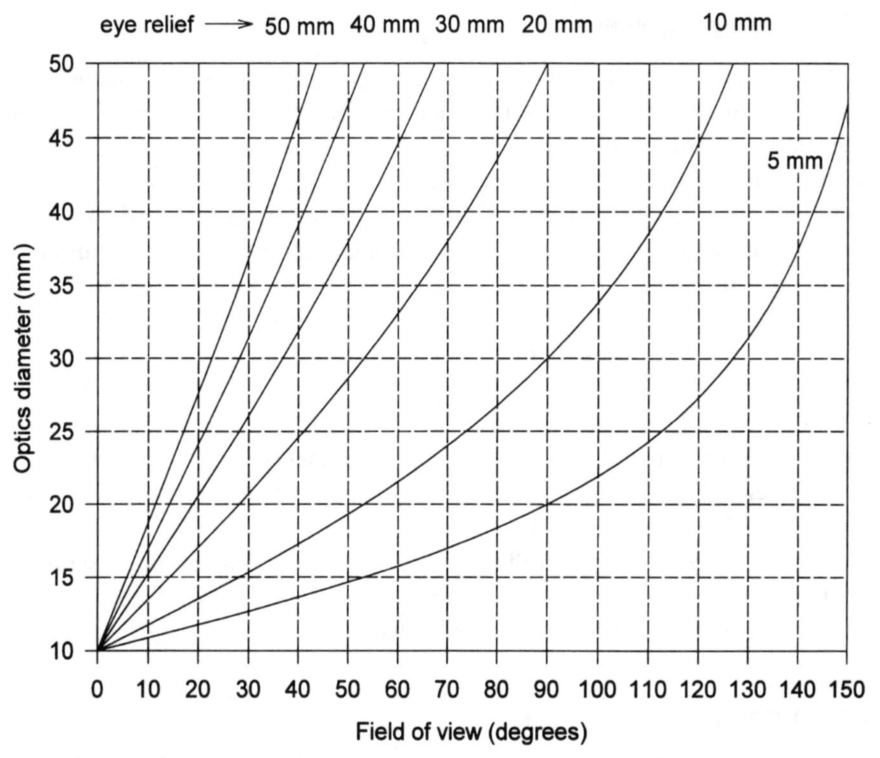

Figure 3.6 The relation between the needed viewing optics diameter and the FOV for various values of eye relief.

eye relief. For typical eye relief of 25 mm, if the FOV is to be increased from 40 to 60 degrees, the viewing optics size will increase by almost 10 mm. Obviously, larger optics means higher weight of the display system and is accompanied by degraded optical performance as more of the outer periphery of the optics is utilized. HMDs with very short eye relief are inconvenient and uncomfortable to use. Wearers of eye glasses require a minimum eye relief of 25 mm to see the entire display. In fighter aircraft, in which the pilot is exposed to the highest risk of injury during crashes or ejection, the helmet visor often is used as the image combiner, thus providing complete face protection for the wearer of the HMD.

3.2.6 Design Eye Point

Design eye position is the point in space at which the eye is considered to be positioned. The line connecting the eye point and the sight reticle center defines the pilot's line of sight. The optical design is executed to position the center of the exit pupil at the design eye point.

3.2.7 Display Brightness

Display brightness needs to be sufficient for good image visibility against the real world in the environment illumination conditions. Displays with poor brightness are prone to display an image that is weak with poor contrast or even washed out in bright illumination conditions. The brightness of the image depends on the image source luminosity, the optics efficiency, and the combiner characteristics. Because the image source has a finite amount of luminous energy distributed across the FOV of the display, the image brightness for a given display source decreases as the FOV becomes wider.

The range of luminance of an HMD must accommodate for conditions varying from background illumination of 10,000 fc (foot-candles), when looking at a cloud illuminated by direct sunlight at an altitude of 10,000 ft, to 0.02 fc during complete darkness. Meeting the highest levels of luminance currently is beyond technology capabilities, so sun visors having low transmittance are used to block the excess ambient light.

Fortunately, during high-illuminance conditions, only graphics symbology usually is needed. This type of symbology requires only two shades of gray, thus a better-contrast ratio may be achievable.

3.2.8 Display Contrast

When viewing an HMD, the viewer sees the combined image of the display sources with the outside scene (see Figure 3.7). The HMD image contrast is the

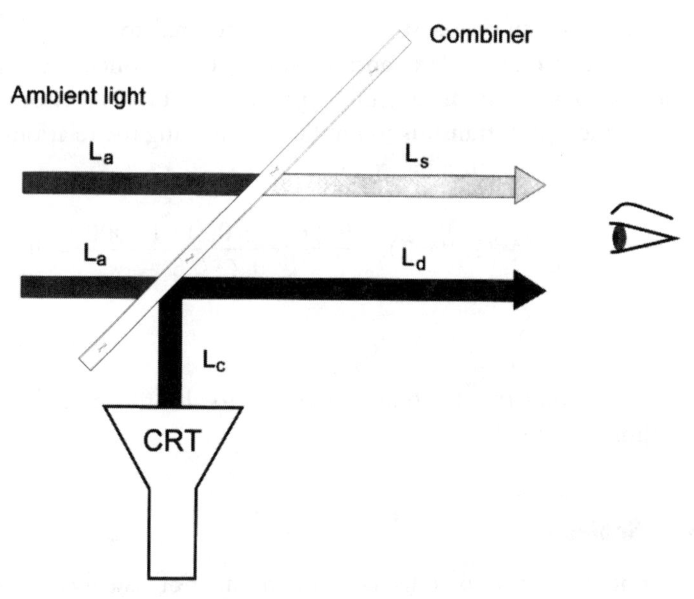

Figure 3.7 Fusion of the display image and the apparent-world image by the combiner. The display image contrast is the ratio between the luminances of the displayed image and the apparent world scene.

ratio of the luminance of the virtual display (L_d) to the apparent real world luminance (L_s):

$$C_d = \frac{L_d}{L_s} \tag{3.4}$$

The luminance of the virtual image is the sum of the luminance of the reflected image from the combiner and the outside-scene illuminance transmitted through the combiner:

$$L_d = r_c t_o L_c + L_s \tag{3.5}$$

where L_c is the image source luminance, t_o is the transmittance of the optics, and r_c is the reflectance of the combiner. The apparent world brightness is

$$L_s = t_v t_c L_a \tag{3.6}$$

where t_v is the transmittance of the visor, t_c is the transmittance of the combiner, and L_a is the ambient luminance.

An HMD operating in a stroke mode is required to be viewed against 10,000 ft-L ambient light with a contrast ratio of 1.2:1, which is adequate for comfortable viewing. Assuming a combiner reflectivity of 20%, visor transmittance of 90%, and optics transmission of 80%, and using the relations in (3.4) to (3.6), the image source luminance must be

$$L_c = (C_d - 1)\frac{t_v t_c L_a}{r_c t_o} = (1.2 - 1) \cdot \frac{0.9 \cdot (1 - 0.2) \cdot 10{,}000}{0.2 \cdot 0.8} = 9{,}000 \text{ ft-L} \tag{3.7}$$

It is assumed here that the combiner losses due to absorption and scatter are negligible; thus, $t_c + r_c = 1$.

3.2.9 Gray Scales

The number of gray shades of a display is the number of visually distinct luminous steps from black to white that the display can reproduce. Each gray scale step is defined as $\sqrt{2}$ times brighter than its predecessor [10]. The number of the gray shades obviously depends on the luminance range of the display, which in turn is characterized by the modulation contrast of the display. The modulation contrast, or the Michelson contrast, is defined as

$$m = \frac{L_{max} - L_{min}}{L_{max} + L_{min}} \tag{3.8}$$

where L_{max} is the maximum luminance of the display, and L_{min} is the minimum luminance. In displays for which $L_{max} \rightarrow L_{min}$, m $\rightarrow$ 0, the modulation contrast is poor; in displays in which $L_{max} \gg L_{min}$, m $\rightarrow$ 1, the modulation contrast is high. The modulation contrast is a common figure of merit used to quantify the image quality of the display [11].

The number of gray shades in CRT displays is related to the modulation contrast (m), using the above definition and according to the following formula:

$$\begin{aligned} G &= 1 + \frac{\log(L_{max}/L_{min})}{\log\sqrt{2}} \\ &= 1 + \frac{\log[(1 + m)/(1 - m)]}{\log\sqrt{2}} \end{aligned} \tag{3.9}$$

Figure 3.8 shows the number of gray shades as function of the modulation contrast.

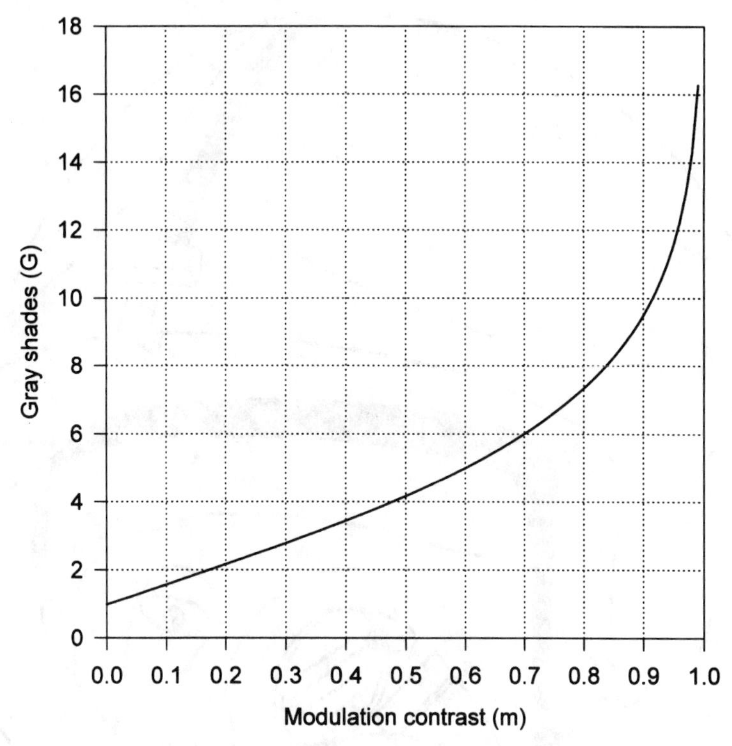

Figure 3.8 The relation between gray shades and the modulation contrast (*m*).

3.2.10 Monocular, Biocular, and Binocular Presentations

HMDs may be constructed in one of three forms: (1) monocular, in which the display is viewed only by a single eye (left or right); (2) biocular, in which the same image is presented for both eyes; and (3) binocular, in which two distinct images (generally generated by two separated image sources) are presented independently to each eye.

Monocular display has a single image source and a single set of optics and present the image to one eye. It has the advantage of low weight and compact volume. It typically is used for presentation of symbology similar to the symbology used in HUDs, although in older systems, such as the Honeywell IHADSS, it has been used to display imagery of thermal-imaging sensors mounted on the nose of AH-64 Apache attack helicopter (Figure 3.9). The disadvantages of monocular displays are the potential introduction of perceptual problems such as binocular rivalry, the lack of stereopsis, and discomfort due to the brightness differences seen by the two eyes [12].

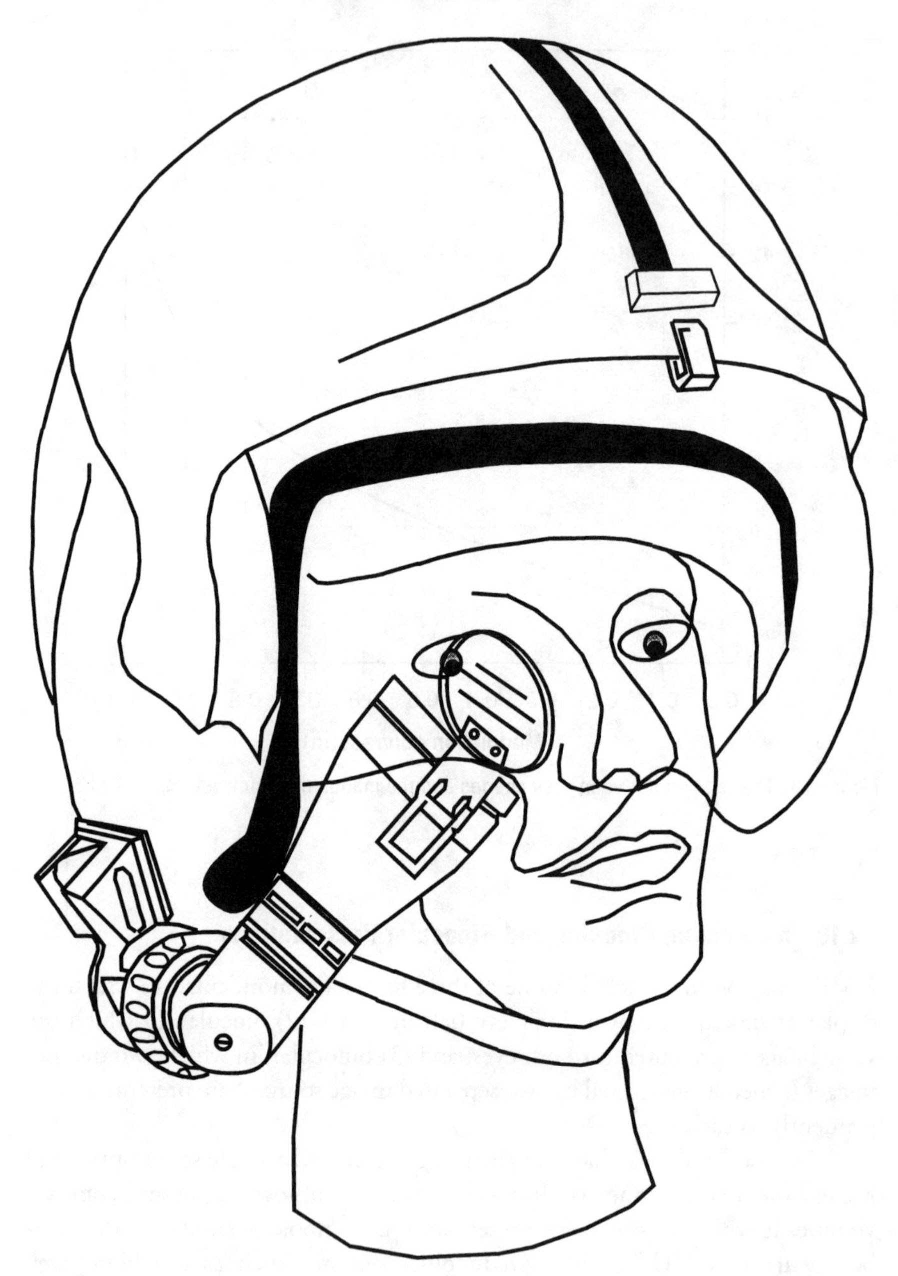

Figure 3.9 The IHADSS used in the AH-64 Apache helicopter is an example of a monocular display.

Biocular and binocular displays present images to both eyes. Biocular display uses one image source and either a single set or double sets of optics and thus have larger weight and volume than monocular systems [8,13,14].

Binocular display employs two image sources and two sets of optics and thus has large weight and volume. There are many advantages to binocular displays. Beyond their capability to provide stereoscopic cues and depth perception [15–17], they can be used to extend the FOV by presenting partially overlapped images [18]. Binocular displays also present a brighter image to the viewer.

For cases in which the image source is an NVD, binocular display also can, at least theoretically, increase the image quality. In NVDs, the display presents an image of the scene that is corrupted by a noise with the nature of scintillations. The noise is temporarily changing and can be regarded as statistically uncorrelated for the two images. The summation of the images by the two eyes tends to cancel out or reduce the noise.

Another advantage of binocular display is added reliability because of the redundancy inherent in the use of two independent displays: if one fails, the other can still function [19]. The main disadvantages, though, are in larger weight and volume, higher cost, added complexity, and the need for extra alignments and mounting precision.

The three types of displays are shown schematically in Figure 3.10.

3.2.11 Real-World Transmission and Obscuration

An HMD can be constructed as having either an opaque or a semitransparent combiner. The opaque combiner arrangement enables a higher contrast of the image, because it completely blocks the high outside ambient light in the area of the display [12]. Therefore, more gray levels can be obtained for an image source with lower luminous. Obviously, this arrangement can be implemented only with monocular displays. Biocular and binocular displays must have some see-through capability to allow vision of the surroundings. When occluded displays are used, it must be ensured that they will cause no perceptual problems, such as binocular rivalry or brightness disparity.

Besides the combiner, the mounting structure and the relay housing may obscure some peripheral or direct vision, which might be vital to the activities and spatial awareness of the user. Such obstructions should be avoided or at least minimized.

In see-through displays, two types of visors generally are used. For high ambient light conditions, visors with typical transmittance of about 30% are used; for low-light conditions, visors with 80% to 95% are common.

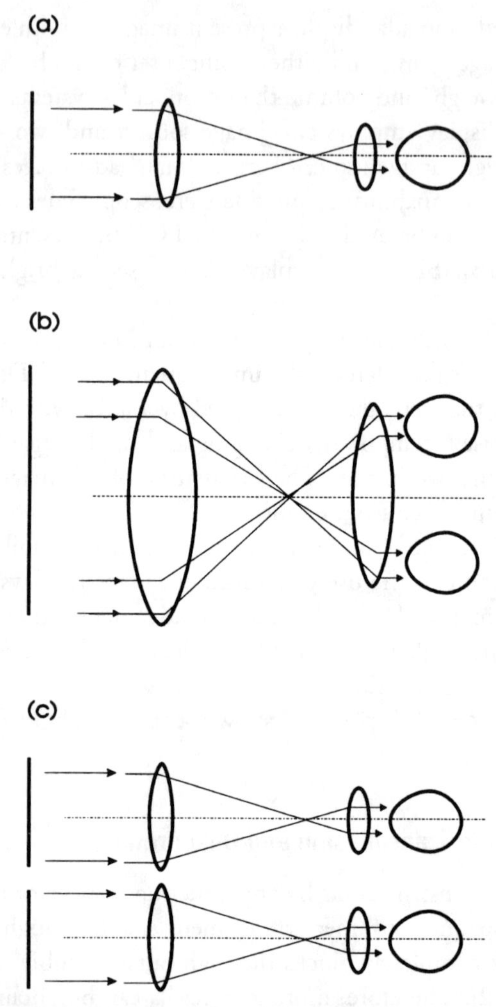

Figure 3.10 Schematic drawings of (a) monocular, (b) biocular, and (c) binocular display systems.

3.2.12 Helmet Size, Weight, and Balance; Head-Supported Weight

The addition of an HMD obviously increases the head-supported weight. The weight increase is reflected in several ways: the associated increase in the head inertia results with lesser capability for fast head motions, thus impairing the pilot's agility during combat missions. Prolonged use of heavy headgear results in fatigue of the neck muscles, which in turn cause sluggish head movements and longer reaction time. Eventually, the slowing of the reaction time may contribute to decreased maneuvering accuracy and pilot performance degradation. This becomes dangerous in operational conditions.

The effect of the addition of head-supported mass is aggravated during high-g maneuvers, in which the mass is multiplied by the g factor. The muscles of the head may no longer be able to endure the large forces, and the pilot may face the risk of injury. Similarly, during aircraft crash the risk even becomes higher, due to the large forces acting on the head.

If the head-supported weight is an important parameter in the design of HMDs, the location of the center of gravity is even more critical. Whereas the higher weight of the helmet can be endured in cases where the wearer is not subjected to high-acceleration conditions, helmet unbalance cannot be tolerated at all. Furthermore, the location of the display is dictated by the obvious need to relay the image to the eyes, so the display must be mounted either on top of the helmet or on its side. This necessarily shifts the center of gravity of the helmet by a large amount.

The task of the designer is to build a helmet-mounted system with the best optical properties in terms of FOV, resolution, eye relief, exit pupil, and so on, without compromising too much the center of gravity location and achieving that almost impossible task without adding any dead-weight counterbalance.

In modern HMD systems, the helmet shape often is modified to reduce excessive weight, to enable embedding the display, the NVD, and the head tracker in the helmet's shell. That results in systems that are more compact and, consequently, less heavy and better balanced. Although optimal integration of the HMD components may require several different helmet sizes, most manufacturers offer two or three standard helmet sizes, to avoid significant cost increase.

3.2.13 Helmet Comfort and Fit

In general, an aviator's helmet is fitted to the individual by molding the liner of the helmet to the exact shape of the user's head. That provides maximum comfort during all operational conditions and ensures that the helmet will not move excessively during impact, high-g maneuvers, or aircraft vibrations.

The HMD introduces another problem that must be considered in the fitting process: the helmet with the display mounted on it must be fitted so the design eye point of the optical system and the user's LOS are aligned. Wrong fitting may impair either the FOV or the exit pupil size. Another aspect that might affect the comfort of wearing an HMD is the thermal comfort of the helmet. As integration of the displays and the NVDs within the helmet becomes more common, attention should be paid to the heat dissipated in the devices, which are close to the head. Proper air ventilation and good thermal insulation definitely are needed.

3.2.14 Accuracy, Rate, and Resolution of Head-Position Measurement

A key element in HMDs is the helmet tracking sensor. Its two main functions are to implement the sight part of the system (i.e., measure the pilot's LOS to the target) and to measure the head orientation for correct alignment of the display imagery according to the orientation of the head. Typical head-position tracker measures the six degrees of freedom (6-DOF) of the head, that is, the three angular orientations (roll, pitch, and yaw) and the three-dimensional linear positions (x, y, z). Some head trackers measure only the three-dimensional angular orientation, while some measure only the LOS direction, namely, elevation and azimuth alone.

Most head trackers are implemented using one of the following key technologies: magnetic, electro-optical, and ultrasonic. The three technologies enable noncontact measurements. Older systems employed mechanical arrangements for head tracking and required mechanical coupling to the helmet.

Helmet-position tracking sensors are characterized by the following key measures: range of measurement, accuracy and resolution, responsiveness, and robustness.

3.2.14.1 Range of Measurement

Range of measurement defines the angular range within which measurements of the head position can be obtained. It also specifies the velocities and accelerations for which the system can maintain tracking.

3.2.14.2 Accuracy and Resolution

Accuracy is the range within which the measured position of the helmet is correct. Accuracy requirements range from 2 mrad, to comply with the accuracy of a HUD and thus enable weapon aiming, to 10 mrad, which is considered sufficient for launching air-to-air missiles.

Resolution is the smallest change in the angular linear position of the head that can be detected by the sensor. The resolution in most head-tracking sensors is better than the accuracy.

3.2.14.3 Responsiveness

The responsiveness of a helmet tracker is determined by the sample rate of the sensor, its data rate, the update rate, and the overall lag.

Sample rate is the rate at which the sensors that constitute the helmet tracking system are sampled. Data rate refers to the rate at which the position of the helmet is computed using the sensors' sampled data, while update rate refers to the rate at which the data is transmitted to the head-coupled device system. All these contribute to the lag of the system, which is the time delay, or latency, between the actual movement of the head and the new position value reported to the system.

3.2.14.4 Robustness

Robustness refers to the immunity of the tracking system to external interferences. Most helmet trackers are sensitive to some interference source, which degrades their accuracy. For example, alternating current (ac) magnetic systems are highly susceptible to the presence of nearby metal objects, and ultrasonic systems are sensitive to the ambient pressure and temperature. Robust systems are devices with low sensitivity to possible disturbances.

3.2.15 Ejection Safety Considerations

The prime role of the helmet is to protect the pilot from the effects of windblast and impact during ejection. The helmet must exhibit a high degree of stiffness to spread concentrated impact loads and pressures and prevent parts of the helmet from tearing apart or becoming deformed in a way that would hurt the pilot. The attachment of the HMD to the helmet and the structure of the HMD parts should be such that the pilot is protected. The dangers of large helmet deformations often lead HMD designers to avoid any optical elements in front of the pilot's eye and instead use the visor as the only optical element that is placed in front of the face of the pilot.

3.2.16 Head Motion Box

The head motion box is the rectangular box defined in space at given dimensions and usually is related to the head pivot point. It represents the limit at which the head-tracking system is capable of measuring the head position and the angular orientation with a specified accuracy. Out of the motion box, the accuracy of the head tracker degrades beyond the specified accuracy, but head tracking does not fail.

In most cases, the head motion box depends on the particular installation of the tracker in the aircraft. Figure 3.11 shows the head motion box for the Honeywell IHADSS system.

3.2.17 Electro-Optical Weapon Protection

In the near future, pilots of modern aircraft may have to face nonconventional weapons that inflict damage to the eye vision. That may vary from temporary visual incapacity to permanent eye damage. The HMD must provide protection against such threats and certainly not aggravate their effects by

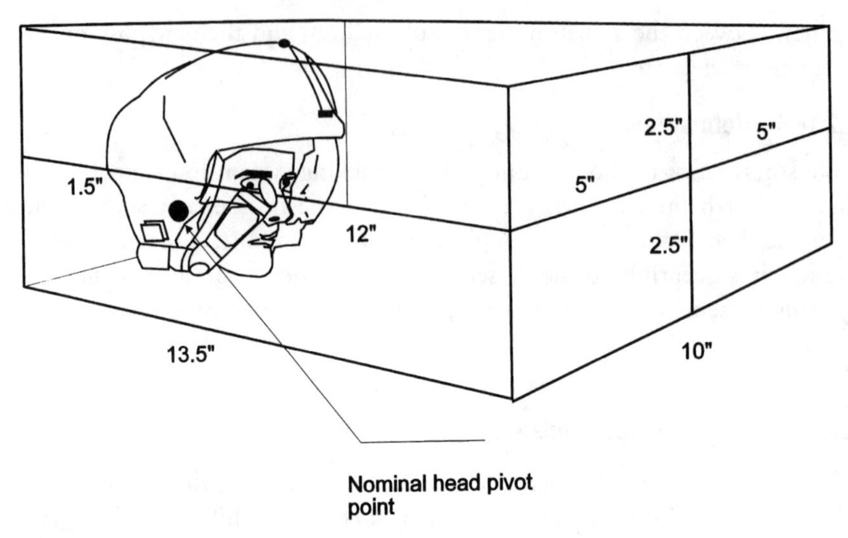

Figure 3.11 The head motion box for the IHADSS system.

having optical elements that might collect and concentrate damaging radiation at the pilot's eyes. A possible approach is to have a special visor with narrow band filters at the appropriate wavelengths. Another approach would be to use some fast-response threat-detection sensor that will activate a shielding device.

3.2.18 Installation Constraints

The HMD/HMS system eventually has be installed in the cockpit of the aircraft. Without exception, the space in a cockpit is confined and always poses stringent constraints on the location of the various elements of the system. The main problem is to accommodate the head-position tracker sensor. The head motion box, which defines the volume in which head-position measurements can be obtained, is certainly dependent on the relative locations between the aircraft-installed and the head-mounted parts of the HMS. Lack of sufficient room or poor location of the sensor highly constrains the usability of the head tracker. Electromagnetic sensors, due to their sensitivity to metal objects and electromagnetic radiation, must be located at particular locations. The allowed distance from the radiator to the detector also limits possible locations.

Electro-optical head trackers require direct LOS between the emitters and the detectors and sufficiently large FOV of the helmet; otherwise, loss of tracking would occur for large head angles.

3.3 Testing and Performance Evaluation of HMD/HMS Systems

Before any head-coupled device system can be flown in an aircraft and certainly before entering active service, it must be thoroughly tested for safety and performance. Testing and performance evaluation can be divided into four main categories: (1) safety of flight in terms of direct risk to the pilot during typical mission conditions; (2) safety during emergencies, such as ejection or crash; (3) comfort and integrity with other pilot equipment; and (4) system functional performance.

3.3.1 Safety of Flight

The weight, center of gravity, and moments of inertia of the complete system must be measured to ensure that the aviator is not exposed to potential risk during high-g maneuvers. The measurements must be validated further in centrifuge testing to include sustained, gradual, and rapid changes in acceleration. Other safety parameters include explosive atmosphere tests to guarantee the safety of the system operation in the cockpit environment without causing an explosion [20] and electromagnetic interference tests to ensure that operation of the system does not interfere with the operation of other aircraft instruments and is immune to their interferences as well [21].

3.3.2 Emergency Safety

Ejection testing is a complex process that involves several stages. Windblast testing that simulates the airstream effects on the HMD after canopy jettison are used to predict head and neck forces during the initial stage of ejection. The next stage of ejection simulation is performed by vertical-drop tower testing and is followed by parachute-deployment tests. In all the tests, the forces acting on the pilot's head are measured, the structural integrity of the HMD and possible contact of any parts of the display system with the face of the wearer are determined. Also, no interference with the parachute harness is verified.

3.3.3 Comfort and Integrity With Pilot Equipment

Testing is performed to verify that the use of the HMD does not interfere with the aviator's ability to perform any mission-essential task. Tests verify the self-ability of the pilot to put on or take off the helmet comfortably and with reasonable speed before and after the mission. Further testing ensures compatibility of the HMD with other personal life support and mission-essential equipment, such as life vest, oxygen mask, communication microphone, and spectacles. Other tests determine sound attenuation by the helmet and speech intelligibility.

3.3.4 System Functional Performance

All the previously mentioned tests are used to ensure that the introduction of a head-coupled device does not adversely affect the existing capabilities of the pilot. System functional performance tests assess the performance and the added value of the device. These tests check the quality of the display and displayed image in terms of FOV, resolution, image uniformity, brightness, optical alignments, and exit pupil; identify potential perceptual conflicts; and confirm the accuracy of the head-tracking system, display, and head tracker lags, and so on. All these parameters eventually determine the benefits and drawbacks of using the device.

3.4 Integrated HMDs Versus Modular Systems

In early versions of HMDs, the display and the sight were constructed with the intent of being mounted on a standard pilot helmet. That design obviously was nonoptimal in terms of weight, since its weight directly added to the helmet's weight and for the same reason had a large volume. As long as the HMD/HMS system comprised only a monocular display and a small and lightweight helmet tracking sensor, that was acceptable. However, as more electro-optical devices, such as image intensifiers, were added to the helmet and binocular displays, simply mounting them on the helmet made the helmet bulky and awkward.

The solution for restoring compactness and low weight was to integrate the added devices, as well as the display itself, within the helmet [22–24]. The integration is performed by designing custom helmets with the display and NVDs embedded in the helmet. However, the newly designed systems still weighed more than desired.

It was suggested that lightness could be achieved by constructing the helmet-coupled device as a modular system. In that approach, some components of the helmet were removed and reinstalled according to the specific need. For instance, a dark visor was used during high-illumination conditions and removed at night. Alternatively, an NVD could be removed in daylight and stowed in the cockpit. One example of such a modular system is an HMD of the early family from GEC, a representation of which is shown in Figure 3.12. Different modules, which can be called mission-configured display modules, such as the NVG module for night use and shown in Figure 3.12, can be

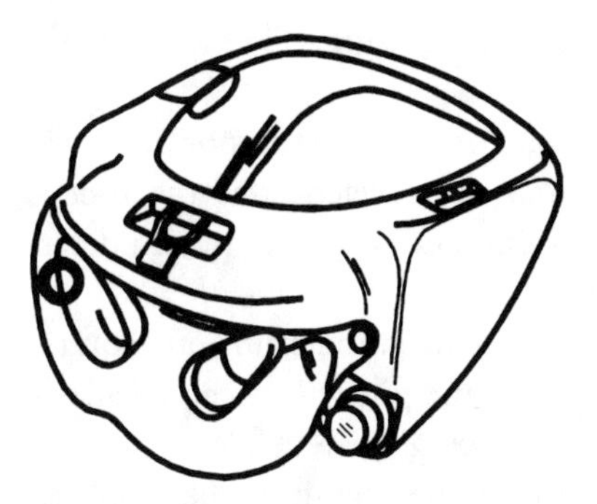

Figure 3.12 The modular HMD approach of GEC. Shown are the basic helmet and a night-use module with NVG and displays. The HMD module can be substituted with various other modules, each of which is optimized for a specific task.

attached and clipped on the basic helmet or unclipped and replaced with a similar in-shape module for day use but with a brighter visor and lighter weight.

Modular systems have the advantage of enabling replacement of the expensive complex electro-optic assembly while retaining the personalized shell. Moreover, if the HMD malfunctions when the pilot is already strapped in the cockpit, it is easier to acquire another module from another pilot unless a spare HMD is retained for each pilot.

Arguments against this approach include the inconvenience during extended-hours flight of the in-flight changeover of equipment under stress conditions and in the highly confined space in the cockpit. Also, the necessary provision of mechanical and electrical interfaces between the modules introduces structural weakness and degrades alignment accuracies.

3.5 Tailoring an HMD/HMS for a Specific Application

Ideally HMDs/HMSs would have characteristics that would satisfy the desires of all users for all types of missions and tasks. However, the requirements from the HMD/HMS depend on the purpose for which it is intended. For instance, a system made for a fighter pilot will not necessarily fit a tank commander's or helicopter pilot's needs. Systems aimed at the fighter pilot must be lightweight, have high resolution, have high brightness stroke symbology, and track head orientation with high accuracy [25]. On the other hand, helicopter pilots and armored vehicle crew require systems with a wider FOV and raster imagery–oriented systems. The first decision to be made in the design or selection of such a device would be to define the use of the system. Obviously, the better the operational conditions of the system can be defined or narrowed down, the easier the design task. Eventually a better design will emerge that is better suited for the genuine needs.

The operational task defines by itself some of the more important parameters of the head-coupled system. It specifies the accelerations the user is expected to experience, defines the ambient illuminance conditions, and determines if withstanding of ejection is required.

If the system is intended for a nonejection vehicle, the stringent requirements on placing the optics between the visor and the pilot's face can be alleviated or lifted. If so, FOV can be increased by reducing eye relief.

Tailoring of the HMD to the specific task can be exemplified by looking at two major parameters of the HMD: the FOV and the HMD weight. Because the increase of FOV usually involves some weight penalty due to larger optics elements, it is worthwhile to look at the needed FOV for each type of aircraft, together with the expected g factor for the same aircraft. Figure 3.13 shows the expected g load for each aircraft type and the required FOV expressed in square degrees. The plots show that moving from nonejecting aircraft to ejecting ones decreases the FOV requirements while increasing the expected g load.

The type of task helps in determining what display arrangement to use: monocular, binocular, or perhaps biocular. The task type also helps in selection of the display type.

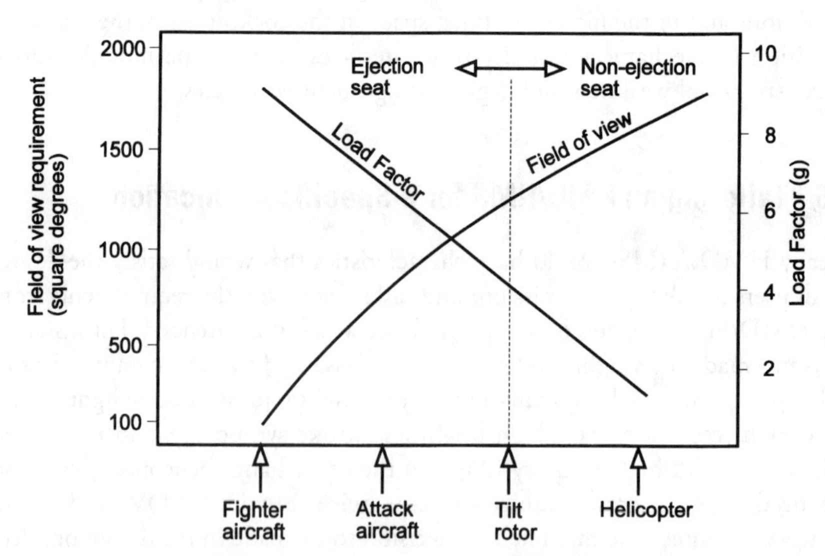

Figure 3.13 The required FOV of the HMD typically needed for each aircraft type. Also shown is the expected g-load factor for each aircraft type.

3.6 Integration of HMD/HMS With Sensors

Contemporary systems integrate night vision sensors with the HMD. Currently available NVDs can be divided into two categories: IITs and thermal imagers, or FLIRs. IITs amplify reflected or emitted light so the user can more readily see poorly illuminated scenes. Obviously, they require some minimum light to produce a usable image. Thermal imagers, on the other hand, do not depend on the ambient light since they sense temperature differences in the scene.

Currently, FLIRS are too big and bulky to be integrated in helmets. For that reason, they are always mounted on a steerable pod outside the cockpit. Images are relayed and displayed on the CRT of the HMD. IIT NVDs, on the other hand, have a small size and easily can be mounted on the helmet. Their images also are presented, together with the symbology, on the HMD.

The fusion of the IIT image with the symbology can be realized in one of two ways [3]. The common way is to blend the images of the two sources optically using a partially transparent combiner. The drawback of that method is the inevitable reduction of brightness of both images by the combiner. Another way is to combine both images electronically by first converting the IIT image to a video signal using a CCD sensor and then mixing both video signals using a video mixer (Figure 3.14). Such a conversion, however, ultimately impairs the quality of the IIT image.

A common use of an HMD/HMS is in conjunction with a VCS. It is used for steering a FLIR system mounted on a steerable gimbal in the nose of the aircraft. VCS enables the pilot to view images generated from multiple waveband sensors ranging from visible to IR. Since they are mounted on a steerable gimbals slaved to the pilot's head motion rather than mounted on the helmet itself, the sensors enable use of large focal length optics, thus providing images with higher resolution. Furthermore, because the pilot points at the target using the sensor rather than the helmet sight, thus giving a magnified view of the target, the pointing accuracy is superior to helmet pointing and can be used for precision targeting. The sensor is located far away from the pilot and can introduce problems of apparent motions, parallax, and incorrect distance estimation [26].

3.7 Summary

Head-coupled devices are characterized by many attributes. They can be categorized by the optical properties of the display system, the image quality, the fidelity of the head-position tracking system, and the sociability of the system.

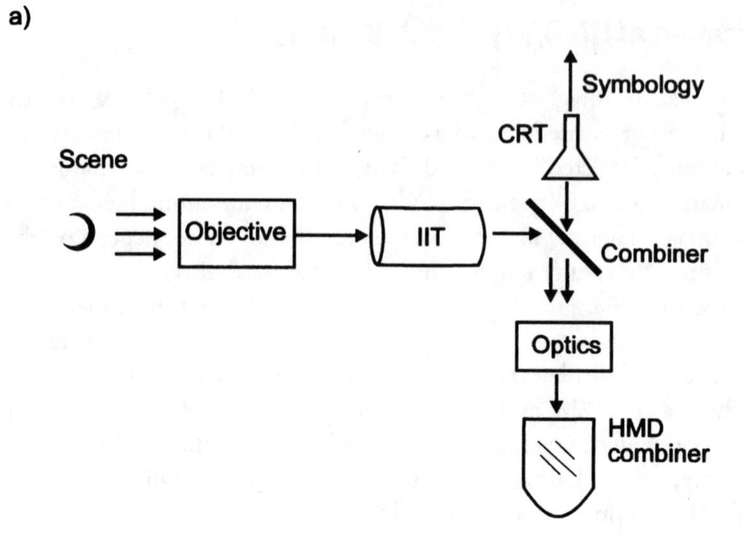

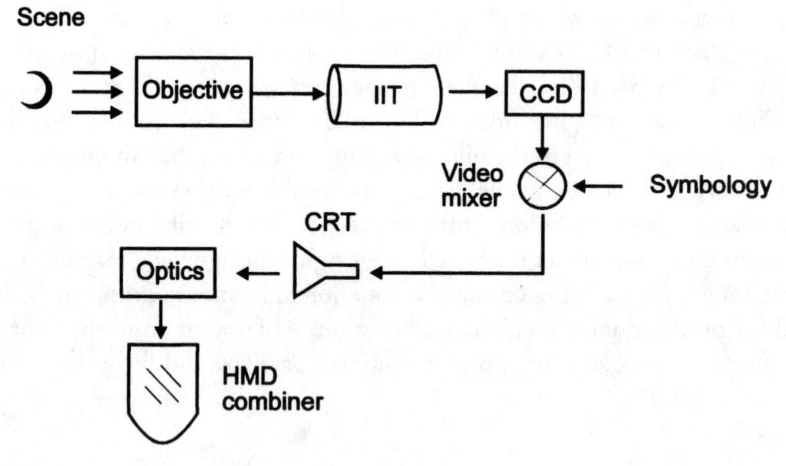

Figure 3.14 IITs and HMD image fusion: (a) optically by use of a combining mirror and (b) electronically by mixing of both video signals.

Each category includes many parameters that must be considered in the design and selection the proper system.

Because of the vast amount of design parameters of head-coupled displays and because it is beyond the practical capability of the designer to construct an ideal system that will meet all those conflicting parameters, the main task is to determine which attributes are more important than others and what trade-offs should be made. Naturally, in that judgment process, the designer or, alternatively, the user must examine how each characteristic of the system affects the task or mission for which the system is intended to be used. The process eventually narrows down the extent of all parameters and enables concentration on the really vital ones. Still, even with that approach, some compromises must be made to meet the essential requirements, and the final solution will not be perfect. The only area where no compromise can be accepted is pilot safety.

References

[1] Thomas, R. M., "Visually Coupled System Integration," in *Helmet-Mounted Displays*, SPIE, Vol. 1116, 1989, pp. 33–36.

[2] Berry, J., et al., *PNVS Handbook*, Fort Rucker, AL: Directorate of Training and Doctrine, 1984.

[3] Bohm, H.-D. V., and Schranner, R., "Requirements of an HMS/D for a Night-Flying Helicopter," in *Helmet-Mounted Displays II*, SPIE, Vol. 1290, 1990, pp. 93–107.

[4] Wood, R. B., "Holographic Head-Up Displays," in *Electro-Optical Displays*, New York: Marcel Dekker, 1992, pp. 337–415.

[5] Bull, G. C., "Helmet Displays Options—A Routmap," in *Helmet-Mounted Displays II*, SPIE, Vol. 1290, 1990, pp. 81–92.

[6] Fischer, R. E., "Optics for Head-Mounted Displays," *Information Display*, Vol. 10, Nos. 7–8, July–August, 1994, pp. 12–16.

[7] Clapp, R. E., "Field of View, Resolution and Brightness Parameters for Eye Limited Displays," in *Imaging Sensors and Displays*, SPIE, Vol. 765, 1987, pp. 10–18.

[8] Kocian, D., and H. L. Task, "Fundamentals and Optics of Helmet-Mounted Displays," Short Course Notes, *SPIE's Internat. Symp. on Optical Engineering in Aerospace Sensing*, April 1994.

[9] Awwal, A. A. S., "Standardization of Nondiscrete Displays," in *Electro-Optical Displays*, New York: Marcel Dekker, 1992, pp. 447–474.

[10] Verona, R. W., et al., "Dynamic Sine Wave Response Measurements of CRT Displays Using Sinusoidal Counterphase Modulation," in *Helmet- and Head-Mounted Displays and Symbology Design Requirements*, SPIE, Vol. 2218, April 1994, pp. 105–114.

[11] Task, H. L., "An Evaluation and Comparison of Several Measures of Image Quality for Television Displays," Wright-Patterson AFB, Ohio, AMRL-TR-79-7-9, 1979.

[12] Jacobs, R. S, T. J. Triggs, and J. W. Aldrich, "Helmet Mounted Display/Sight System Study," Volume 1, USAF Flight Dynamics Laboratory, Wright-Patterson AFB, Ohio, Tech. Report No. AFFDL-TR-70-83, March 1971.

[13] Farrell, R. J., and J. M. Booth, "Design Handbook for Imagery Interpretation Equipment," Seattle: Boeing Aerospace Co., February 1984.

[14] Rogers, P. J., and M. H. Freeman, "Biocular Display Optics," in M. A. Karim, ed., *Electro-Optical Displays*, New York: Marcel Dekker, 1992.

[15] Patterson, R., et al., "Depth Perception in Stereoscopic Displays," *J. Society for Information Display*, Vol. 2, No. 2, 1994, pp. 105–112.

[16] Patterson, R., and W. L. Martin, "Human Stereopsis," *Human Factors*, Vol. 34, December 1992, pp. 669–692.

[17] Rash, C. E., R. W. Verona, and J. S. Crowley, "Human Factors and Safety Considerations of Night Vision Systems Flight Using Thermal Imaging Systems," in *Helmet-Mounted Displays II*, SPIE, Vol. 1290, 1990, pp. 142–164.

[18] Klymenko, V., et al., "Convergent and Divergent Viewing Affect Luning, Visual Thresholds and Field-of-View Fragmentation in Partial Binocular Overlap Helmet Mounted Displays," in *Helmet- and Head-Mounted Displays and Symbology Design Requirements*, SPIE, Vol. 2218, April 1994, pp. 82–96.

[19] Leger, A., et al., "Binocular HMD for Fixed-Wing Aircraft: A Trade-Off Approach," in *Display Systems*, SPIE, Vol. 1988, 1993, pp. 160–168.

[20] Stiffler, J. A., and L. Wiley, "I-Night and Beyond," in *Helmet-Mounted Displays III*, SPIE, Vol. 1695, 1992, pp. 13–20.

[21] Benedict, C. P., and R. G. Gunderman, "Helmet-Mounted Systems Test and Evaluation Process," in *Helmet-Mounted Displays III*, SPIE, Vol. 1695, 1992, pp. 8–12.

[22] Whitcraft, R. J., "Helmet Integration: An Overview of Critical Issues," in *Helmet-Mounted Displays*, SPIE, Vol. 1116, 1989, pp. 122–125.

[23] Jarrett, D. N., and A. Karavis, "Integrated Flying Helmets," *Proc. Institute of Mechanical Engineers*, Vol. 206, 1992, pp. 47–61.

[24] Cameron, A. A., and D. G. Steward, "The Viper HMD—From Design to Flight Test," in *Helmet- and Head-Mounted Displays and Symbology Design Requirements*, SPIE, Vol. 2218, April 1994, pp. 137–148.

[25] Lohmann, R. A., and A. Z. Weisz, "Helmet-Mounted Displays for Helicopter Pilotage: Design Configuration Tradesoffs, Analyses, and Test," in *Helmet-Mounted Displays*, SPIE, Vol. 1116, 1989, pp. 27–32.

[26] Brickner, M. S., "Helicopter Flights with Night Vision Goggles: Human Factors Aspects," NASA Technical Memorandum 101039, Moffett Field, CA, 1989.

4

Image Sources

The image source is the core element of the HMD. It has the largest impact on most of the attributes of HMDs: the brightness of the displayed image is determined almost solely by the image source, the gray levels or colors, the resolution, and the attainable FOV. Finally, the size, volume, and weight of the HMD are directly affected by the image source device.

The requirements from the HMD image source are perhaps the most difficult that may be demanded from any other display application. It must be very small and lightweight to be tolerated by the HMD wearer; it must have high luminous efficiency to provide the brightness needed to be seen clearly during the highest ambient illumination conditions; it must have very high resolution so that even after being magnified by the HMD optics its resolution match the resolution of an equivalent full-size display; and it must provide the safety for the user because of its proximity to the user's head.

The image sources can be divided into two main categories:

- *Emissive displays,* which generate the image illumination;
- *Passive displays,* in which the illuminance is generated by some independent light source and the image is generated by selectively blocking and passing the light in a way that eventually forms the image.

Emissive displays include all types of image sources that use phosphor, the best known being the CRT. Other emissive displays are vacuum fluorescent displays, electroluminescent displays, plasma display panels, and field emission displays. One other well-known emissive device is the LED. The best known image source in the category of passive displays is the liquid crystal light valve.

For many years—and as is still the case today—miniature CRTs have dominated in HMD applications, because of their capability to provide high resolution together with high luminous efficiency. No other type of image source device rivals the luminance, resolution, and cost of miniature CRTs, but in recent years several types of alternative flat-panel displays are closing the gap.

4.1 Display Parameters

4.1.1 Luminance

Because HMDs are required to be used at the highest ambient lighting conditions, the luminance of the image source must be very high. The luminance of the HMD is determined mainly by the luminance of the image source. In see-through HMDs, which are intended to be used in daylight conditions, the luminance of the image source has to be in excess of 5,000 ft-L [1], although luminance of 8,000 ft-L is more appropriate for HMDs and a target of 12,000 ft-L has been sought. Currently, this level of luminance can be achieved only with CRTs operating at stroke mode, and thus, capable of presenting symbology only.

4.1.2 Resolution

The resolution of the image source is considered differently for the various types of the image source. That is because of the different principles of operation of the image sources. For CRTs operating in raster mode, the vertical dimension resolution is determined by the scan line width, while in the horizontal dimension it is the discernible cycles or line pairs per scan line.

Fundamentally, the resolution of the CRT is determined by the spot size of the display, which is the footprint of the electron beam on the screen. There is an obvious relation among the spot size, luminance, resolution, and line rate of the beam, since a larger and "slower" moving beam activates the phosphors more and thus produces more luminous energy.

For flat-panel displays, which are arranged at a matrix form, the vertical and horizontal resolutions are determined by the vertical and horizontal sizes of the discretely addressed picture element, or pixel.

One resolution limit common to all flat-panel displays technologies is the interconnect density required around the perimeter of the display. Flat-panel displays for HMDs are being developed at sizes of 20 to 50 mm capable of addressing 1,000 to 2,000 pixels [2], thus indicating a discernible pixel size on the order of 15 to 30 μm.

4.1.3 Contrast Ratio

The contrast ratio of the image source is the ratio of the luminance of the brightest and dimmest regions of the display. A high contrast ratio is required to achieve sufficient gray levels while operating in a raster or video mode. The contrast ratio also affects the perceived vividness of the displayed image. The contrast ratio is increased by using spectral filters and fiber-optic faceplates, which better trap the light of the individual light spot and concentrate it on the specific picture element. Current image sources have contrast ratios between 16:1 and 1:50.

4.1.4 Gray Shades and Color

Most military applications of HMDs use a monochrome image source because no high-brightness and high-resolution miniature image sources have yet been produced to support full color. Moreover, all electro-optic sensors are monochrome. The rationale for color in a military HMD is only to enhance target information or warnings or to duplicate head-down displays when operating in shut-down mode with synthetic video.

The color of the image source has been dictated mainly by the properties of the phosphor used. The selection of the phosphor is made to satisfy the display brightness by selection of phosphors with the highest luminous efficiency. The gray shades, which are required to present video images, are dictated mainly by the contrast ratio of the display. Typically, HMDs are produced with 10 or more gray levels.

4.1.5 Size and Weight

Naturally, the image source is expected to be as small and as lightweight as possible. Its size and volume directly affect the size and weight of the entire HMD.

Today, miniature CRTs, the most commonly used image sources, have a length of 80 to 130 mm and weight ranging from 35g for a 1/2-in CRT to about 90g for a miniature 1-in tube [3]. The usable screen area for the 1/2-in display is 11 mm in diameter; for a 1-in CRT, it is 19 mm. The CRT has an overall diameter close to that of the usable area: its outside diameter is 20 mm for the 1/2-in tube and 27 mm for the 1-in tube. The overall CRT dimension and weight are increased because of the need to add MU-metal shielding to isolate the display from external magnetic fields that exist in the cockpit and may affect the operation of the display.

The flat-panel display has a larger unusable area with integrated drivers, for example, and connections around the perimeter.

Ideally, the image source should have a length not exceeding 15 mm or a weight less than 25g. It is expected that some flat-panel technologies will approach this goal without compromising the other attributes of the display.

4.2 CRT Displays

The CRT is an electron vacuum tube that has an electron gun in one end and a phosphorous screen in the other. The electron gun is composed of three main elements: the emission system, the focusing grids, and the deflecting electrodes. The electron gun generates a narrow focused beam of electrons. The electrons generated by thermionic emission from a cathode then are accelerated by application of an electrostatic field.

The cathode is enclosed in an electrode structure, or triode, that has two grids: the control grid, G1, and the accelerating grid, G2. The accelerating grid is biased at a positive potential with respect to the cathode and thus accelerates the beam away from the cathode. The control grid is placed between the cathode and the accelerating grid. By biasing the control grid with negative voltage, the electron beam current and, consequently, the CRT spot brightness are controlled. For certain cutoff voltage, the beam is shut off. The structure of a miniature CRT used for HMD applications is illustrated in Figure 4.1.

The two grids produce an electric field that causes the trajectories of the electrons to cross over near the grids. If the beam current becomes high, the high electron density in the crossover region produces repulsion forces, which tend to increase the spot size and hence reduce the CRT resolution. In some high-resolution CRTs, a laminar flow gun approach is used, which eliminates the beam crossover and thus enables smaller spot sizes to be achieved [4].

As the electron beam trajectories diverge from the crossover, a focusing lens system is used to focus the beam into a spot on the CRT screen. The focusing lens is either electrostatic or magnetic. The electrostatic lenses exhibit, similar to optical lenses, aberrations that impair the ability to produce finely focused spots. These astigmatisms result in an elliptical beam cross section and coma, which produces unevenly distributed spots [5].

Magnetic focusing enables higher-resolution displays. The magnetic field produces forces that are always perpendicular to the electron velocity vector. Therefore, it affects only the direction of the beam without affecting the kinetic energy of the electrons, resulting in a more effective lens.

Finally, the electron beam is deflected to the desired position on the screen. In HMD CRTs, a magnetic deflection system is used. The magnetic deflection yoke, which is mounted externally to the CRT envelope, generates a uniform magnetic field perpendicular to the electron beam. The yoke is con-

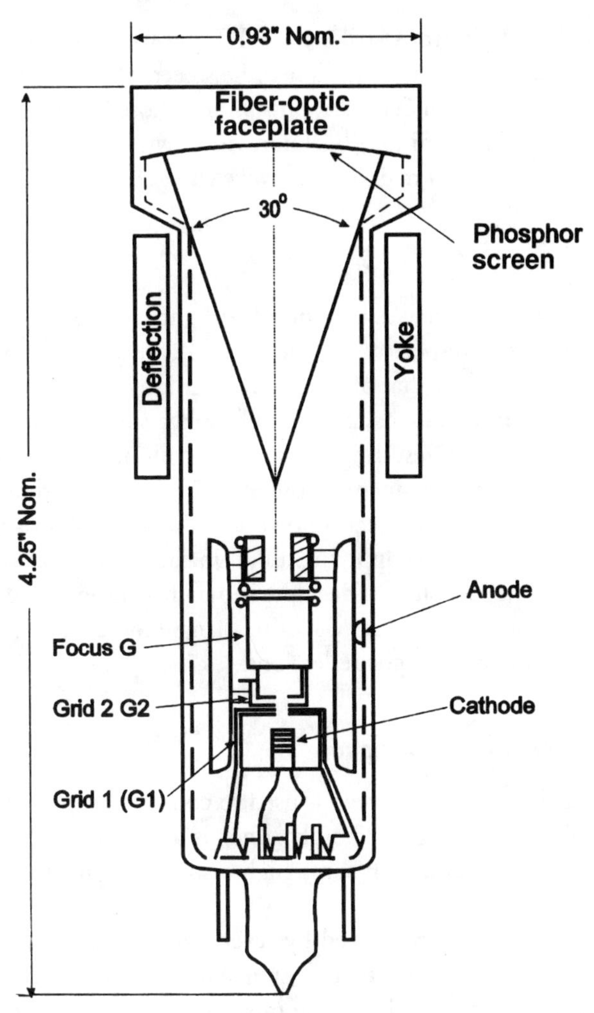

Figure 4.1 Structure of the CRT for HMD applications. The dimensions are typical to a 1-in tube.

structed of copper windings and a high-permeability core. The yoke has a separate set of windings for the lateral and vertical deflections. The beam deflection is steered by the intensity of the magnetic field, which is controlled by the current introduced in the coils' windings.

There are two basic types of deflection coils: saddle coils and toroid coils [4]. The saddle coils have their ends raised up for optimum sensitivity and look like a saddle. They are simpler and less expensive than toroid coils. Toroid coils are wound toroidally on a ferrite core and have greater precision in the winding distribution. The main disadvantages of electromagnetic deflection coils, which are significant for HMD applications, are their higher weight and volume.

The deflected electrons strike the CRT phosphorus screen, causing the phosphors at that location to emit light. The screen of the CRT is coated with a thin layer, about 5 μm in depth, of phosphor enclosed between the glass faceplate and a very thin aluminum backing with a width of about 0.1 μm. The aluminum backing is used mainly to prevent charge buildup on the screen [5].

4.2.1 Halation

When the electron beam strikes the phosphor, light rays enter the glass faceplate at various angles with respect to the faceplate. Because the index of refraction of the glass is greater then that of the air, rays striking at angles above a critical angle are reflected internally back to the phosphor surface and again scattered by the phosphor surface, resulting in concentric illuminated rings around the illuminated spot. The rings increase the effective spot size, thus reducing the resolution of the CRT. The effect, known as halation, is shown in Figure 4.2. Halation can be reduced by applying an antihalation film between the phosphor screen and the faceplate. The film, if matched properly to the emission spectrum of the phosphor, causes the critical angle to occur in the film and to some extent eliminates the light reflections.

4.2.2 Characteristics of Phosphor

The phosphor coating of the screen is used to convert the kinetic energy of the electron beam to visible light. The phosphor is an inorganic crystal with fine particles sized 3 to 10 μm, with some high-resolution miniature CRTs having particles below 2 μm [3].

The phosphor is characterized by several attributes, including the emission spectra, luminous efficiency, persistence or decay time, and particle size. Table 4.1 lists the characteristics of the most common phosphors.

Most HMDs use a P43 or P53 phosphor, with the latter dominating because of its high resistance to burning and because it contains most energy in a narrow band, which permits the use of frequency-selective optical elements such as combiners or holographic elements, as well as optimizing the optical

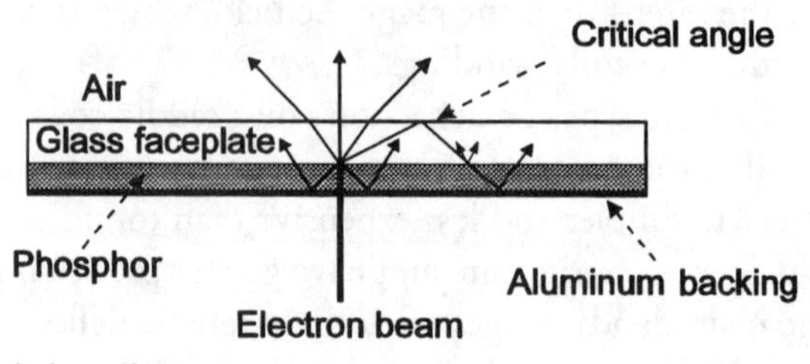

Figure 4.2 The halation effect.

Table 4.1
Characteristics of the Major Phosphors Used for HMDs

Type	Phosphor	Peak Wavelength (nm)	Color	Luminous Efficiency (lm/W)	Persistence*	Decay Time to 10% Brightness
P1	Zn_2SiO_4:Mn	525	Yellow-green	30	Medium	24 ms
P31	ZnS:CU	520	Green	35–45	Medium-short	40 μs
P43	Gd_2O_2S:Tb	544	Yellow-green	40	Medium	1.2 ms
P45	Y_2O_2S:Tb	Multiple peaks	White	20	Medium	1.7 ms
P53	$Y_3Al_3Ga_2O_2$:Tb	544	Yellow-green	30	Medium	7 ms

*Phosphor persistence classification is based on the time required to rise to 90% or decay to 10% of peak luminance, as follows: very long = 1 sec or more; long = 100 ms to 1 sec; medium = 1 ms to 100 ms; medium-short = 10 μs to 1 ms; short = 1 μs to 10 μs; very short = less than 1 μs.

such as combiners or holographic elements, as well as optimizing the optical relay design to the specific wavelength.

4.2.3 Fiber-Optic Faceplates

The plain-glass faceplate of the CRT can cause spurious screen illumination due to internal reflections, such as halation, and chromatic aberrations [6]. The fiber-optic faceplate, on the other hand, acts as an image plane device with a virtual zero thickness and as such captures most of the light that otherwise would be wasted due to the internal reflections. As a result, the fiber-optic faceplate improves the display contrast and the resolution and improves the light transmission.

The fiber-optic faceplate is a coherent array of millions of optical waveguides, each having a diameter as small as 3 μm [7]. It is made by fusing many fiber-optic bundles into a boule (Figure 4.3). The boule then is sliced and polished to the required shape of the faceplate. The fiber-optic faceplate is curved on the inside to the deflection angle of the tube but is flat on the outside. That configuration helps to eliminate the need for dynamic focusing, which compensates for the increase in spot size caused by the oblique electron incidence at the edge of the screen.

4.2.4 Line Width

Line width is the attribute that most characterizes the CRT operating at a stroke mode. The required display line width is on the order of 15 to 30 μm, which is needed to obtain a line width of 0.5 to 1 mrad in the complete HMD. Those line-width requirements can be achieved only with a very fine grain phosphor screen, with particle size smaller than 2 μm and limited final anode voltage. Line width or resolution, luminance, and line rate are closely related. Luminance, for example, can be increased by sacrificing line width or by increasing writing line rate, and finer lines can be obtained by reducing luminance.

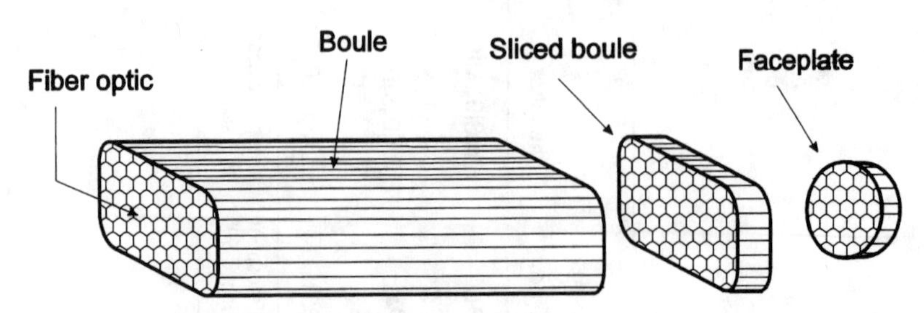

Figure 4.3 The production process of the fiber-optic faceplate.

4.2.5 Final Anode Voltage

The final anode voltage controls the beam electrons' energy that bombards the screen phosphor and thus affects the luminance of the CRT. The anode voltage in practice was limited to 7 kV for operation in aircraft because of safety of the HMD wearer in the flammable environment in the cockpit.

Increasing the final anode voltage, although increasing the CRT luminance, also enlarges the beam spot size and thus increases the line width and reduces the display resolution. Use of finer phosphor particles enables an increase of the CRT luminance by increasing the final anode voltage with smaller loss of resolution. In high-performance HMDs, the final anode voltage of the miniature CRT is in the range of 5 to 20 kV, with 7 to 10 kV being the standard.

4.2.6 Raster and Stroke Presentation

The image on the CRT is generated by one of two methods: raster mode and stroke mode. In raster mode, the screen is scanned sequentially line by line. Commonly, raster scanning is performed in one of two ways: noninterlaced and interlaced (see Figure 4.4). Interlaced scanning enables reducing the video bandwidth requirements of high-resolution displays. In interlaced displays, the complete raster, called a frame, is subdivided into two fields consisting of the odd and even raster lines, respectively. The fields are scanned alternately at a rate above the critical flicker frequency, thus refreshing a complete frame at a rate of one-half the critical frequency. However, because of the averaging effect of the human eye, the perception of flicker is minimized. Since raster presentation in HMDs is mainly used for presentation of video imagery, which mainly consists of areas with continuous brightness distribution, flicker is not noticeable. This is not the case, however, for high-resolution graphics and symbology; therefore, it is preferably presented in a stroke mode. The video standard, RS-170, used for raster presentation in the United States defines an interlaced scanning of 525 lines, refreshed at a field rate of 60 Hz and a frame rate of 30 Hz. For each scan line, the RS-170 standard allocates about 58.5 μs. For a typical 0.75-in useful scan line, that dictates a line-writing rate of about 13,000 in/s. For higher display resolutions, the RS-343 standard allocates about 25.6 μs for each line, so the required line-writing rate is about 30,000 in/s.

The European standard, CCIR, defines 625 lines refreshed at a field rate of 50 Hz and a frame rate of 25 Hz. In HMD applications, which typically are characterized by higher-resolution image generation, the RS-343 standard, which specifies resolutions up to 1,023 scanning lines, is often employed.

Stroke presentation has several advantages, including improved brightness and higher symbology sharpness. The higher brightness is achieved by using a slower writing speed than in raster presentation. If the stroke information is pre-

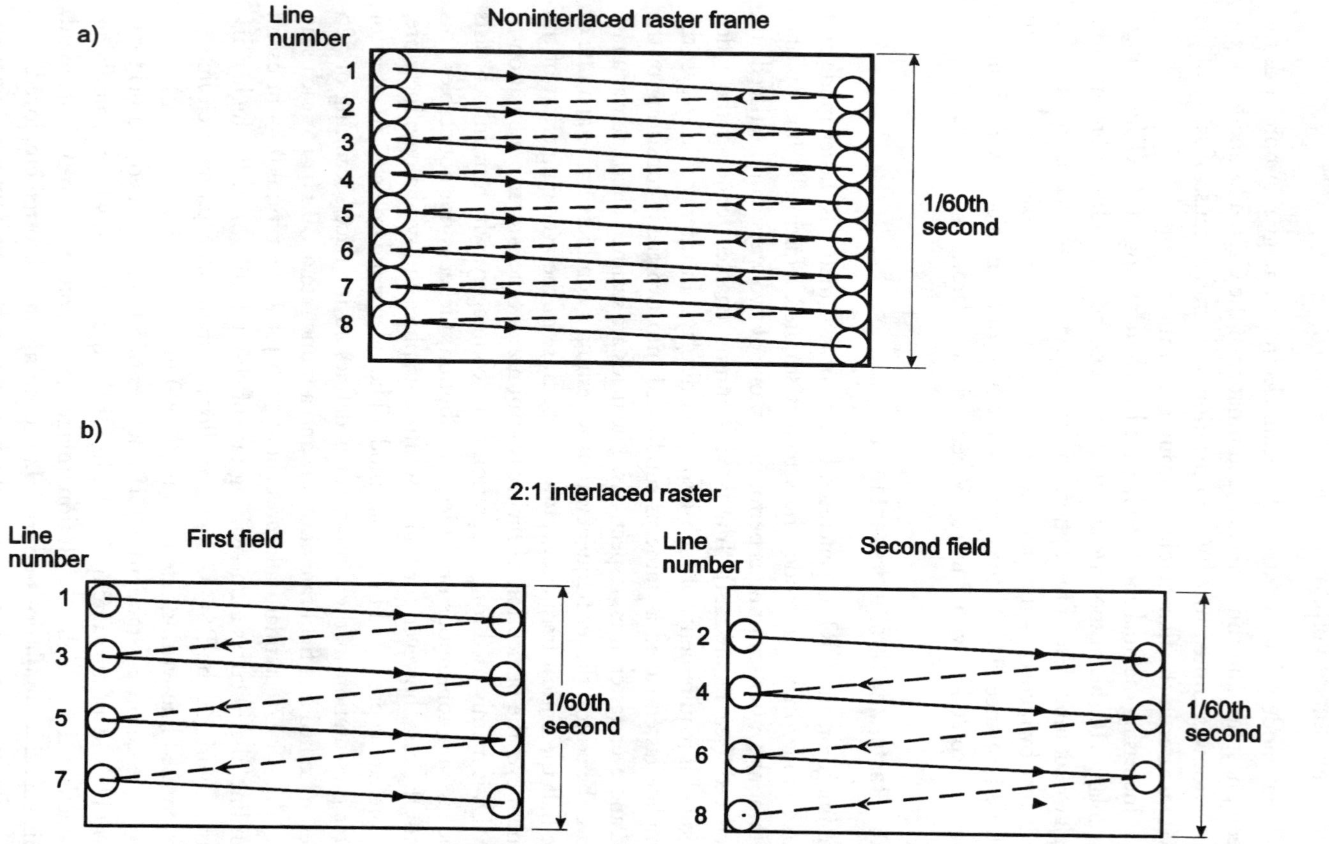

Figure 4.4 Raster scanning pattern: (a) noninterlaced frame and (b) 2:1 interlaced scanning, with the odd-lines field (left) and even-lines field (right).

sented during the vertical retrace (defined by the RS-343 standard as 1.25 ms) and assuming total line length of 5 in, then the required writing line rate is 4,000 in/s. Obviously, that is an important attribute for HMDs, which are required to be viewed against a very bright ambient background.

The better sharpness of the symbology is due mainly to the way the symbols are generated using line segments rather than the combination of dots or pixels used in all other displays. The use of straight-line segments eliminates the jaggedness that occurs in the raster mode when a line in the image crosses two or more raster lines at a shallow angle.

One important feature of the miniature CRT is its capability to present both a video image in a raster format and symbology in a stroke format. Video, or raster information, is needed to present imagery from sensors such as FLIRs or TV cameras. The video image needs to be presented with sufficient gray scales. Stroke symbology, on the other hand, is needed to display very high resolution symbology, such as cross-hair, or aiming reticle; numerals; and indicators. That symbology is needed to be seen clearly at the highest ambient illumination conditions.

Most HMDs, thus, use a mixed or hybrid mode of operation. In this approach, the same quantity of symbology as written in the normal cursive mode is written during the field flyback period of the raster. The writing speed required to do so is about 10 times faster. A typical slow cursive rate is about 1,000 in/s, and a fast cursive rate is about 10,000 in/s.

4.2.7 HMD Drive Electronics

The high resolution and the need to recombine stroke and raster presentation requires special-purpose and fast electronics of the CRT circuitry [8].

Figure 4.5 shows a block diagram of the electronic circuitry of an HMD system capable of displaying both raster image and stroke symbology during the vertical retrace. Standard composite video signal together with stroke symbology, which is obtained from a symbol generator circuitry, are fed to the display electronics unit. A synchronization separation network strips the synchronization from the video signal and provides synchronization to the horizontal and vertical sweep generators. The outputs of the sweep generators are sawtooth waveforms for controlling the x and y positions of the CRT beam, so the CRT is scanned with raster lines across the phosphor faceplate of the CRT. The deflection signals are amplified by linear deflection amplifiers and coupled to the deflection coils in the CRT.

The video signal is amplified by a video amplifier and coupled to the final video driver, which in turn excites the control grid of the CRT and modulates

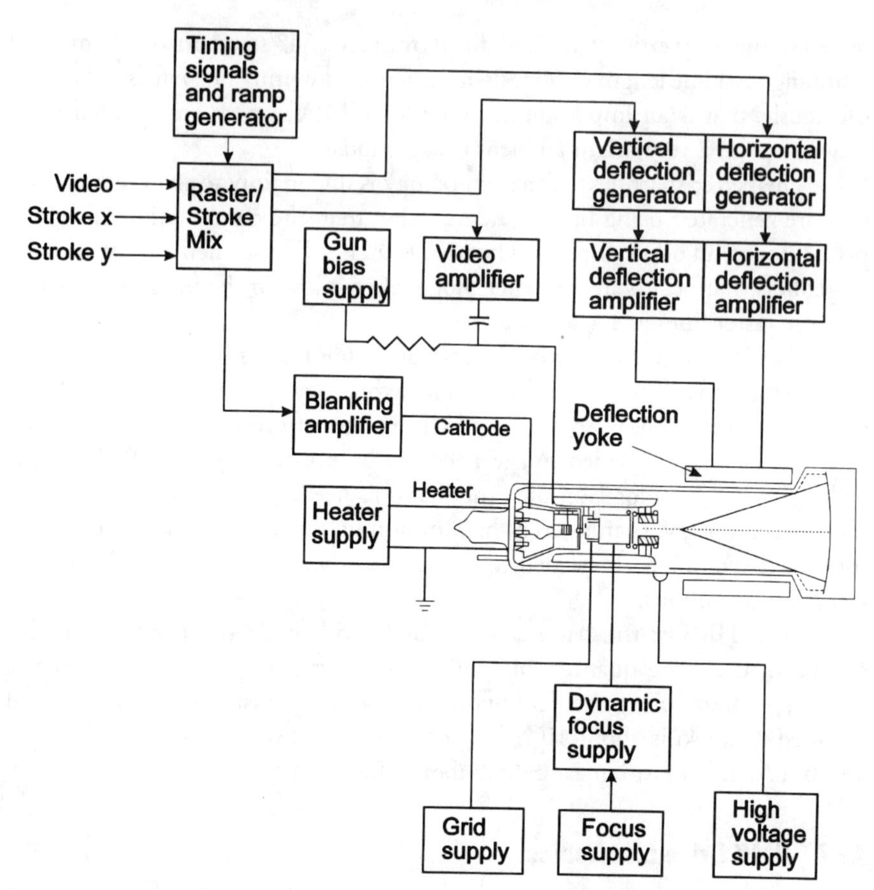

Figure 4.5 Schematic block diagram of a CRT electronic circuitry for an HMD.

the intensity of the electron beam. During horizontal and vertical retrace, the beam is blanked except during the instances of drawing of the stroke symbology.

The stroke symbology is drawn during the vertical retrace. The switching between the two modes, raster and stroke, is controlled by the timing circuitry of the display. The stroke electronics is designed in a way that the beam position on the screen is linearly proportional to a direct current (dc) voltage command, when 0V corresponds to the center of the display.

To present the CRT image correctly into the FOV, it usually is necessary to provide compensation in the form of barrel distortion to compensate for the pin-cushion effects created in the optic relay. Distortion corrections are added as an analog function comprising nonlinear corrections. The analog nature of the CRT drive allows real-time correction.

4.3 Flat-Panel Displays

Flat-panel displays offer several important advantages over CRTs for use in HMDs. They are thin and lightweight, have high resolution, and mostly consume little power [9]. Flat-panel displays are found in both the categories of emissive displays and passive displays. Typically, flat-panel displays are a discrete type of displays, with their pixels arranged in matrix form. Each pixel is located at the intersection of a row and a column of electrodes.

4.3.1 LCDs

Liquid crystal is a liquid organic compound that, over a specified temperature range, exhibits in an intermediate state of matter. In that state, called liquid crystal state or mesophase state, the compound features characteristics of both liquid and crystalline materials. While in the mesophase, the liquid crystal molecules have long-range ordering properties, similar to the solid crystalline state, but exhibits some freedom of movement, as in the isotropic liquid state.

In the basic LCD type, the twisted-nematic (TN) LCD, the nematic crystals are aligned in a spiral configuration that twists polarized light by 90 degrees. The liquid crystals are placed between two transparent conductive electrodes and crossed linear polarizers, as shown schematically in Figure 4.6. A light source is placed at the back of display. As the randomly polarized light from the light source enters the display, only vertically polarized light is able to pass through

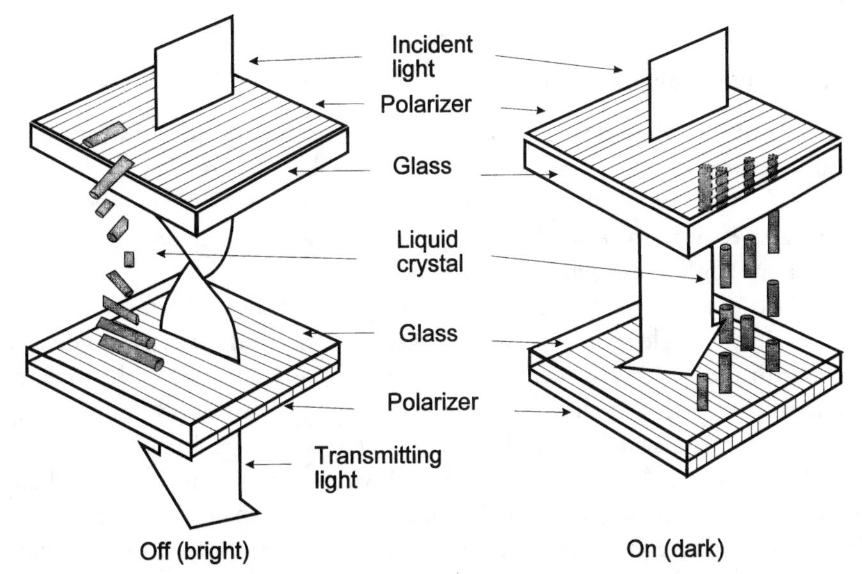

Figure 4.6 Schematic structure of the TN-LCD.

the front polarizer. The polarization of the light is rotated by 90 degrees as it passes through the rear, horizontal polarizer.

By applying voltage to the transparent electrodes, an electrical field is established in the liquid crystal layer. Because the molecules have some degree of freedom of movement and an electric dipole moment, they align with the electric field. In that state, the liquid crystal passes the incoming light but no longer rotates its polarization. The vertically polarized light is not transmitted by the horizontal rear polarizer, so the light is extinguished. The individual pixels that transmit or block light are formed at the points where row and column lines intersect on the separated glass plates.

One disadvantage of the TN-LCD is the cross talk between adjacent pixels. Turning on row and column lines partially turns on adjacent pixels, producing partial lighting of the adjacent pixels and thus reducing the contrast of the display. Other disadvantages are sensitivity to extreme temperatures and limited response time. Also, the presence of the polarizer reduces the optical transmission substantially. Typically, a color LCD has a transmission of 5% to 8%, while a monochrome panel is nearer 17%. With such poor transmission, the necessary level of illumination to achieve comparable performance as a CRT is some 40,000 ft-L.

Higher-resolution LCDs are obtained by using the active-matrix approach, in which each pixel is equipped with a thin-film transistor (TFT) or diode switch. Active-matrix LCDs (AMLCDs) have improved response time, resolution, and contrast. The structure of an AMLCD is shown in Figure 4.7.

4.3.2 Spatial Light Modulator

Spatial light modulator (SLM) image sources are passive devices in which light is reflected from a backplane electrode that acts like a mirror. The device consists of a single crystal silicon backplane that contains CMOS addressing circuitry, a layer of ferroelectric liquid crystal, and a glass substrate coated with a transparent conducting oxide.

Each pixel of the device contains a polysilicon row and a metal column address line, a single transistor, and a large metal electrode. The electrode acts as a mirror that reflects incident light. The electrode is attached to the drain of the transistor so that in the "off" state no voltage is present and the overlying liquid crystal is unswitched. In the switched state, a voltage is applied, and the liquid crystal switches and acts as a half-wave plate so the incident light reflected by the mirror goes through a $\pi/2$ polarization change and is viewed as "on." The device requires, however, polarizing optics, which reduces the overall luminous efficiency.

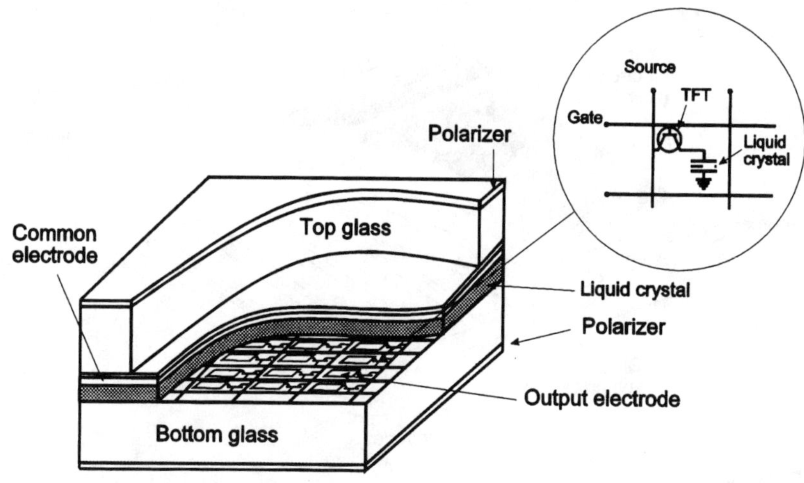

Figure 4.7 Cross section of a TFT-LCD.

The advantage of the electrically addressed SLM (EASLM) device is that all the complex addressing circuitry is placed on the backplane, which is fabricated by standard CMOS processes. Small fabrication geometries are used so very high resolutions of more than 1,000 lines/in can be realized [10]. The display is illuminated by a laser source and thus can produce high luminance levels, making it attractive for HMD applications. Gray levels are obtainable by temporal dithering, and color can be achieved by addressing the display with three colored lasers.

4.3.3 Electroluminescent Displays

An electroluminescent (EL) display is a solid state device with a thin-film luminescent layer sandwiched between transparent dielectric layers and a matrix of row and column electrodes deposited on a single-glass substrate. The column electrodes are transparent, usually indium tin oxide (ITO), and are deposited first in front. The aluminum row electrodes are deposited last in the back. Both electrodes are separated from the luminescent layer by insulating layers. The insulating layers act as current limiters to the capacitive luminescent layer and also store charge, which increases the internal electric field and thus significantly increases the luminous efficiency of the display. The structure of an ac thin-film EL (TFEL) display is shown in Figure 4.8.

The solid state nature of the EL display allows for a low profile and rugged display with a wide operating temperature. When ac voltage is applied between a column electrode and a row electrode, the phosphor thin film between them emits light, which passes through the transparent electrode and through the faceplate and forms an illuminated pixel. The luminance of the TFEL display is

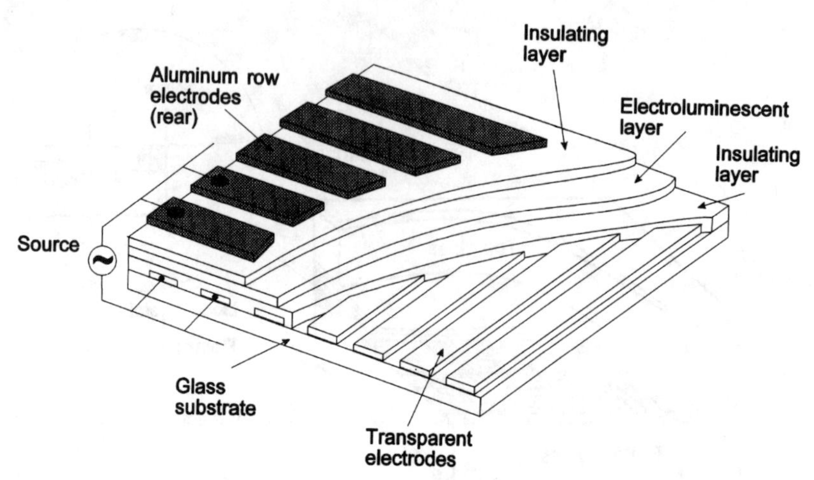

Figure 4.8 Structure of an ac TFEL.

directly proportional to the refresh rate of the display. Increasing the refresh rate enables quite high display brightness, exceeding 300 ft-L, to be achieved. Also, placing a light-absorbing black layer between the rear electrodes and the insulator enhances the contrast ratio of the display in high ambient brightness conditions. Monarchie, Budzilek, and Cupero [11] reported contrast ratios of 1:43 at 2,000 fc conditions and 1:9.4 at 10,000 fc.

An improvement to the basic EL display is the active-matrix EL (AMEL) display, which is accomplished by the addition of data storage and high-voltage control of each pixel. That capability of independent control of a single pixel leads to higher brightness, better power efficiency, and expanded gray scale. Prototype AMEL displays were constructed having luminance of higher than 500 ft-L with contrast ratio greater than 1:100. AMEL displays with resolutions of 1,280 by 1,024 pixels are expected.

4.3.4 Field Emitter Cathode Displays

A field emitter device (FED) is an emissive flat-panel display. It comprises a base plate on which electron emission sites, or emitters, are located. The emitters, which are micron-sized sharp cones, produce electron emission in the presence of an intense electric field [12]. The electrons bombard a phosphor-coated faceplate, which is separated from the base plate by a vacuum gap, similar to the faceplate of a CRT. Each pixel of the display is composed of several cones. Figure 4.9 illustrates the structure of the FED. A cross section of a FED is shown in Figure 4.10.

The addressing of each pixel is achieved by addressing the specific emitter row simultaneously with a specific extraction grid column.

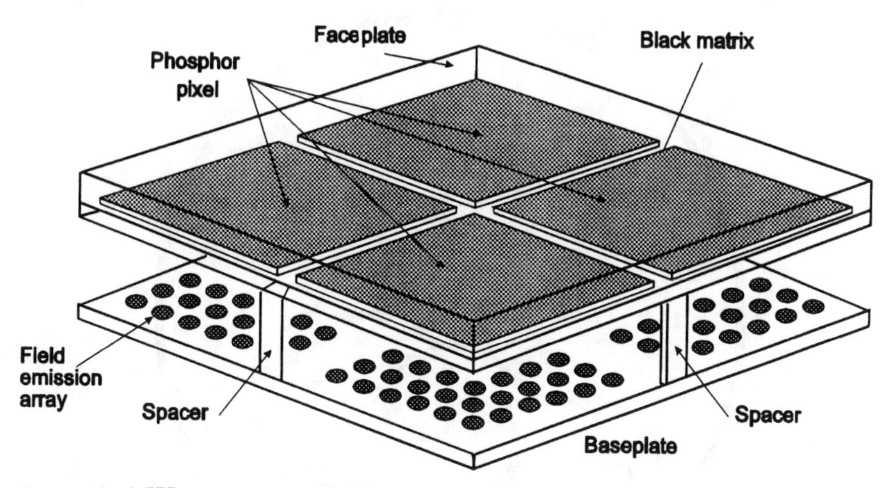

Figure 4.9 A FED.

The luminance of the FED is controlled by either varying the time the emitter is activated or by varying the emitter's current. High-luminance devices (10,000 ft-L) are feasible but require high voltage of about 10 kV anode and thus have the same deficiency of requiring high voltage on the head similar to the common CRT.

4.3.5 LEDs

Image sources based on LEDs are used in applications where a simple, low-density, and almost fixed-pattern information display, for example, a simple aiming reticle, is sufficient [13]. In those cases, the LED's image source is a good

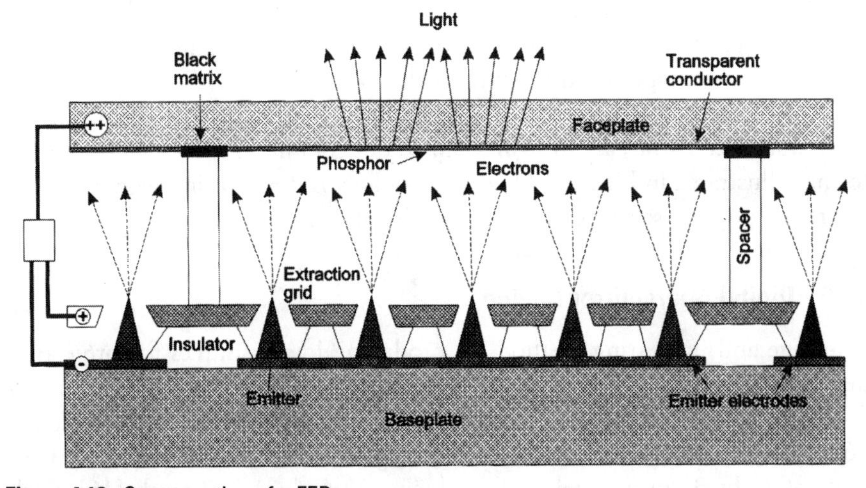

Figure 4.10 Cross section of a FED.

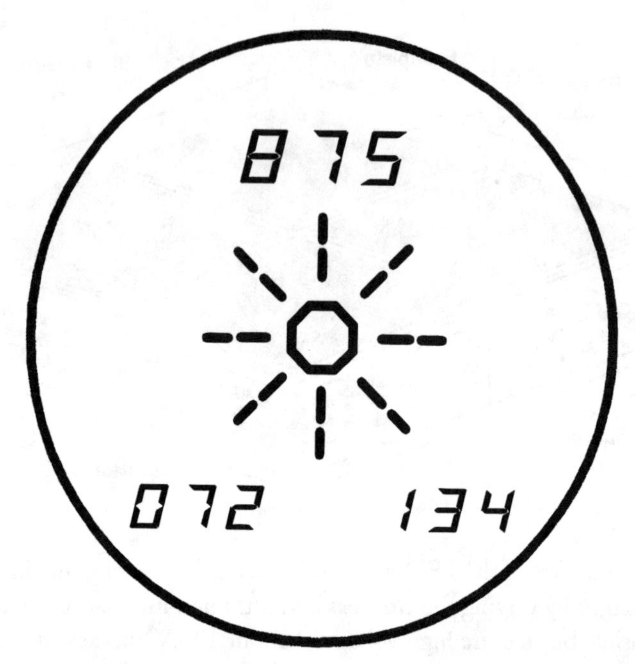

Figure 4.11 Fixed-pattern typical LED-based display.

substitute for the other alternative of a CRT operating in a stroke mode. It is ideal when small, compact, and very low weight displays are sought.

LEDs are single crystal semiconductor devices that emit light when a p-n junction is forward biased. LEDs are fabricated using a range of semiconductor materials but are based mainly on gallium arsenide (GaAs) or related crystals, such as gallium arsenide phosphide (GaAsP) and gallium aluminum arsenide (GaAlAs). LED displays are produced in colors of red, green, and yellow, and in recent years blue LEDs have become available. The LED is a low-voltage but high-current device.

LED-based displays are used in the following arrangements: fixed-format array, dot-matrix array, and mirror-scanned dot column [14]. In the fixed-array format, illustrated in Figure 4.11, the symbology is drawn using segments similar to the popular seven-segment numerals pattern.

4.3.6 Digital Micromirror Device

A unique and interesting method for producing image sources for projection displays developed by Texas Instruments has some potential advantages useful for HMD applications. It uses a large number of micromirrors, each as small as 16 by 16 μm arranged in a matrix array (Figure 4.12(a)) on a small chip with a size of 37 by 32 mm, capable of producing 2,048 by 1,152 pixels [15]. Each

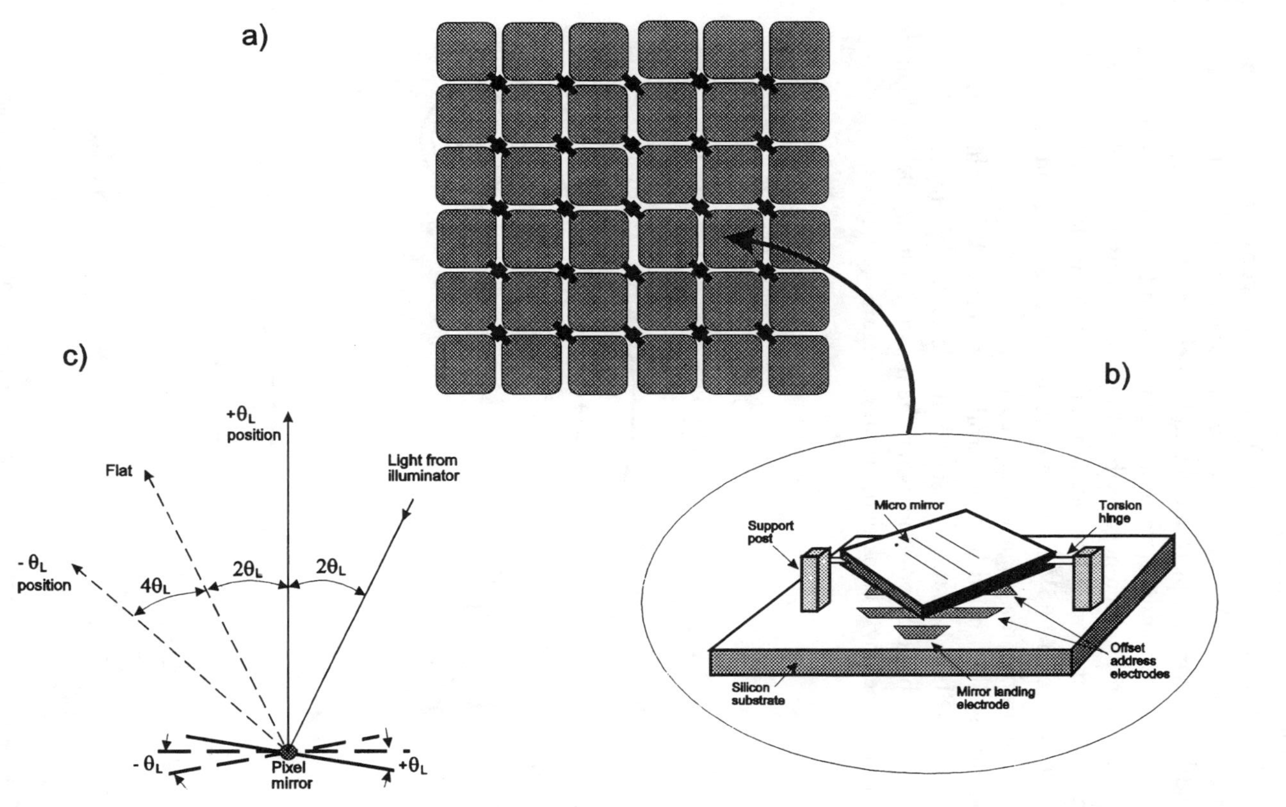

Figure 4.12 (a) The DMD chip is a matrix array of micromirrors suspended on torsional hinges. (b) A schematic structure of the individual micromirror. (c) The principle of operation of the DMD: at positive angular tilt, the mirror is "on" and the light is reflected onto the screen. At flat and negative angles, the light is reflected away from the projection screen and is absorbed.

micromirror is suspended on a torsion support and is capable of either +10- or −10-degree tilt (Figure 4.12(b)).

The digital micromirror device (DMD) chip is constructed over a random access memory (RAM) on which two address electrodes and two landing pads are added. The mirror plate and the address electrodes form capacitors. By applying +5V to one address electrode (digital 1) and 0V (digital 0) to the other address electrode and a negative bias to the mirror itself, the electrostatic charge in the capacitor created between the mirror and the electrode causes the mirror to tilt toward the +5V electrode until it hits the landing pad. It remains at that position until a reset signal is applied.

The chip is illuminated by a powerful light source. When the mirror is tilted at the $+\theta_L$ tilt, it is in the "on" state and reflects the light toward the projection screen. In the "off" position, when the mirror is tilted $-\theta_L$, the light is reflected and trapped by a black-light absorber (Figure 4.12(c)).

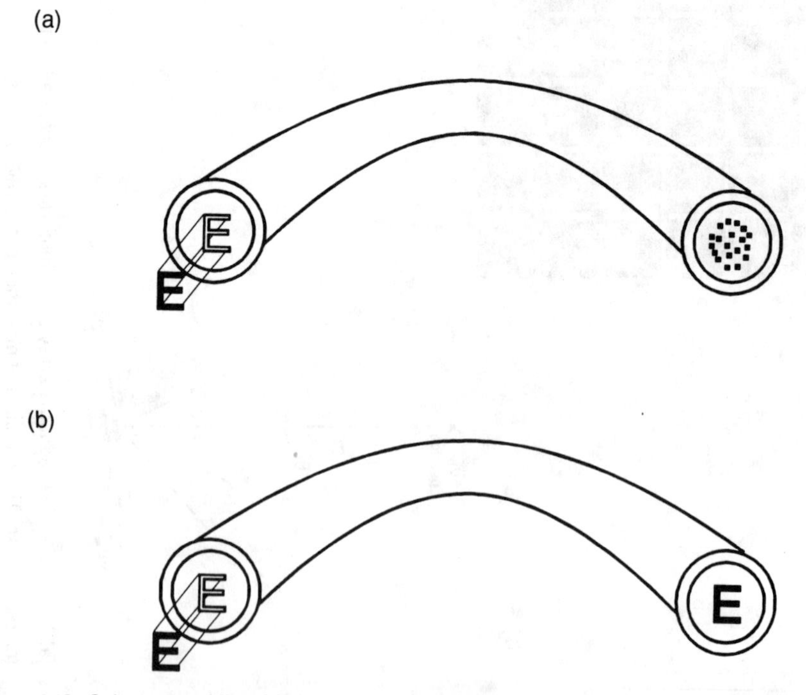

Figure 4.13 Coherent and illumination fiber-optic cables. In the coherent cable, the fibers retain their order along the cable; for the illumination fiber, such order is not required since it is required to transmit illuminating light only.

4.4 Fiber-Optic Image Guide Displays

The limitation on the luminance and resolution of most miniature helmet-mounted image sources can be circumvented partially by using a bright and large image source located off-helmet and conveying the image to the helmet by means of fiber-optic image guides [16,17].

The basic idea behind the arrangement is simple. The image is generated on a large, high-luminance, and high-resolution CRT. Through use of conventional optics, the image is reduced to a smaller size, fed to the fiber-optic image guide, and transmitted to the helmet, where it is enlarged back to the original size. In that way, the final image exhibits the high luminance and resolution of the larger display (less the losses in the fiber and the resolution reduction in the optics and in fiber).

The fiber-optic image guide is a coherent fiber that retains the spatial correspondence of the incoming and the outcoming images. It basically is the same fiber bundle of which the fiber-optic faceplate of the miniature CRT is produced. Figure 4.13 shows the difference between the image guide coherent fiber and the ordinary illumination fiber.

The quality of an image transmitted through the fiber-optic bundle can be seriously degraded by the structure of the fiber bundle. The fiber-optic cable can be regarded as a light tube with a light-blocking mesh made of the cladding of the individual fiber bundles.

A common practice to increase the resolution of the fiber-optic image guide display is the wavelength multiplexing method [18]. In this method, a prism is placed at the input end of the cable to spread each pixel of the image into a linear spectrum covering several multifibers. An identical prism is placed at the other end of the fiber, which acts inversely and collects the spread light back into a single pixel. In that manner, the image of each pixel is distributed along several neighboring multifibers, and the information is not lost in the cases when a single bundle is broken. The principle of the wavelength multiplexing method is illustrated in Figure 4.14.

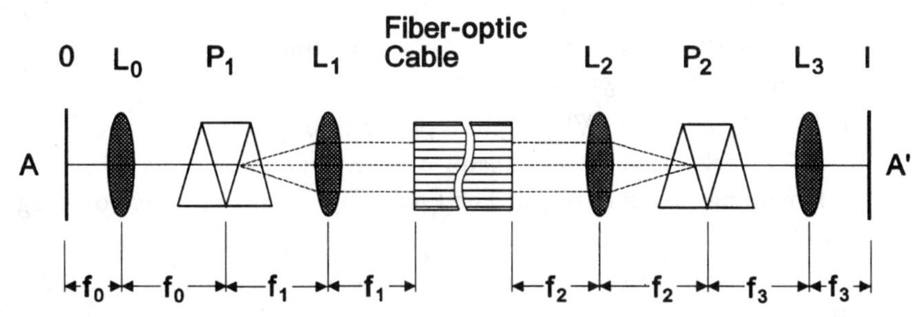

Figure 4.14 Schematic diagram of the wavelength multiplexing method.

4.5 Color Displays

The effectiveness of inclusion of color, at least for color coding, has been long recognized [19]. The use of color is even more justified in HMDs, which condense a large amount of graphics and symbology in a small display area, and the symbology usually is crucial to the user. Yet, most HMD applications still use monochrome displays.

The main reasons why color is being slowly introduced for HMDs are the inevitable reduction of image brightness in the color display and the almost unavoidable loss of resolution.

Although most HMDs still are monochromatic, two major exceptions, which almost always employ color, exist: displays for simulation applications and HMDs used in virtual reality systems.

Almost every technology of monochrome displays also is used in color displays. As such, there are color CRTs and various types of flat-panel displays. Regardless of the technology employed, there are four basic ways to generate the color image:

- *Spatial integration:* Each image pixel is composed of a triad or a quad of smaller pixels, each one with a different primary color. As long as the groups subtend a sufficiently small angle at the eye, they are perceived by the eye as a single full-color pixel. Spatial integration is the most common method of color displays and is used in most television and commercial CRT monitors and color LCDs.
- *Temporal integration:* Red, green, and blue spatially registered full-screen images are presented in rapid succession at a sufficiently high rate so they are integrated by the visual system and perceived as a single full-color image.
- *Spatial addition:* Red, green, and blue images are generated by three separated image sources and combined optically on a single screen, where they are summed to produce a full-color image. The spatial addition method is hardly appropriate for HMDs because the need for three separated image sources has a large impact on the size and the weight of the image source.
- *Spatial subtraction:* White light is passed through three transmissive displays, each of which modulates one primary color while passing the remaining two unchanged. The final image results in full-color image.

4.5.1 Color CRTs

The most common way to produce a color CRT is through use of a metal mask and three electron guns, each corresponding to a primary phosphor granule: red, blue, and green. The electron guns are positioned in either a delta or an in-line configuration. The faceplate is covered with triads of primary-color phosphors. The metal mask is placed so to ensure that each electron beam coming from each color gun, after passing the particular hole in the mask, hits the corresponding color phosphor. This method is common in commercial television sets and monitors, but is impractical in miniature CRTs, mainly because of the small size and manufacturing accuracies.

An alternative method for producing color on a miniature CRT is based on a voltage-dependent penetron phosphor screen [3]. The phosphor is layered with a fine P1 green phosphor, over which a vapor phase deposited silica barrier and a fine red P22R layer of phosphor are added. The red phosphor is excited at low electron beam energy, 6 kV, giving red color, and at high energy of 12 kV. Both phosphors are excited, with the P1 dominating, giving a green color display. The introduction of color, however, reduces the luminance of the display.

4.5.2 Color LCDs

LCDs are the most common image sources in color displays, especially in displays used for virtual reality systems. Most displays use the spatial integration method, although that limits the attainable resolution [20].

A high-brightness color display based on spatial subtraction was demonstrated by Honeywell [21]. It uses a high-luminance xenon arc lamp as a white-

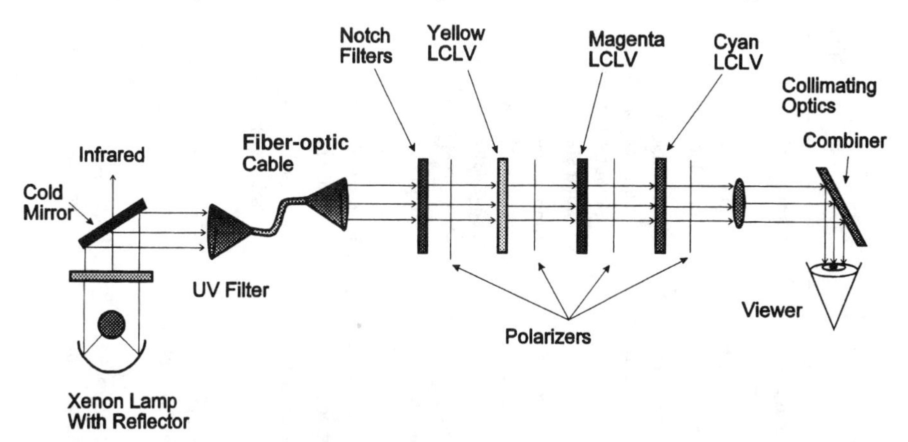

Figure 4.15 Schematic drawing of a spatial subtraction color display.

Table 4.2
Summary and Comparison of Image Source Technologies

Technology	Strengths	Weaknesses	Status	Main Applications
Cathode ray tube (CRT)	High resolution, high luminance	High voltage on head; large, heavy	Mature technology	Military (avionic) HMDs
Electroluminescent (EL) display	High resolution, high contrast, low power, ruggedness, solid state	Low luminance	In development	Military (nonavionic) HMDs; high-resolution commercial applications
Liquid crystal display (LCD)	High resolution, color readily available	Low-light transmission, slow temporal response	Emerging technology	Training and simulation, virtual reality, consumer products
Spatial light modulator (SLM)	High resolution, high luminance, solid state	Polarizers reduce efficiency relative to DMD	In development	Military (avionic) HMD training and simulation
Field emission device (FED)	High resolution, solid state	High luminance requires high voltage	In development	Consumer products
Light emitting diode (LED)	High luminance, low voltage, solid state	High power consumption, limited number of elements	Mature technology	HMSs, vehicular applications
Digital mirror device (DMD)	High luminance, high contrast and resolution, solid state	Unproved reliability and endurance in harsh environmental conditions	In development	Military (avionic) HMD training and simulation
Fiber-optic image guide	High resolution, high luminance	Heavy and bulky fiber-optic bundle requires special QDC	Emerging technology	Simulators; past attempts for avionic HMDs unsuccessful

light source. The light, coming from the off-helmet-located lamp, is filtered to remove IR and ultraviolet light, to avoid damaging the LCD due to excessive heat, and then is fed into an illumination fiber-optic cable that carries the light to the helmet. The light is passed through three active-matrix liquid crystal light valves (LCLVs) incorporating yellow, magenta, and cyan color polarizers. The outcoming light exits as a full-color image. A schematic illustration of the display is shown in Figure 4.15.

4.6 Summary

Numerous technologies are used for image sources for the HMD. The unique and demanding requirements from the image source have led to the use of almost every potential technology. Table 4.2 summarizes and compares the strengths and weaknesses of each technology as well as its status.

Miniature CRTs offer high resolution, high luminance, and luminous efficient emissive displays. Their main drawback is in their relatively high volume and weight. Also, the need for a high-voltage final anode imposes some risk when used in inflammable environments coupled with the requirement for quick disconnect in cases of ejection. Although a quick disconnect coupler (QDC) is required regardless of the type of image source used, the high-voltage connections compound the problem. Moreover, the lack of CRTs with full color limit their use.

Emerging flat-panel technologies, such as AMLCDs, TFEL displays, and FEDs, offer great promise for meeting the demanding requirements for HMD applications in terms of resolution and luminance. The need for low-cost, full-color, and high-resolution consumer products such as camcorders, for which flat-panel displays are suitable, is pushing the advancement of these technologies. For the same reasons, new technologies such as micromirror devices are promoted.

References

[1] Sauerborn, J., "Helmet-Mounted Tubes for Displays," *Information Displays*, Vol. 6, 1995, pp. 10–13.

[2] Spitzer, M. B., and J. P. Salerno, "High Resolution Miniature AMLCDs for Projection and Head-Mounted Systems," *IEEE AES Mag.*, Vol. 10, No. 4, April 1995, pp. 33–35.

[3] Leyland, J. D., F. Walters, and D. G. Etherington, "Developments in CRTs for HMD Applications," in *Helmet-Mounted Displays II*, SPIE, Vol. 1290, 1990, pp. 30–40.

[4] Masterman, H., C. Johnson, and M. Silverstein, *How to Select a CRT Monitor*, Millwood, VA: Beta Review, 1990.

[5] Karim, M. A., and A. F. M. Yusuf Haider, "Intensifier and Cathode-Ray Tube Technologies," in M. A. Karim, ed., *Electro-Optical Displays*, New York: Marcel Dekker, 1992, pp. 1–17.

[6] Cook, L., and S. Patterson, "Fiberoptics for Displays," *Information Display*, 1991, pp. 14–16.

[7] Brain, J., "Fiber Optics for Displays," *Information Display*, 1994, pp. 14–15.

[8] Blow, B. A., "A Drive Electronics System for Helmet-Mounted Displays Using Mixed-Mode Integrated Circuits for Video Signal Processing," in *Helmet-Mounted Displays II*, SPIE, Vol. 1290, 1990, pp. 30–40.

[9] Sobel, A., "Flat Panel Displays," in M. A. Karim, ed., *Electro-Optical Displays*, New York: Marcel Dekker, 1992, pp. 121–185.

[10] Worboys, M. R., et al., "Miniature Display Technologies for Helmet and Head Mounted Displays," in *Helmet- and Head-Mounted Displays and Symbology Design Requirements*, SPIE, Vol. 2218, 1994, pp. 17–24.

[11] Monarchie, D., R. Budzilek, and F. Cupero, "Sunlight Viewable Electroluminescent Displays for Military Application," *IEEE Systems Mag.*, Vol. 10, No. 8, August 1995, pp. 21–25.

[12] Cathey, D. A., Jr., "Field-Emission Displays," *Information Display*, 1995, pp. 16–20.

[13] Marconi Avionics, "Helmet Mounted Display and Optical Position Sensing System," 1979.

[14] Becker, A., "Design Case Study: Private Eye," *Information Display*, Vol. 6, No. 3, March 1990, pp. 8–11.

[15] Younse, J. M., "Mirrors on a Chip," *IEEE Spectrum*, November 1993, pp. 27–31.

[16] Naor, D., O. Arnon, and A. Avnur, "A Lightweight Innovative Helmet Airborne Display and Sight (HADAS)," in *Display System Optics*, SPIE, Vol. 778, 1987, pp. 89–95.

[17] Thomas, M. L., et al., "Fiber Optic Development for Use on the Fiber Optic Helmet Mounted Display," in *Helmet-Mounted Displays*, SPIE, Vol. 1116, 1989, pp. 90–101.

[18] Koester, C. J., "Wavelength Multiplexing in Fiber Optics," *J. Optical Society of America*, Vol. 58, No. 1, January 1968, pp. 63–70.

[19] Melzer, J. E., and K. W. Moffitt, "Color Helmet Display for the Tactical Environment: The Pilot Chromatic Perspective," in *Helmet-Mounted Displays III*, SPIE, Vol. 1695, 1992, pp. 47–51.

[20] Leinenwever, R. W., L. G. Best, and B. J. Ericksen, "Low-Cost Color LCD Helmet Display," in *Helmet-Mounted Displays III*, SPIE, Vol. 1695, 1992, pp. 68–71.

[21] Post, D. L., "Miniature Color Display for Airborne HMDs," in *Helmet- and Head-Mounted Displays and Symbology Design Requirements*, SPIE, Vol. 2218, 1994, pp. 2–6.

5

HMD Optics

The function of the optical system is to relay the image from the image source, which is usually placed on the top of or to the side of the helmet, to the front of the eye. The constraints of packaging require that the image source be as small as possible. As a result, the miniature image of the image source is too small to be directly viewed by the user with the unaided eye. For that reason, a further function of the optics is to magnify the image to increase its angular subtense to cover a wider FOV. Also, the image must be perceived farther than its actual position so the viewer is able to accommodate it comfortably. If the image is to be seen superimposed on the outside world, it should be collimated so the user is not forced to switch focus between the display and the external scene and so both are seen in focus simultaneously.

The image source and the optics system are placed as close as possible to the viewer's head to reduce helmet unbalance and to maximize the FOV. The image should be seen with the highest resolution and brightness and extend the largest attainable FOV. To comply with all those requirements the display optics must contain several optical elements and yet must be lightweight and take up little volume. To meet compactness without compromising optical quality, the multi-element optical system must be wrapped around the helmet. In that manner, the essential optical path is obtained while the center of gravity of the helmet is shifted only minimally.

The HMD has a functional resemblance to the HUD. They both use similar miniature image sources, and they both relay a collimated image superimposed on the outside world scenery and project it on a combiner in front of the user's eye [1]. Finally, both displays share many system requirements, such as a wide FOV and high brightness, and many similar constraints, such as space and weight restrictions. As a result of this parallelism, the basic optical design of the HMD originally was derived from the design of the HUD.

5.1 The Basics of Optics

5.1.1 Physical and Geometrical Optics

By its nature, light is a wave motion. Physical optics describe the phenomena related to the wave motions, such as diffraction, interference, and scattering, that determine the fine structure and light distribution within an image. In the design of an optical system, it is more convenient to treat the image formation by ray tracing or by so-called geometrical optics.

When an incident light wave reaches a surface separating two media in which the wave propagates, a wave is transmitted into the second medium, and another wave is reflected back to the first medium. These waves are termed refracted and reflected waves, respectively. Figure 5.1 illustrates incident, reflected, and refracted rays.

The phenomena of reflection and refraction are portrayed by the basic law of optics, Snell's law, which describes the change in the direction of light wave propagation when it crosses an interface between two materials having different refractive indexes, at angles other than 90 degrees to the interface. The relation between the angle of incidence and the angle of refraction is given by

$$n_1 \sin\theta_1 = n_2 \sin\theta_2 \tag{5.1}$$

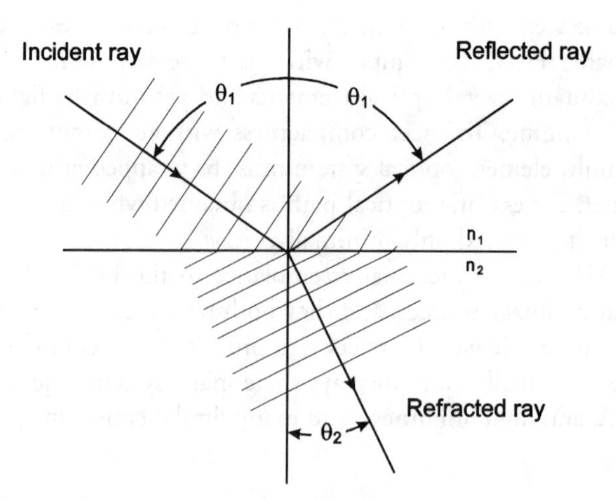

Figure 5.1 Incident, reflected, and refracted plane waves.

where n_1 and n_2 are the refraction indexes of the two materials, and θ_1 and θ_2 are the incident ray angle and the refracted ray angle, respectively.

Snell's law is exploited by the lens, which uses the refraction at a curved surface to converge or diverge a beam of light. Snell's law remains valid when neither the wave surface nor the interface material is a plane, because each point on both surfaces can be regarded as infinitesimally small planes [2].

5.1.2 Lenses

Lenses are the most common elements used in optical systems. The lens is a transparent medium bounded by either two spherical curved surfaces or one curved and one flat surface. The lens acts by refraction to converge or diverge a beam of light. Each ray from the object is mapped by the lens to a corresponding point on the image plane, thus forming an image of the object. This process is called imaging. The procedure of constructing the image is called ray tracing [3].

Consider a convex thin lens whose thickness is very small compared with the radii, as shown in Figure 5.2. If the lens is illuminated with a collimated beam of light parallel to the optical axis, the beam comes to a focus at a distance f to the center of the lens. In the same fashion, if a small point source is placed in the focal point of the lens, a collimated light is produced. If the point source is moved farther away from the focus to point q, the rays no longer are parallel to the axis but instead cross the axis at a finite distance p farther from the lens (Figure 5.3). The relation between those two points and the focal length of the lens is given by the Gaussian lens equation or Descartes's formula for a thin lens:

$$\frac{1}{p} - \frac{1}{q} = \frac{1}{f} \tag{5.2}$$

The lens focal length f is given by the so-called lensmaker's equation:

$$\frac{1}{f} = (n - 1)\left(\frac{1}{r_2} - \frac{1}{r_1}\right) \tag{5.3}$$

where r_1 and r_2 are the two lens surface radii, and n is the index of refraction.

Equations (5.2) and (5.3) are derived by assuming a very small inclination of both the incident and the refracted rays. Such rays are termed paraxial rays. In this case, Snell's law can be approximated by $n_1\vartheta_1 = n_2\vartheta_2$, and the treatment of the optical system is called the first-order theory or Gaussian optics.

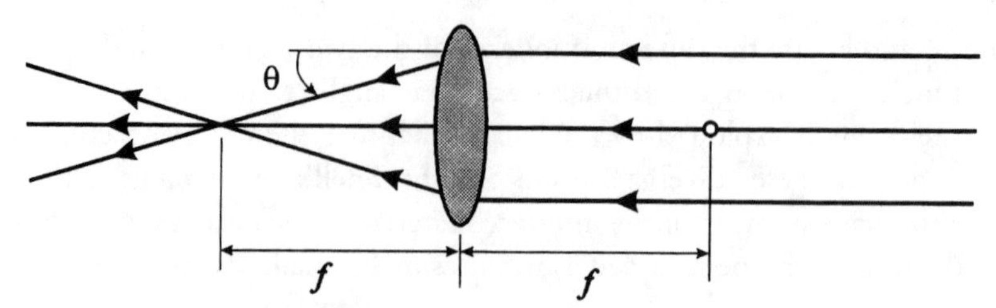

Figure 5.2 Collimated light converges at the focal point of a positive thin lens.

The magnification produced by the lens depends on the position where the object is placed and is given by

$$M = -\frac{q}{p} \tag{5.4}$$

The negative sign in (5.4) indicates that the image is inverted if the object and the image are in positive positions. The magnification may be greater or less than unity, depending on the ratio of the two distances.

If the lens focal length f is positive, the lens is called convergent; if f is negative, it is called divergent. Figure 5.4 illustrates the principal rays for convergent and divergent thin lenses.

The amount of light that passes through a lens is given by its numerical aperture (NA), which is defined by the cone of light rays that pass through the lens:

$$NA = n \sin\left(\frac{\phi_{max}}{2}\right) \tag{5.5}$$

where ϕ_{max} is the cone angle. Alternatively, a lens system often is characterized by the f-number, which is the ratio between the lens focal length f and the aper-

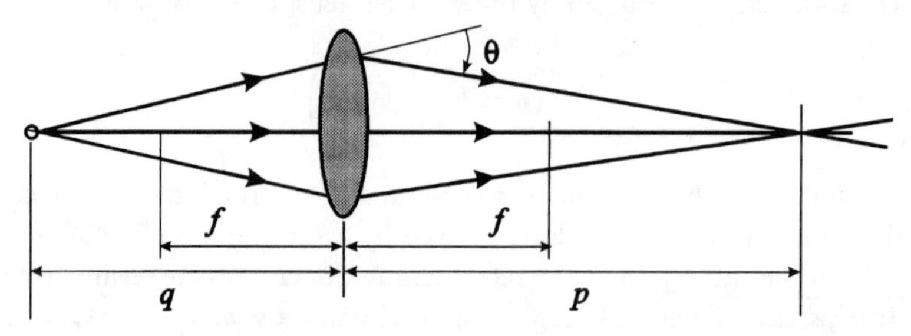

Figure 5.3 Image generation of a point light source by a positive lens. The image position is determined by the Gaussian lens equation (5.2).

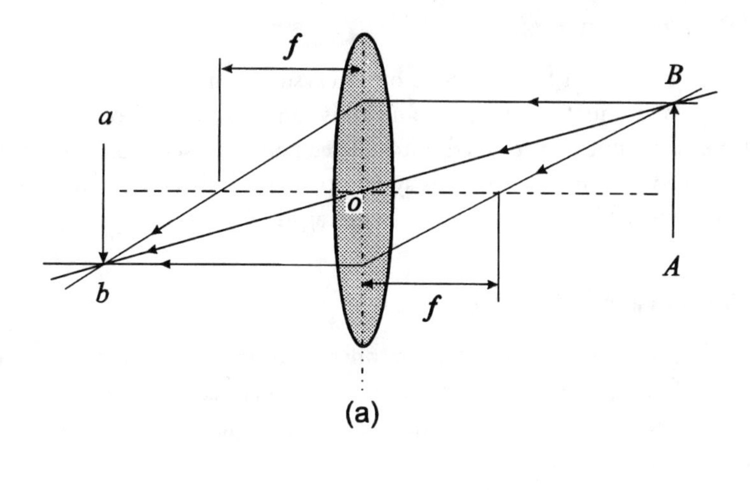

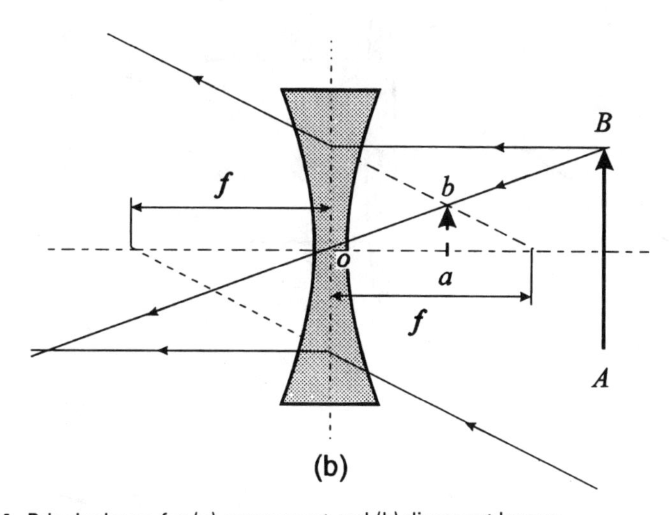

Figure 5.4 Principal rays for (a) convergent and (b) divergent lenses.

ture D. For small cone angles, the relation between the f-number and the numerical aperture is

$$f\text{-number} = \frac{f}{D} \approx \frac{1}{2NA} \tag{5.6}$$

Lenses can be combined in succession to construct a lens system capable of performing a variety of tasks. The property of a lens system is found by successive application of the Gaussian lens equation in which the object to the next lens is the image of the previous lens.

5.1.3 Real and Virtual Images

The image of an object formed by an optical system can be real or virtual. A real image is formed outside the system and thus can be projected onto a screen. A virtual image cannot be projected onto a screen and can serve only as an object to be reimaged by a subsequent lens system or viewed directly. Figure 5.5 shows the real and virtual images of a real object imaged by lenses.

5.1.4 Mirrors and Reflection

The simplest reflecting objective is the spherical mirror. If an object is placed in the center of curvature of a mirror, its image also appears at the center of curvature, and the image has no aberrations of any kind. If the object is moved

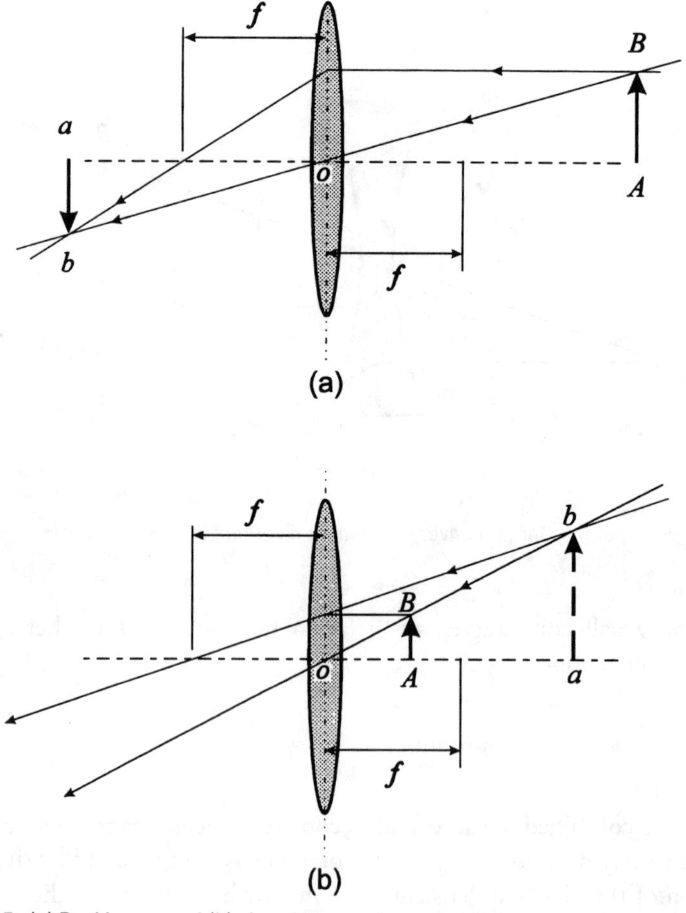

Figure 5.5 (a) Real image and (b) virtual image of a real object.

away from the center of curvature, the image moves, according to the Descartes's formula for reflection at a spherical surface, as in a lens:

$$\frac{1}{p} + \frac{1}{q} = \frac{1}{f} \tag{5.7}$$

The focal length is at the midway point between the center of curvature and the mirror center; that is, $f = r/2$. The reflection of a ray by a spherical mirror is illustrated in Figure 5.6. The angles of incidence and reflection of the ray at the mirror are equal. P is the source point, Q is its image, and C is the curvature center.

Among other types of curved mirrors, parabolic mirrors often are used. The parabolic mirror has the property that all rays entering the mirror parallel to the axis converge at its focus. Off-axis parabolic mirrors frequently are used in wide-FOV HUDs. The disadvantage of a parabolic mirror is the presence of large amount of coma aberration due to the increasing focal length of rays that are more distant from the mirror axis (coma aberration is explained in Section 5.2). Figure 5.7 shows an off-axis parabolic mirror.

Reflective or mirror optics have the advantages of freedom from chromatic aberrations (described in Section 5.2), and reduced weight compared to equivalent refractive optics. Those advantages are significant when wide-FOV optics and, consequently, large optical components are required.

5.1.5 Catadioptric Systems

A catadioptric system is an optical system that employs both reflective and refractive components to achieve its optical power [4]. Usually, most of the

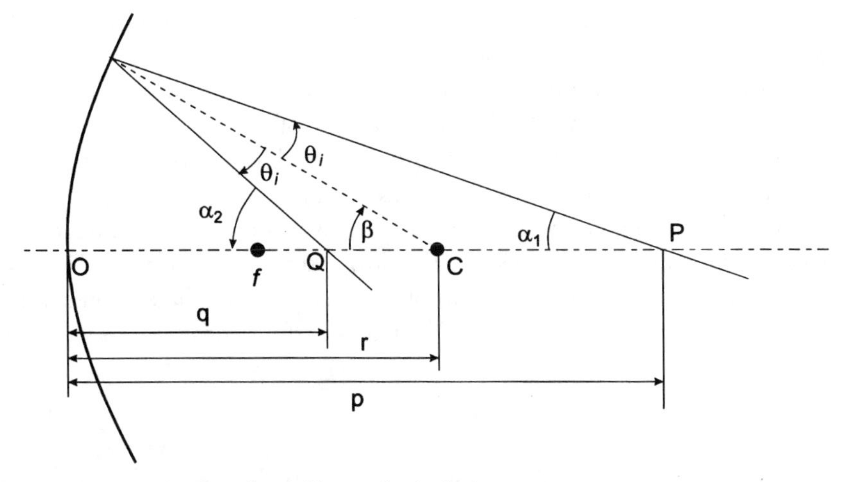

Figure 5.6 Path of a ray reflected by a spherical mirror.

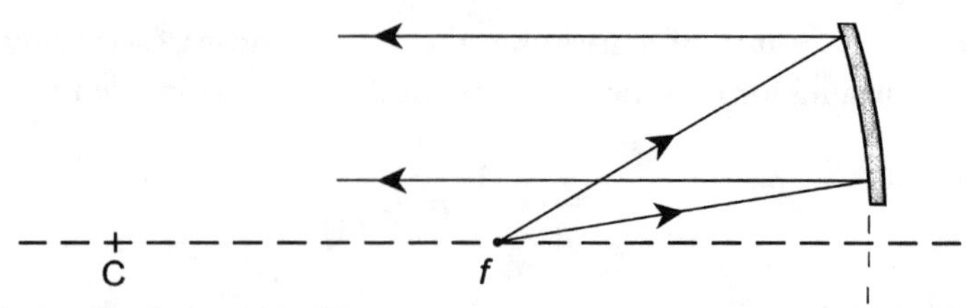

Figure 5.7 An off-axis parabolic mirror.

optical power is obtained from the reflective mirror. The refractive lens has little or zero power and is used to correct the system's aberrations [5].

5.1.6 Prisms

Prisms are used in optical systems to deflect or deviate beams of light. Prisms are blocks of optical material with flat polished sides that are oriented at precise angle with respect to each other. The prism functions in most applications as a set of flat mirrors that are precisely aligned. In that regard, they are superior to flat mirrors, which are sensitive to mechanical structure deformations. The main use of prisms is to fold the optical path to make the system more compact, although at the expense of higher weight. Prisms may or may not invert the image, and they cannot produce a real image, only a virtual image. Also, they must be used only in collimated beams; otherwise, they introduce aberrations. Figure 5.8 show three common types of prisms.

5.1.7 Diffraction Effects and the Airy Disk

Diffraction phenomena are the consequence of the wave nature of light. In geometrical optics, it is assumed that light travels in straight-line rays. Light, however, propagates in a wave motion, and each point on the wave front can be regarded as being a source of spherical wavelets that in turn form a new wave front. When the wave front passes through a small aperture with a size on the order of the wavelength, which limits the passage of rays of close vicinity only, and is focused on a screen, the wavelets reinforce each other to produce a bright area or interfere with each other to produce a dark zone [3].

A single point in the object field is projected as a set of concentric rings of light having rapidly diminishing intensities. This image structure is called an Airy disk. The disk, shown in Figure 5.9, consists of a central core of light surrounded by a series of dark and bright rings. The angular radius of the first dark ring is given by

$$\alpha = \frac{1.22\lambda}{D} \tag{5.8}$$

where λ is the light wavelength, and D is the lens diameter.

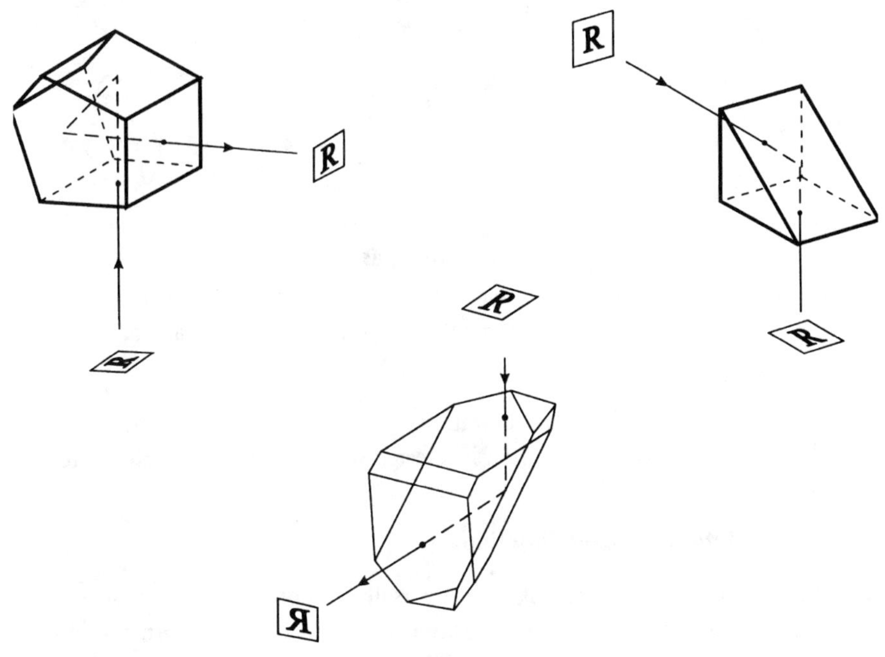

Figure 5.8 Three common prisms used to turn a beam by 90 degrees.

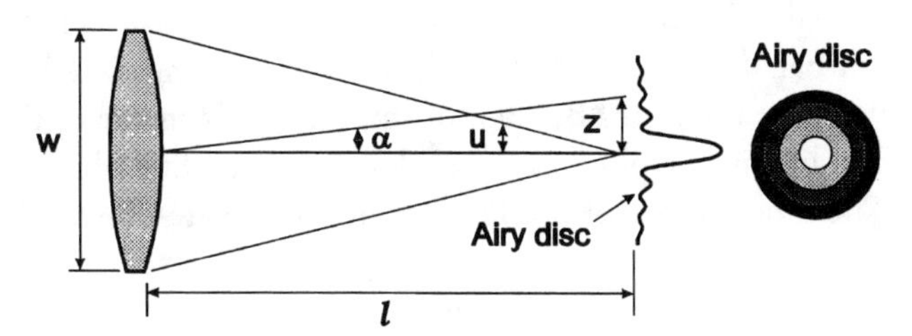

Figure 5.9 The distribution of illumination at the image point and the appearance of an Airy disk.

Refraction, evident by the Airy disk, constitutes a limit on the resolution, or resolving power, of an optical system. If two source points, that are close to each other are imaged by the lens as bright light disks, their diffraction patterns may overlap fully or partially. The resolution limit is the minimal separation in which the images still can be distinguished by the optical system.

The common criterion for resolution limitation is the Rayleigh criterion [3], shown in Figure 5.10, in which the first dark ring of one image point coincides with the other point bright spot. The dashed lines represent the diffraction patterns of two point images at various separations, while the solid lines describe the combined pattern.

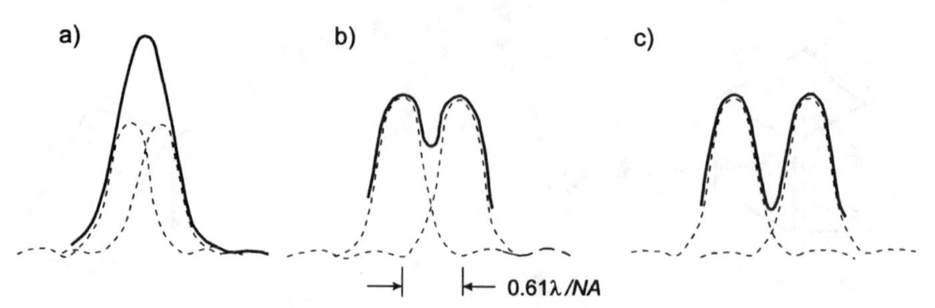

Figure 5.10 The criterion for limiting resolution of an optical system: (a) an unresolved pattern, (b) the Rayleigh criterion, and (c) a clearly resolved pattern.

A lens system in which the aberrations are so small that the image of a point is no larger than the Airy disk, often is referred to as being diffraction limited.

5.1.8 Modulation Transfer Function

The MTF of an optical system is the universally accepted measure for the quality of the system. It implies to what extent the optical system forms a faithful image of the object. The MTF describes how the contrast of an object is affected by the lens system at various spatial frequencies. The concept of the MTF is illustrated in Figure 5.11 for a bar target with alternating light and dark bars. The light bars have a specified brightness and spacing. The bars target is imaged by the optical system. Because of imperfections in the optical system, each geometric line in the object is imaged as a blurred line; thus, its contrast is reduced. The modulations of both object and image are expressed by

$$M = \frac{I_{max} - I_{min}}{I_{max} + I_{min}} \tag{5.9}$$

where I_{max} and I_{min} are the maximum and minimum illuminances of the target or image bars. The modulation is computed for both object and image for various spatial frequencies $1/N$. The MTF is computed by the ratio of the two modulation functions of the object and the image:

$$\text{MTF}(\nu) = \frac{M_i(\nu)}{M_o(\nu)} \tag{5.10}$$

where $M_i(\nu)$ and $M_o(\nu)$ are the modulations of the image and the object, respectively, and ν is the spatial frequency, $\nu = 1/N$. Figure 5.11 shows that, as the spacing between the bars decreases, the contrast of the image increases and thus the MTF is reduced.

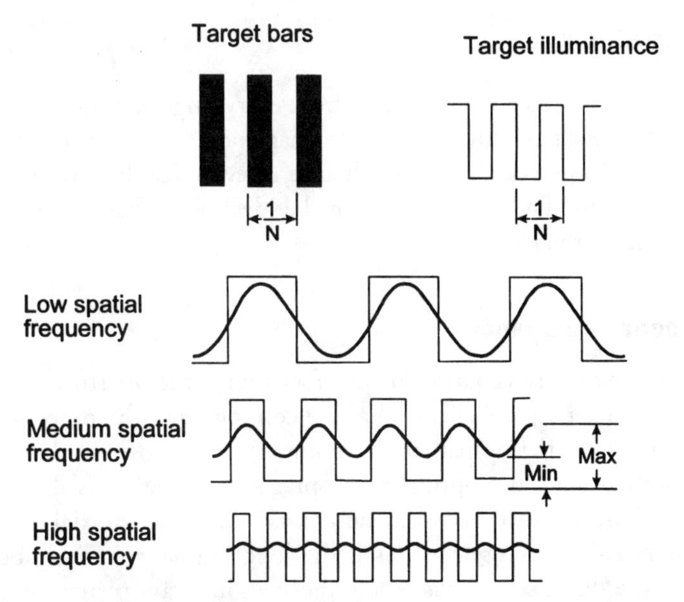

Figure 5.11 The concept of modulation and MTF as reflected by the imagery of a bar target. The optical system round off the sharp edges of the square-wave target. As the test pattern becomes finer, the image contrast is reduced.

Figure 5.12 shows a typical plot of the MTF of an optical system. The upper curve shows the diffraction limit of the system, and the other two curves show the modulations at the center and the peripheral areas.

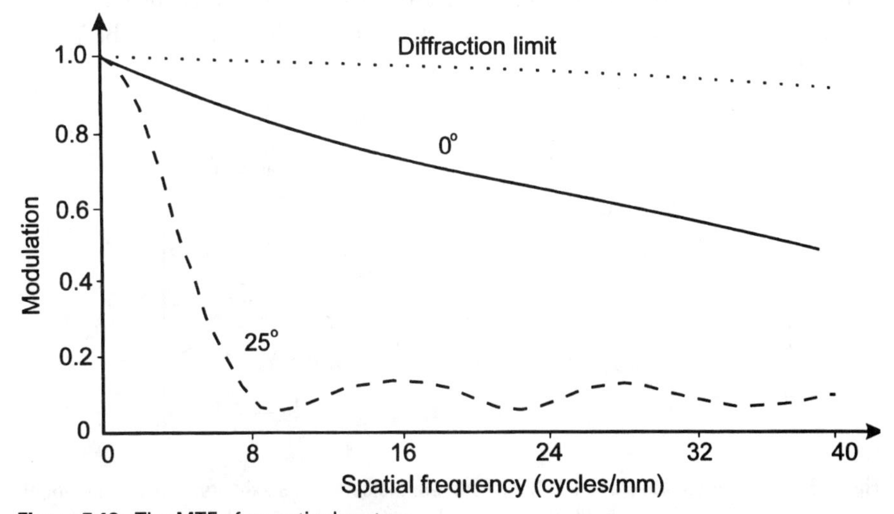

Figure 5.12 The MTF of an optical system.

5.2 Aberrations

In a perfect optical system, all light rays from every single point of an object on one side of the lens pass through a single image point on the other side of the optical system. In practice, the performance of every real lens system departs from the ideal lens. The deviations from the ideal performance are termed as the aberrations of the lens.

5.2.1 Spherical Aberration

Spherical aberration is variation of the focus with the aperture. For lenses with large diameters, parallel rays that are coming from the outer periphery of the lens intersect the optical axis at points that are closer to the lens than rays that are closer to the optical axis. Spherical aberration is illustrated in Figure 5.13. The distance between the points where the paraxial focal plane and the marginal rays cross the axis is the longitudinal spherical aberration (LSA), and the distance off axis where the marginal rays pierce the paraxial focal plane is the transverse spherical aberration (TSA). Spherical aberrations can be minimized in several ways. One way is to block the marginal rays by employing aperture stop or, better, to use a smaller-diameter lens. Different orientation of the lens also may reduce spherical aberration, for example, a plano convex lens, when oriented with its flat surface toward a collimated beam, exhibits more aberration than the same lens with its curved surface oriented toward the collimated beam. Finally, using a two-lens combination that minimizes the outer-ray bending reduces the spherical aberrations but with a penalty of increased weight. Spherical aberrations must be corrected before other aberrations can be reduced. A lens corrected for spherical aberration and coma is aplanat.

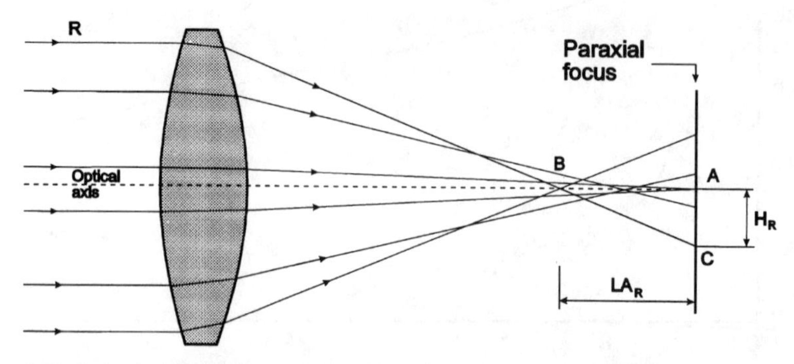

Figure 5.13 Spherical aberration. When reversed, the same lens exhibits a different amount of spherical aberration.

5.2.2 Coma

An off-axis aberration in which the image of a point appears as a comet-shaped blur (*coma* is Latin for "comet"). It is the result of the variation of the lens magnification with aperture. The coma is illustrated in Figure 5.14.

5.2.3 Astigmatism

The plane in the optical system that includes the optical axis and the object point is called the meridional plane. The plane that contains the object point but is perpendicular to the meridional plane is called the sagittal plane. Rays in the meridional plane come to focus at a point closer to the lens than rays in the sagittal plane. Due to astigmatism, the image of a point is not a point but appears as a pair of focal lines, as illustrated in Figure 5.15. Between the two lines, the bundle of rays have a minimum diameter, called the circle of least confusion.

In a simple lens, astigmatism can be reduced by judicious use of stops or apertures that limit the size of rays passing through the lens. Full correction of astigmatism, however, requires fabrication of multi-element lenses. A lens corrected for astigmatism is called anastigmat.

5.2.4 Field Curvature

Object points that lie in a plane are not imaged in a plane but rather on a paraboloidal surface called the Petzval surface. In positive lenses, the Petzval surface

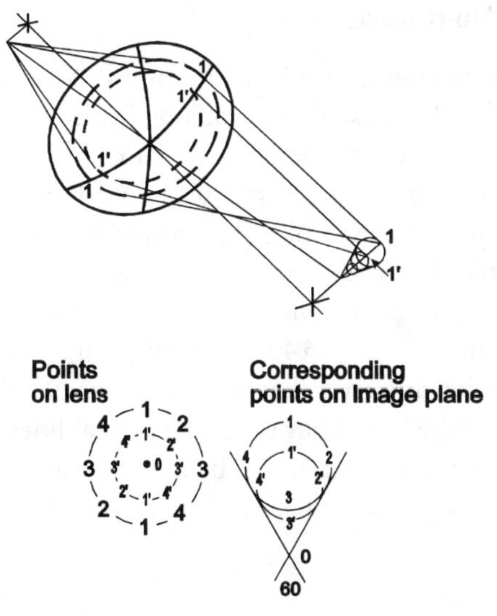

Figure 5.14 A coma results as rays coming from the periphery of the lens focus at points higher above the optical axis than paraxial rays.

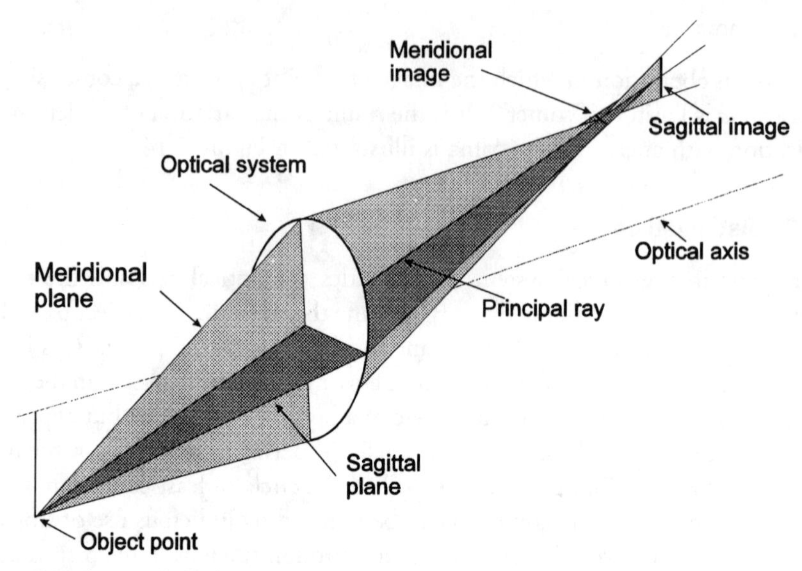

Figure 5.15 Lens astigmatism.

is curved toward the lens; in negative lenses, it is curved away from the lens. The field curvature is corrected using a combination of positive and negative lenses. When the lens is corrected for astigmatism, its meridional and sagittal planes are flat and coincide with the Petzval surface (Figure 5.16).

5.2.5 Chromatic Aberrations

Axial chromatic aberration is the focus variation along the longitudinal axis with the wavelength. The cause for the focus variation is the dispersion of light when it passes through the lens. Because the index of refraction is different for each wavelength, each wavelength is focused at a different point. The effect of chromatic aberration can be exemplified by considering a ray of light in air incident on a piece of flat flint glass at $\theta_i = 25°$, with the refractive index for red n = 1.6658, green n = 1.673, and blue n = 1.6808. Using Snell's law (5.1), the refraction angles are θ(red) = 14.6967°, θ(green) = 14.6321°, θ(blue) = 14.5627°. The angular difference between the extremes (i.e., red and blue) is 0.134° or 2.3 mrad, which is about twice the typical linewidth of an HMD.

Chromatic aberration is minimized by the use of achromatic doublets composed of a positive and a negative lens, each having a different refraction index.

5.2.6 Distortion

Distortion is the change in transverse magnification with the distance of the image point off axis. When the distortion is positive, the magnification increases

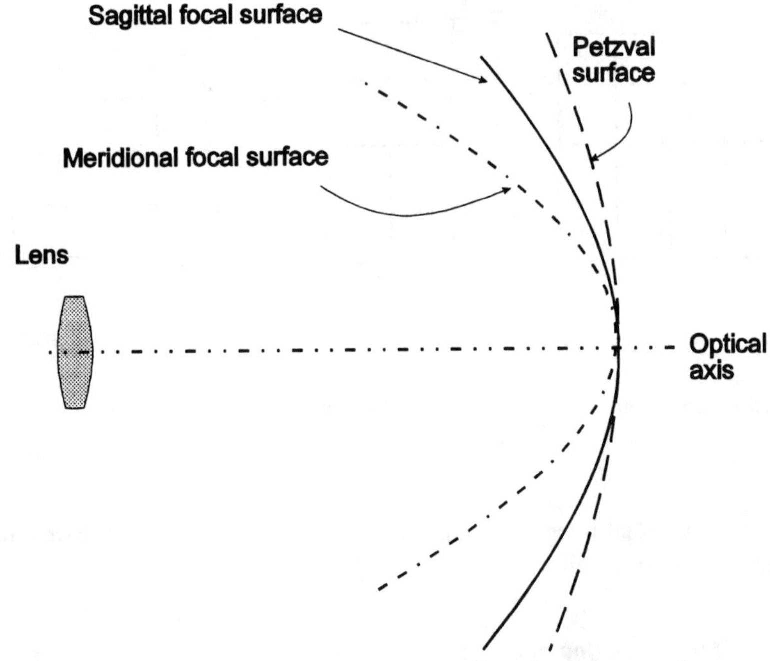

Figure 5.16 The sagittal and meridional images of an object. As the image point moves away from the axis, the amount of astigmatism increases.

and results in pincushion distortion. In negative distortion, the magnification decreases, resulting in barrel distortion (Figure 5.17). Distortions smaller than 1% are not noticed by most people.

5.3 Fundamentals of HMD Optics

The main role of an HMD optical system is to provide the user a clear, sharp, and bright image of the CRT superimposed on the real world. Because the user needs to maintain constant visual contact with the outside world, both the real world scene and the display image must be in focus simultaneously, so the image must be projected at optical infinity. The process of presenting to the viewer a close image that is perceived as projected at infinity is called collimation.

It is almost always desirable that the image of the display be brought as close to the eye of the wearer as possible, for two main reasons: (1) the closer the image is to the eye, the closer the optics elements are to the head, so the center of gravity shift is minimal, and (2) a closer image enables achievement of a wider FOV for a given image size. However, for distances closer than 25 cm, the image cannot be pleasantly viewed by the user. Therefore, the optical system is required

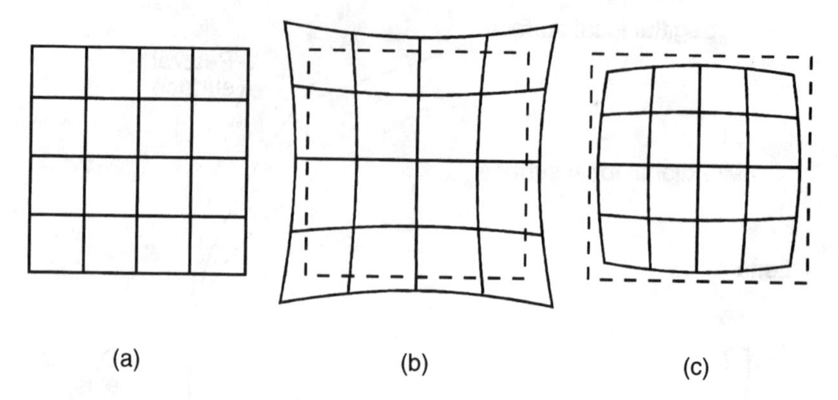

Figure 5.17 Distortion: (a) undistorted image; (b) positive, or pincushion, distortion; (c) negative, or barrel, distortion.

to generate a virtual image so the eye will perceive the image as if it were more distant or even at infinity.

5.3.1 HMD Optics Components

Almost all HMD optical systems comprise several or all of the following basic optical elements:

- An *eyepiece,* which is a lens system with external exit pupil. The exit pupil is designed to have its location at the user's eye.
- A *field lens,* which bends the ray bundles toward the optical axis to redirect them into the eye lens. The use of a field lens allows the use of a small-diameter eye lens.
- A *relay lens system,* which conveys an image along a relatively long distance without increasing the diameter of the lens.
- A *field flattener,* which is a plano lens used near the image plane to correct aberrations, mainly field curvature, by reshaping the image surface without drastically affecting the effective focal length, image size, and location.
- *Folding prisms,* which are used for beam deflection and deviation. Together with flat mirrors, folding right-angle prisms are used to fold the optical path to reduce the volume of the optical system and achieve more compact optical systems.
- *Beam splitters,* which are used to combine two images. Beam splitters, or combiners, are semitransparent mirrors and are used mainly to combine the display image with the direct view scene.

5.3.2 Occluded Versus See-Through Displays

The simplest way to view a close image is by using a magnifying lens. In that approach, the image is placed between the lens and the focus of the lens, so the image falls at a distance equal to or larger than the minimal distance of distinct vision. This approach is adopted in many low-cost occluded displays.

However, for see-through displays, in which the image is to be superimposed on the real world scene, the image must be perceived at much larger distances. Otherwise, when the viewer accommodates on distant objects, the display would appear blurred. Therefore, the image must be produced at a significantly larger distance, so it will be sharp when accommodating on the external scene. Consequently, in that type of display, the optical system is much more complex [6,7].

The fusion of the display image with the outside scene view is achieved by insertion of a combiner in front of the viewer's eye. A combiner is a semi-transparent mirror that enables a direct view of the scene and also reflects the display image to the eye. In most cases, the combiner is optimized for the specific wavelength matching the wavelength of the image source. As a result, most of the display source illuminance is transmitted to the viewer, while ambient light at that wavelength is blocked. In that manner, the image source illuminance is attenuated minimally and the display contrast is enhanced.

Most existing HMDs are basically in the form of a microscope erecting eyepiece, which consists of an eyepiece and a relay lens system with a field lens imaging the aperture of the relay to form the exit pupil at the eye [8]. The principle of the erecting eyepiece is shown in Figure 5.18.

5.3.3 Refractive Versus Reflective Optics

Two basic types of optical design can be used in HMDs: refractive and reflective. Refractive optical systems are simpler in construction, have a lower cost, and are characterized by low aberrations, higher weight, and limited FOV. Reflective systems typically have lower weight, are more complex, and employ

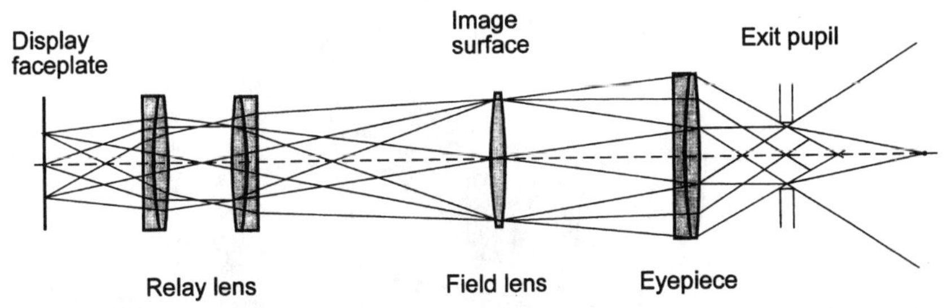

Figure 5.18 The microscope erecting eyepiece, which is the basis for most HMD optical designs.

higher-cost optical elements with higher aberrations; they require special corrective measures but offer a wider FOV.

5.3.4 Refractive Optics Design

In refractive systems, a refractive lens is used for collimation and a flat dichroic or wavelength-selective combiner reflecting the particular wavelength of the display image. The optical performance of the system approaches diffraction limits over most of the FOV [9,10]. The objective of such a display has a large diameter, and the image location is close to the combiner, to maximize the FOV; thus, the eye relief is relatively small.

An example of a refractive type of optical system is the IHADSS display, shown in Figure 5.19.

5.3.5 Reflective Optics Design

Reflective optics design, or more specifically catadioptric optics design, enables achievement of a significantly larger FOV. Reflective designs use curved combiners that perform image collimation. Often, the helmet visor is utilized as the

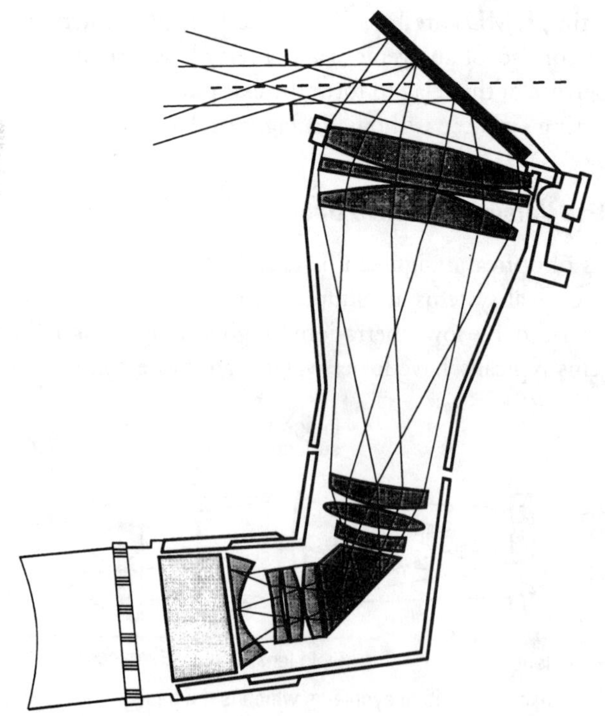

Figure 5.19 The optical system of the IHADSS HMD.

curved combiner [11–13]. There are two basic types of reflective systems: on-axis, or rotationally symmetric geometry and off-axis, or rotationally nonsymmetric geometry [11,12].

On-axis systems use a flat beam splitter, ideally with 50% reflection at the wavelength of the CRT phosphor, and a spherical semitransparent combiner. The image from the optic relay system is projected onto the flat beam splitter reflected on the combiner, which in turn generates an image in its center of curvature where the eye is placed. The role of the beam splitter is to fold and reproduce the reflector's center of curvature in such a way that in one point an aerial image is generated and in the other point the eye is placed.

The advantage of this method is its capability to provide a display with a wide FOV, large exit pupil, low distortion, and low weight [14]. The complexity of the optical system is comparable to the refractive systems because only spherical elements are utilized. The disadvantage of such a design is the reduction in the brightness of the displayed image as it passes twice through the flat beam splitter. Typically, such systems transmit 15% to 20% of the CRT luminance [12].

A further increase of the eye relief can be achieved by tilting the beam splitter and relay optics. That arrangement allows moving the beam splitter, which is the closest optical element to the eye, farther from the wearer. Typical on-axis catadioptric design is shown in Figure 5.20.

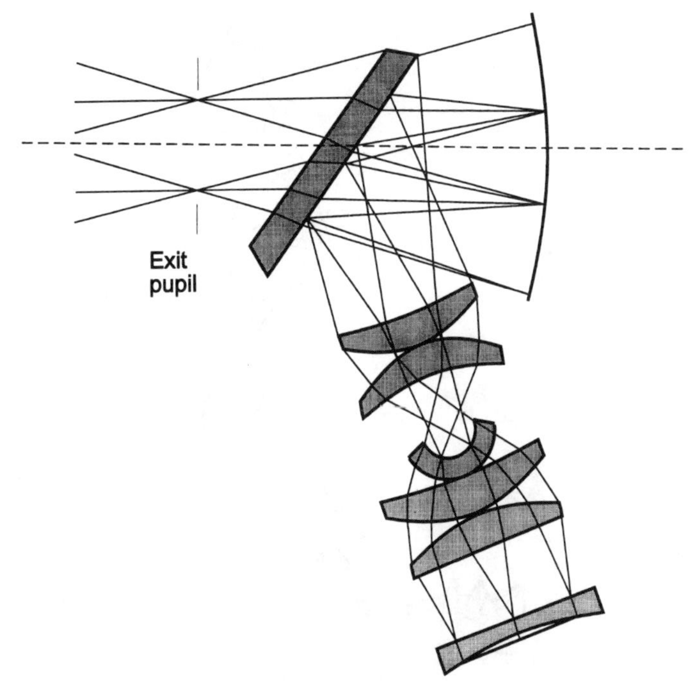

Figure 5.20 On-axis catadioptric HMD optical system.

Off-axis displays offer the capability of providing a very high FOV of 50 to 60 degrees. Together with the wider FOV, the tilted systems make it easier to move any obstructing elements from the viewer's FOV. That is achieved by tilting the combiner with respect to the optical axis of the display. The principal ray at the center of the field reflects at an offset or fold angle. The need to keep the visor close to the wearer's face while mounting the optics on the helmet usually requires fold angles of 40 to 60 degrees. That results in aberrations in the form of astigmatism and distortion.

Astigmatism can be corrected by using tilted or decentered surfaces. Other alternatives require using ellipsoidal or toroidal surfaces for the combiner. At larger fold angles, the nonspherical combiner must be used. In cases when the distortion exceeds about 5%, it commonly is compensated electronically on the display device by generation of a reciprocally distorted image.

Figure 5.21 shows an off-axis tilted catadioptric lens system with spherical combiner. The display optics include the following elements: combiner with collimating optics, relay lens with turn prism, tilted field lens, and second relay lens system with turn prism. The combiner includes a beam splitter with angle sensitive coatings.

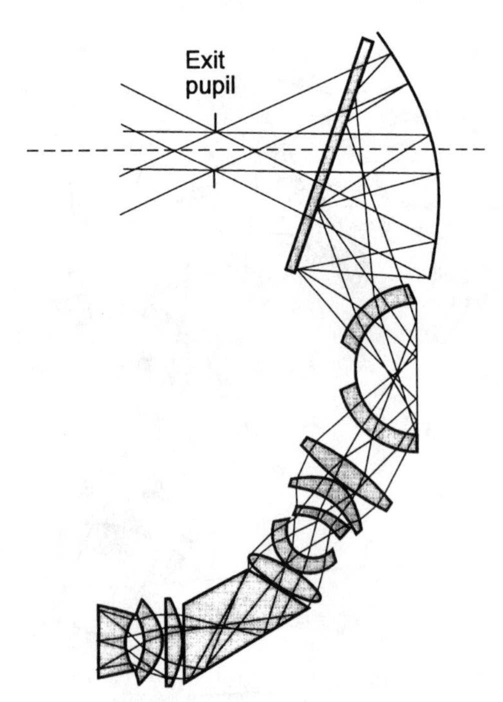

Figure 5.21 Off-axis tilted catadioptric lens system with spherical combiner.

An alternative for a very wide FOV of about 80 degrees, high-resolution display is the Pancake Window™, which was adopted from the flight simulator field [15–17]. The Pancake Window, described in Figure 5.22, is composed of spherical and flat elements. The spherical part includes a spherical beam splitter mirror enclosed between a linear polarizer and a quarter-wave plate [15,16]. The flat part comprises a front flat beam splitter, a quarter-wave plate, and a back linear polarizer. The spherical mirror in the Pancake Window also is a beam splitter. When an object is placed behind the spherical mirror, it is essentially as if the object is placed at its focus, hence emerging as a collimated image. The polarizers and the quarter-wave plates have the function of blocking the direct vision of the object, which otherwise would be transmitted through the two beam splitter mirrors.

With the Pancake Window, FOVs in excess of 80 degrees are attainable. The main drawback of the Pancake Window is the poor transmission of the display image, typically less than 10%.

5.4 Diffraction Optics and Holographic Optical Elements

Significant improvement of the optical performance of an HMD potentially can be achieved by the substitution of conventional optical elements with holo-

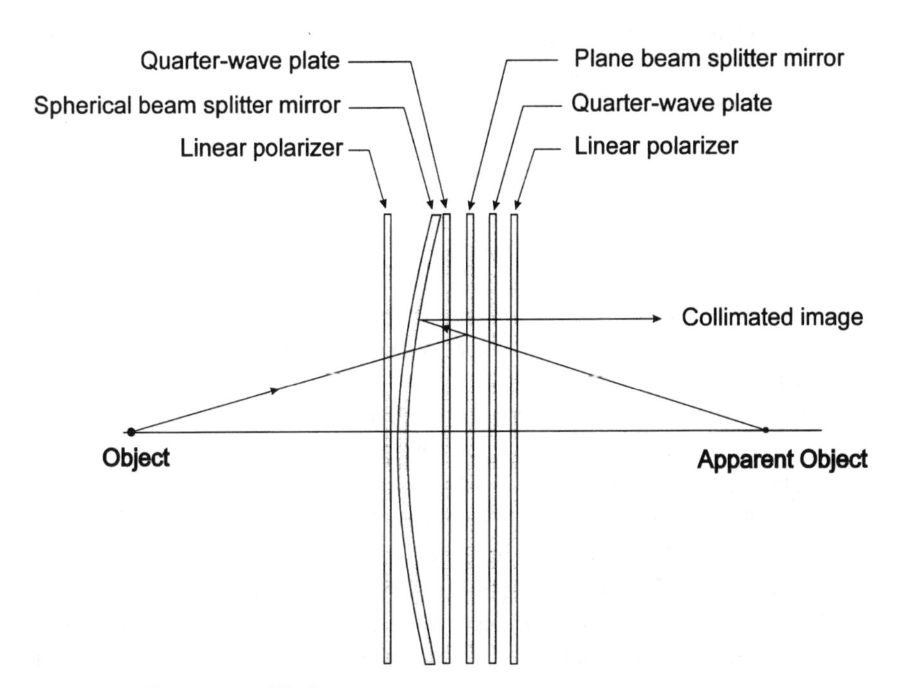

Figure 5.22 The Pancake Window.

graphic elements, mainly holographic combiners. The benefits of holographic combiners are significant. The spectral bandwidth of the hologram can be accurately matched to the spectral bandwidth of the image source, so the outside-world transmission is improved substantially while display brightness is maintained. Other advantages are more flexibility in the design and use of complex optical components, compactness, and lower-weight optical systems.

Holographic optical elements (HOEs), also called diffractive optical elements (DOEs), have many advantages over conventional optical elements. They are lighter and more compact, and, more important, they can perform complex optical operations that cannot be performed by conventional optics [18,19].

Many conventional optical elements can be replaced by equivalent DOEs, including lenses, mirrors, notch filters, and combiners [20–23]. Diffractive elements, in contrast to conventional optical elements such as lenses and mirrors (which use refraction or reflection for imaging functions), have grating-line structure and work by diffraction [24].

The advantage of holography is its ability to reflect a narrowband wavelength efficiently while enabling complete see-through capability.

Diffractive optics can perform many optical functions (including correction of aspheric, chromatic, and higher-order aberrations), tilt, and decenter functions by acting on the incident wavefront. Because most wide-FOV designs dictate the use of optical elements that are decentered, tilted, cylindrically shaped, or aspheric, the possibility of replacing them with holographic elements constitutes a substantial advantage.

Holography is based on the Bragg effect in the thick gelatin. Thick, or volume hologram, is a hologram in which the three-dimensional interference pattern is recorded and used in depth.

Holographic lenses are generated by using two coherent laser beams, one called the reference wave and second the object beam. A photosensitive plate is placed between the beams, and the diffraction gratings are recorded on the plate. The recording process creates a transmission hologram, because both beams are impinging from the same side of the plate. The resulting fringe pattern is such that each fringe surface, which crosses the HOE in the depth dimension, acts as a mirror reflecting the incident ray through the hologram. When the hologram is illuminated by a coherent light source from point *R*, the beam converges at the object point *O*, imitating the action of a lens. The recording and readout processes are shown in Figure 5.23, in which the reference beam is a diverging beam while the object beam is a converging beam.

A reflection hologram, acting like a mirror, is created by the interference of a diverging reference beam and a converging object beam, as illustrated in Figure 5.24. In that recording, the two beams impinge from opposite sides of the photosensitive plate.

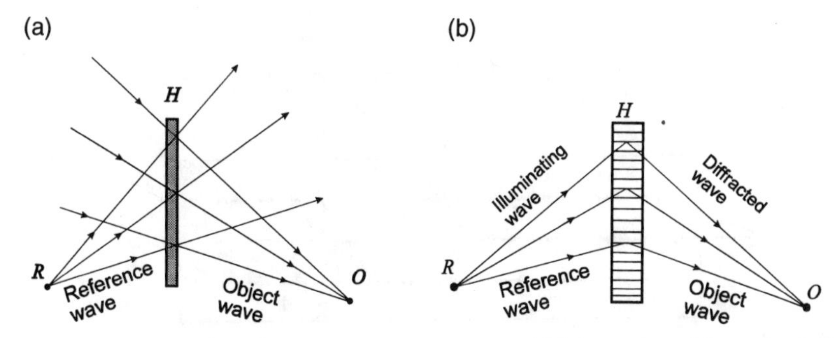

Figure 5.23 (a) Recording of a transmission hologram and (b) the action of the transmission hologram as a lens.

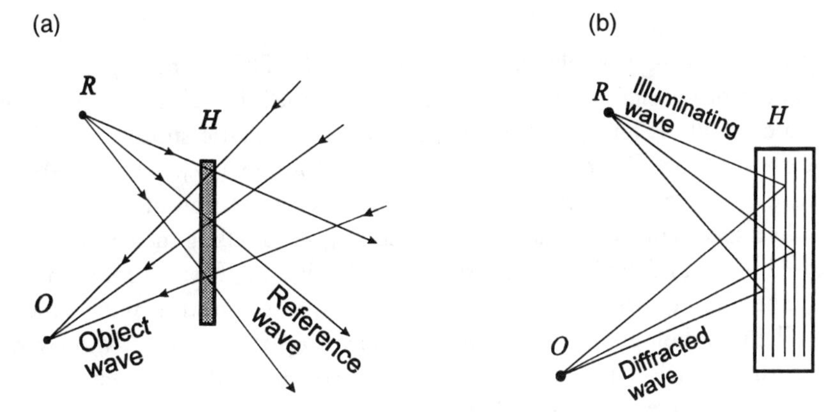

Figure 5.24 (a) Recording of a reflection hologram and (b) the action of the reflection hologram as a mirror.

The fringe pattern in the reflection hologram is now flat and oriented in parallel to the substrate, and the fringe spacing is according to Bragg's law:

$$2d \sin \alpha = K\lambda / n \tag{5.11}$$

where d is the fringe spacing, λ is the wavelength, α is the refraction angle of the ray in the substrate (according to Snell's law), n is the refraction index of the substrate, and k is the diffraction order and a positive integer. The grating structure of the reflection hologram illustrating Bragg's law is shown in Figure 5.25.

Volume HOEs for HMDs are made almost exclusively from dichromated gelatin (DCG). Gelatin is a fibrous protein collagen that comes from animals' skin and bones. It is made sensitive to light by chemical sensitization through the addition of such materials as ammonium dichromate [25,26]. The HOE is

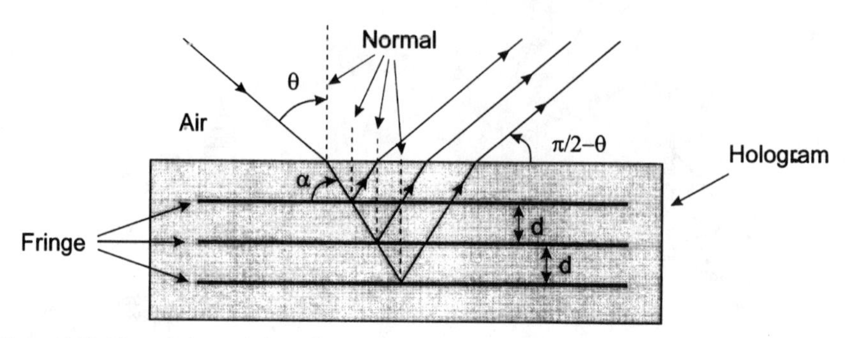

Figure 5.25 The gratings of the reflection hologram and the Bragg's law outline.

prepared by coating the substrate with light-sensitive DCG film, exposing it to the pattern of interference fringes using a laser at the selected wavelength. The substrate usually is an optical quality glass or some plastic substrates such as polycarbonate, acrylic, or CR39 ophthalmic substrate [27]. The pattern is recorded on the gelatin film and transformed to a modulation of the refractive index by a special process. That process is washing and swelling the film in water, followed by rapid dehydrating and shrinking the film in propanol solutions.

The main drawback of DCG is its sensitivity to moisture and temperature. The final hologram must be tightly sealed to prevent contact with water or moisture, which may reduce the diffraction efficiency of the system or alter the optical characteristics of the hologram.

Diffraction efficiency is a measure of the performance of a DOE. It is defined as the ratio of the useful diffracted light flux to the total light flux used to illuminate the hologram. The diffraction efficiency of a DCG thick hologram approaches 90% to 100%.

Holographic elements are used in HMDs in two ways: (1) as wavelength-selective holographic combiners, which are employed in conjunction with refractive optics to enhance the photometric characteristics of the display, and (2) as full-power optical elements capable of combining optical characteristics of large-volume, complex optical systems in a single thin hologram. The optical elements that are replaced by HOEs are mirrors, combiners, and collimators.

HOEs or DOEs function optically to the first order, independently of the physical shape of the element. As such, they can implement complex optical operations that cannot be achieved with conventional optics. Furthermore, holographic elements can be readily produced on arbitrarily shaped surfaces such as helmet visors. Such holograms that are conformal with the substrate curvature are called conformal holograms.

5.5 Summary

The function of an HMD optical system is to provide the user with a high-quality collimated image of the display over the required FOV, while meeting the other system objectives and constraints.

Because the characteristics of HMD optical system are common to HUDs, originally the optical system designs were directly adopted or modified HUD designs. The constant quest for wider FOV and lower weight, however, prompted the introduction of more sophisticated and complex designs based on catadioptric system concepts and the use of multiple holographic elements.

The optical system is constructed by selection among the many available basic optical elements. The number of possibilities of component selection is very large, so there are many potential solutions for the specified requirements of a system.

Almost always, the resulting optical system for an HMD involves trade-offs among several constraints. Those constraints include requirements for minimum aberrations, maximum display brightness, large FOV and exit diameter, minimum obscurations, low weight, compactness, and proximity to the head center of gravity.

The approaches for occluded HMDs usually are based on classical microscope eyepieces. See-through displays employ relay lens systems and semitransparent mirrors, which combine the display image with the outside world view. In wider FOV, catadioptric systems are almost mandatory, and the tendency is to use the helmet visor to function as the collimating and combining element of the optical system.

A significant weight reduction of the HMD can be achieved by incorporating in the optical system HOEs or DOEs, which can replace many conventional elements. Apart from the lower weight and compactness, diffractive optics are capable of performing complex optical operations that cannot be achieved with conventional optics.

References

[1] Wood, R. B., "Holographic Head-Up Displays," in M. A. Karim, ed., *Electro-Optical Displays*, New York: Marcel Dekker, 1992.

[2] Alonso, M., and E. J. Finn, *Fields and Waves*, Vol. II of Fundamental University Physics, Reading, MA: Addison-Wesley, 1967.

[3] Smith, W. J., *Modern Optical Engineering*, New York: McGraw-Hill, 1968.

[4] Smith, W. J., *Modern Lens Design*, New York: McGraw-Hill, 1992.

[5] Kingslake, R., *Optical System Design*, Orlando, FL: Academic Press, 1983.

[6] Jacobs, R. S., T. J. Triggs, and J. W. Aldrich, "Helmet Mounted Display/System Study," Air Force Flight Dynamics Laboratory, Tech. Report AFFDL-TR-70-83, March 1971.

[7] Fischer, R. E., "Optics for Head-Mounted Displays," *Information Display*, Vol. 10, No. 7–8, July–August, 1994, pp. 12–16.

[8] Kocian, D., and H. L. Task, "Fundamentals and Optics of Helmet-Mounted Displays," Short Course Notes, SPIE, *Internat. Symp. on Optical Engineering in Aerospace Sensing*, April 1994.

[9] Faklis, D., and M. J. Hoppe, "Effects of Diffraction on the Performance of Diffractive Relay Optics," in *Helmet- and Head-Mounted Displays and Symbology Design Requirements*, SPIE, Vol. 2218, 1994, pp. 115–119.

[10] Cox, J. A., T. A. Fritz, and T. Werner, "Application and Demonstration of Diffractive Optics for Helmet-Mounted Displays," in *Helmet- and Head-Mounted Displays and Symbology Design Requirements*, SPIE, Vol. 2218, 1994, pp. 32–40.

[11] Rotier, D. J., "Optical Approaches to the Helmet Mounted Display," in *Helmet-Mounted Displays*, SPIE, Vol. 1116, 1989, pp. 14–18.

[12] Droessler, J. G., and D. J. Rotier, "Tilted Cat Helmet Mounted Display," in *Helmet-Mounted Displays*, SPIE, Vol. 1116, 1989, pp. 19–26.

[13] Gilboa, P., "Visor Projected Display Using a Spherical Combiner," in *Display Systems*, SPIE, Vol. 1988, 1993, pp. 193–197.

[14] Rallison, R. D., and S. R. Schicker, "Combat Vehicle Stereo HMD," in *Large-Screen-Projection, Avionics, and Helmet-Mounted Displays*, SPIE, Vol. 1456, 1991, pp. 179–190.

[15] LaRussa, J. A., and A. T. Gill, "The Holographic Pancake Window," in *Visual Simulation and Image Realism*, SPIE, Vol. 162, 1978, pp. 120–129.

[16] Shenker, M., "Visual-Simulation Optical Systems," *Proc. Los Alamos Conf. on Optics '83*, SPIE, Vol. 380, 1983, pp. 22–29.

[17] Shenker, M., "Optical Design Criteria for Binocular Helmet-Mounted Displays," in *Display System Optics*, SPIE, Vol. 778, 1987, pp. 70–78.

[18] Amitai, Y., A. A. Friesem, and V. Weiss, "New Design of Holographic Helmet Displays," in *Holographic Optics: Design and Applications*, SPIE, Vol. 883, 1988, pp. 12–19.

[19] Twardowski, P., and P. Meyrueis, "Design of an Optimal Single Reflective Holographic Helmet Display Element," in *Large-Screen-Projection, Avionic, and Helmet-Mounted Displays*, SPIE, Vol. 1456, 1991, pp. 164–174.

[20] Sweatt, W. C., "Describing Holographic Optical Elements as Lenses," *J. Optical Society of America*, Vol. 67, No. 6, June 1977, pp. 803–808.

[21] Magarinos, J. R., and D. J. Coleman, "Holographic Mirrors," in *Applications of Holography*, SPIE, Vol. 523, 1985, pp. 203–218.

[22] Schweicher, E. J. F., "Review of Industrial Applications of HOEs in Display Systems," in *Progress in Holographic Applications*, SPIE, Vol. 600, 1985, pp. 66–80.

[23] Perrin, J. C., "Wide Angle Distortion Free Holographic Head-Up Display," in *Progress in Holographic Applications*, SPIE, Vol. 600, 1985, pp. 178–183.

[24] Caulfield, H. J., *Handbook of Optical Holography*, New York: Academic Press, 1979.

[25] Evans, R. W., "The Development of Dichromated Gelatin for Holographic Optical Element Applications," in *Applications of Holography*, SPIE, Vol. 523, 1985, pp. 302–304.

[26] Owen, H., "Holographic Optical Elements in Dichromatic Gelatin," in *Applications of Holography*, SPIE, Vol. 523, 1985, pp. 296–301.

[27] De Vos, G., and G. Brandt, "Use of Holographic Optical Elements in HMDs," in *Helmet-Mounted Displays II*, SPIE, Vol. 1290, 1990, pp. 70–80.

6

Head-Position Measurement

Most head-coupled systems incorporate some type of device that measures the pose of the head, that is, the angular orientation and the linear translations of the head. Head or helmet tracking is required to implement a sighting device, to enable the head orientation so displayed information on the HMD is properly presented with respect to the head motions and to allow slaving of sensors or weapons to the head.

Systems that measure head position and orientation can be divided into roughly three categories. The first category is direct-measurement sensors. These sensors are basically mechanical systems attached to the helmet that directly measure angular orientation by angular sensors such as resolvers. The second category, to which most currently used systems belong, includes all remote sensing systems. Typically, these systems employ some type of emitter radiating signals that are sensed by detectors on the helmet or, alternatively, by the radiator on the helmet and the detector in the cockpit. Electromagnetic, electro-optical, and acoustic sensors all belong to this category. The third category includes all self-contained systems, which use a sensor mounted on the helmet to measure a certain global physical property from which the head position and orientation are resolved. These systems require no radiating element in the cockpit.

The head-pose measurement has been mechanized using a variety of technologies, including various electro-magnetic, ultrasonic or acoustic, electro-optic, and mechanical technologies, as well as technologies based on inertial instruments.

6.1 Performance Considerations of Head Trackers

In the consideration of a system that measures head pose, several factors should be examined. The foremost parameter is what degrees of freedom need to be measured. For steering a gimbaled sensor, only the azimuth and elevation angles are essential, since the gimbals are capable of moving in only those two degrees of freedom. That also is true for pointing purposes, since LOS is defined by only these two angles. When conformal display symbology is desired, measurement of the roll angle should be added. Figure 6.1 illustrates the helmet angular degrees of freedom.

In some applications, mainly if the system is to be used in fighter aircraft, the exact position of the helmet in the cockpit is desired. For instance, the aircraft canopy often deviates the LOS due to optical refraction of the canopy curved surface [1]. In such cases, calibration of the LOS deviation due to the canopy refraction is needed to compensate the LOS deviation. That can be achieved only if the head position relative to the canopy is known. A similar mapping requirement exists when an electromagnetic head tracker sensitive to metal objects and to electromagnetic radiation in the cockpit is utilized.

Naturally, the tracking systems are required to measure the helmet angular orientation over a range of azimuth, elevation, and roll angles that cover the expected viewing angles of the user in a particular vehicle. In addition, they are required to provide those measurements within the complete head-motion box. Table 6.1 gives typical values of the angular and head-motion ranges of the head-motion tracker.

Since the ultimate task of these systems is to measure the head pose accurately, accuracy over the complete measurement range obviously is the most

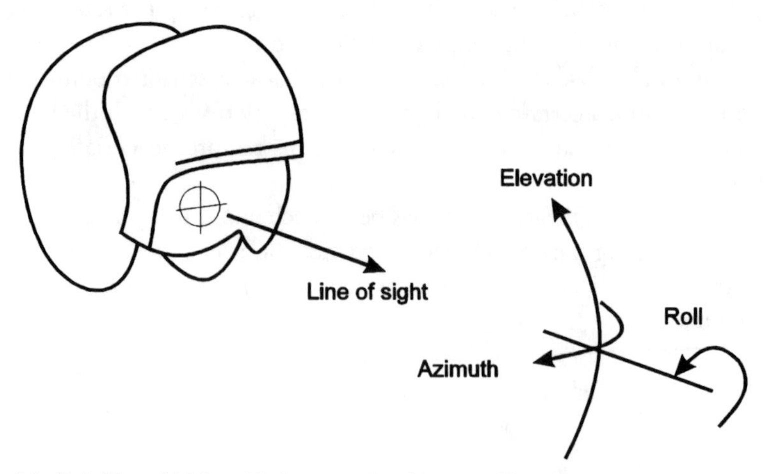

Figure 6.1 Definition of LOS and helmet angular degrees of freedom.

Table 6.1
Typical Required Angular and Translational Ranges of Helmet-Position and Orientation Measurement System

Azimuth	Elevation	Roll	Up/Down	Left/Right	Fore/Aft
±180°	−70° to +60°	±60°	±225 mm	±200 mm	±270 mm

important factor. Commonly, accuracy close to that of a conventional HUD is considered essential, at least in the nominal LOS direction. A HUD can achieve an accuracy on boresight better than 1 mrad, which degrades to around 2 mrad at the extreme of the FOV. An HMD can come close to the HUD on boresight around 2 mrad but degrades to 4 to 6 mrad at 10 degrees off axis. Therefore, current HMDs are not expected in the near future to replace the HUD completely.

Accuracy, however, is only one factor. The measurement range of the tracker also is important. It must retain sufficient accuracy or at least be capable of tracking the head motion for all expected motions. Systems that rely on iterative computation schemes are prone to "break tracking" if the head motion exceeds some value. In such cases, reacquisition or initialization is needed. Head movements may exceed angular velocities of several hundreds of degrees per second, and accelerations up to several thousands of degrees per second squared. Table 6.2 summarizes the maximum values for head motions that may be experienced during operational conditions as measured in two studies.

The process of head-pose measurement adds, naturally, a delay in the combined system. That obviously is undesirable, since information about head position lags behind the true head position. Also, because the process typically is used to display some information overlaid on the true scene, it produces misregistration between the two scenes. To overcome such phenomena, constant attempts to increase the sampling rate of the sensors and to speed up the computation process are made.

The factor that perhaps is most dominant in realizing head-position measurement but that frequent receives less attention or is ignored is installation of

Table 6.2
Peak Head Acceleration and Velocity of Head Motion From Two Different Studies

	Source			
	List [2]		**Wells and Haas [3]**	
	Azimuth	**Elevation**	**Azimuth**	**Elevation**
Peak velocity (deg/s)	437	Not measured	601	344
Peak acceleration (deg/s^2)	5,718	Not measured	4,753	2,452

the system in the cockpit. Most systems function satisfactorily in laboratory conditions, but when they are installed in the real environment, the restricted room and other cabin constraints limit the usefulness of many systems. Some devices require installation on the aircraft canopy, while others suffer from accuracy and coverage deficiencies because of the proximity of the sensor to the helmet. Therefore, the design is not finalized before the proper installation is ensured.

6.2 Mechanical Systems

Mechanical systems measure the angular orientation of the helmet by physically connecting the helmet to a point of reference with jointed linkages. The mechanical system basically is a two-bar linkage, as illustrated schematically in Figure 6.2. The arrangement comprises two rigid bars with precision joints. The first joint between the helmet and the vertical bar allows motion in all three degrees of freedom, that is, rotation around the x, y, and z axes. The second joint between the bars only permits rotation about the x and z axes while the final joint to the airframe has rotation only around the y axis. The whole system thus allows roll, pitch and yaw measurement with some systems also providing lateral displacement. The bars' dimensions are known, and all angular degrees of freedom are measured using angular sensors [4–6]. For the obvious reason of ejection safety, the system is only of use in helicopters and even here rapid egress is compromised.

Mechanical systems usually are precise and responsive because they employ high-accuracy angular position sensors such as resolvers or shaft encoders. They are immune to all external disturbances, and their accuracy is determined almost solely by the accuracy of the angular sensors and installation in the cockpit. In

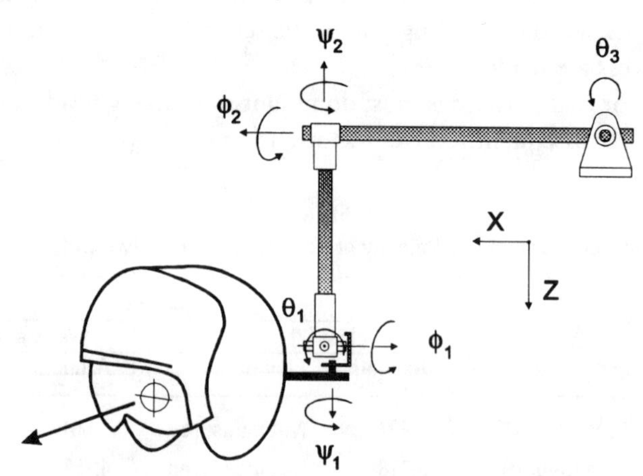

Figure 6.2 Schematic illustration of a mechanical helmet-position sensor.

some cases, the mechanical systems are connected to a sliding carriage to allow some freedom of movement of the head within the motion box.

6.3 Magnetic Sensors

Magnetic sensors use a small transmitter in the vehicle and a similar receiver mounted on the helmet [7,8]. Both units have three mutually orthogonal magnetic coils. Each coil is excited sequentially, producing an electromagnetic vector. The vector is sensed by the three coils of the receiver, each of which senses the respective component of the transmitted vector. The relative position and orientation between the transmitter and the receiver and their respective coordinate frames are depicted in Figure 6.3.

The magnetic field produced by the transmitter has two components, a radial component (Figure 6.4):

$$H_\rho = \frac{NIA}{2\pi\rho^3}\cos\zeta \tag{6.1}$$

and a tangential component:

$$H_t = \frac{NIA}{4\pi\rho^3}\sin\zeta \tag{6.2}$$

where A is the area of the coil loop, N is the number of turns in the coil, and I is the loop current.

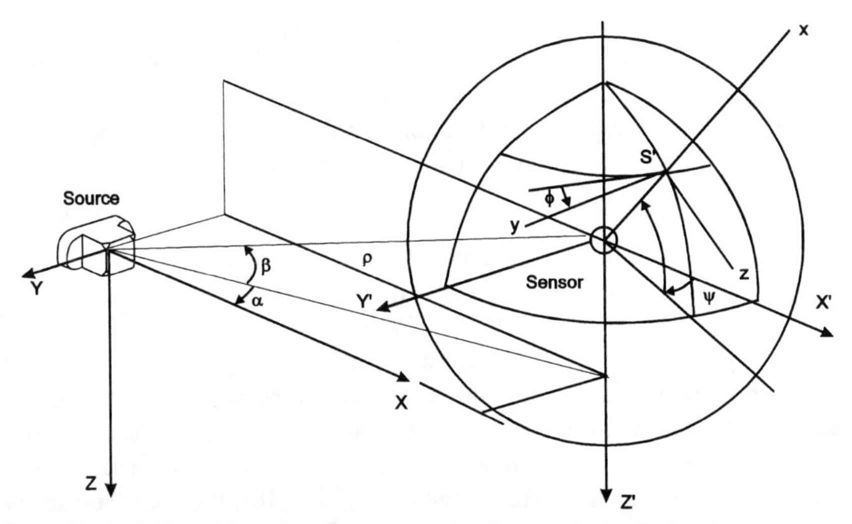

Figure 6.3 Definition of the transmitter and receiver frames of coordinates.

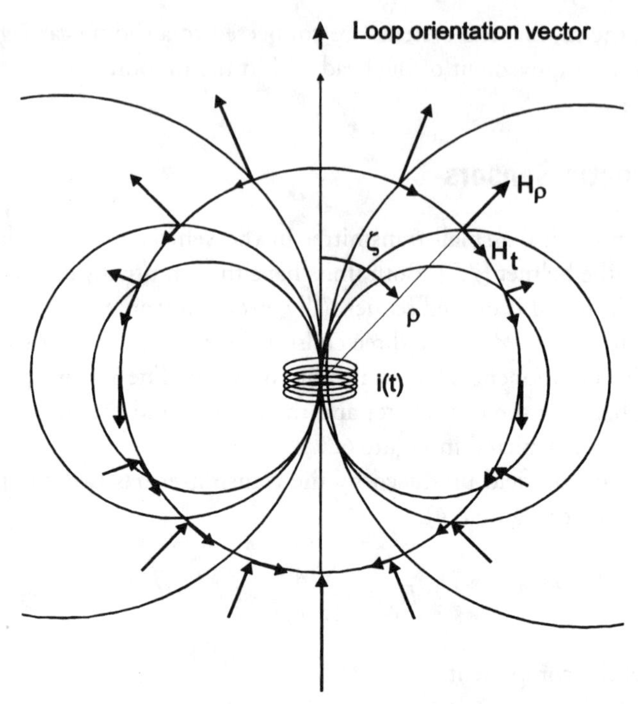

Figure 6.4 The magnetic field produced by each coil in the transmitter unit.

Because the distance between the transmitter and the receiver is significantly larger than the transmitter diameter, it can be considered as a point source and the radiation as a vector. To illustrate the coupling between the transmitter and the receiver, imagine the receiver placed with a parallel orientation to the transmitter and their x axes aligned with the line connecting both units' centers. In that case, the receiver's coil senses the radial magnetic field components, while the two remaining coils sense the two respective tangential components. Using (6.1) and (6.2) and the notation in Figure 6.3,

$$V_{sp} = \frac{C}{\rho^3}\begin{bmatrix} 1 & 0 & 0 \\ 0 & -1/2 & 0 \\ 0 & 0 & -1/2 \end{bmatrix} V_{tp} \tag{6.3}$$

where $C = NAK/2\pi$, and K is the sensor gain.

Next, consider the case in which both the transmitter and the receiver have zero-orientation and the receiver is placed in a remote location described by the distance ρ, by the azimuth angle α_1, and the elevation angle β_1.

If V_t is the excitation vector of the transmitter, then the excitation vector of the equivalent, similarly aligned transmitter with the line connecting the transmitter and receiver is obtained by rotating the vector using the azimuth and elevation rotation transformation:

$$V_{te} = T_{\beta 1} T_{\alpha 1} V_t \tag{6.4}$$

The output of the equivalently aligned receiver now is obtained using the sensor-source coupling relation of (6.3):

$$V_{se} = \frac{C}{\rho^3 S} \cdot V_{te} = \frac{C}{\rho^3 S} \cdot T_{\beta 1} T_{\alpha 1} V_t \tag{6.5}$$

where V_{se} is the output of the equivalent sensor similarly aligned with the source. Now the output of the zero-orientation sensor is obtained by applying inverse position rotations (using the rotation transformation matrices given later in this section, in Table 6.3):

$$V_{so} = T_{-\alpha 1} T_{-\beta 1} V_{se} = \frac{C}{\rho^3} T_{-\alpha 1} T_{-\beta 1} S T_{\beta 1} T_{\alpha 1} V_t = \frac{C}{\rho^3} Q \cdot V_t \tag{6.6}$$

and the output of a sensor with any arbitrary orientation is determined by applying the azimuth, elevation, and roll rotation transformations to the output of the zero-orientation vector:

$$V_s = T_{\phi 1} \mathrm{T}_{\theta 1} \mathrm{T}_{\psi 1} V_{so} = \frac{C}{\rho^3} T_{\phi 1} \mathrm{T}_{\theta 1} \mathrm{T}_{\psi 1}\, Q \cdot V_t = \frac{C}{\rho^3} A \cdot Q \cdot V_t \tag{6.7}$$

Table 6.3
Coordinate Transformation Matrices

	Position	Orientation
Azimuth: rotation around z axis	$T_\alpha = \begin{bmatrix} \cos\alpha & \sin\alpha & 0 \\ -\sin\alpha & \cos\alpha & 0 \\ 0 & 0 & 1 \end{bmatrix}$	$T_\psi = \begin{bmatrix} \cos\psi & \sin\psi & 0 \\ -\sin\psi & \cos\psi & 0 \\ 0 & 0 & 1 \end{bmatrix}$
Elevation: rotation around y axis	$T_\beta = \begin{bmatrix} \cos\beta & 0 & -\sin\beta \\ 0 & 1 & 0 \\ \sin\beta & 0 & \cos\beta \end{bmatrix}$	$T_\theta = \begin{bmatrix} \cos\theta & 0 & -\sin\theta \\ 0 & 1 & 0 \\ \sin\theta & 0 & \cos\theta \end{bmatrix}$
Roll: rotation around x axis	$T_\gamma = \begin{bmatrix} 1 & 0 & 0 \\ 0 & \cos\gamma & \sin\gamma \\ 0 & -\sin\gamma & \cos\gamma \end{bmatrix}$	$T_\phi = \begin{bmatrix} 1 & 0 & 0 \\ 0 & \cos\phi & \sin\phi \\ 0 & -\sin\phi & \cos\phi \end{bmatrix}$

By exciting sequentially the transmitter with three linearly independent excitation vectors, nine nonlinear equations are obtained. Those equations are used to resolve the six unknowns: the three orientation angles and the three positions (the azimuth and elevation angles and the sensor to source distance). Although, the equations can be solved directly, an iterative process commonly is used to simplify and speed up the computation process [7]. That is done by calculating the angular increments between successive measurements. Another frame of reference, the tracking coordinate frame, is defined as that in which a source vector V_o is specified and is related to the true excitation vector by the estimated azimuth and elevation:

$$V_t = T_{-\hat{\alpha}1} T_{-\hat{\beta}1} V_o \tag{6.8}$$

where the difference between the source-sensor position angles are $\Delta\alpha 1 = \alpha 1 - \hat{\alpha}1$, $\Delta\beta 1 = \beta 1 - \hat{\beta}1$.

Substituting (6.8) in (6.7) and (6.6) yields

$$V_s = \frac{C}{\rho^3} A T_{-\alpha 1} T_{-\beta 1} S(T_{\beta 1} T_{\alpha 1} T_{-\hat{\alpha}1} T_{-\hat{\beta}1}) V_o = \frac{C}{\rho^3} A \cdot T_{-\alpha 1} T_{-\beta 1} S \cdot \Delta P \cdot V_o \tag{6.9}$$

The differential position rotation matrix, ΔP, for small changes in the position is

$$\Delta P = T_{\beta 1} T_{\alpha 1} T_{-\hat{\alpha}1} T_{-\hat{\beta}1} \cong T_{\beta 1} T_{\Delta\alpha 1} T_{-\beta 1} T_{-\Delta\beta 1}$$

$$= \begin{bmatrix} 1 & \Delta\alpha_1 \cos\beta_1 & -\Delta\beta_1 \\ -\Delta\alpha_1 \cos\beta_1 & 1 & -\Delta\alpha_1 \sin\beta_1 \\ \Delta\beta_1 & \Delta\alpha_1 \sin\beta_1 & 1 \end{bmatrix} \tag{6.10}$$

The relation in (6.10) exploits the properties sin $\delta \cong \delta$ and cos $\delta \cong 1$, where δ is a small angle, and the order of small angles rotations may be arbitrary. Similarly, the position rotation matrix can be expressed relative to the tracking coordinate frame by rotating the source-sensor in roll, elevation, and azimuth:

$$\Delta P = T_{\beta 1} T_{\alpha 1} T_{-\hat{\alpha}1} T_{-\hat{\beta}1} \cong [(T_{\Delta\gamma 0} T_{\Delta\beta 0} T_{\Delta\alpha 0}) \mathrm{T}_{\beta 1} \mathrm{T}_{\alpha 1}] T_{-\alpha 1} T_{-\beta 1}$$

$$= T_{\Delta\gamma 0} T_{\Delta\beta 0} T_{\Delta\alpha 0} = \begin{bmatrix} 1 & \Delta\alpha_0 & -\Delta\beta_0 \\ -\Delta\alpha_0 & 1 & \Delta\gamma_0 \\ \Delta\beta_0 & -\Delta\gamma_0 & 1 \end{bmatrix} \tag{6.11}$$

where $\Delta\alpha 0$, $\Delta\beta 0$, $\Delta\gamma 0$ are referenced to the tracking coordinate frame. In the same way, the attitude transformation can be reduced to a small angle transformation using

$$V_R = T_{\hat{\beta}1} T_{\hat{\alpha}1} \hat{A}^{-1} V_s = \frac{C}{\rho^3} T_{\hat{\beta}1} T_{\hat{\alpha}1} \hat{A}^{-1} A \cdot T_{-\alpha 1} T_{-\beta 1} S \cdot \Delta P \cdot V_o \quad (6.12)$$

$$T_{\hat{\beta}1} T_{\hat{\alpha}1} \hat{A}^{-1} A \cdot T_{-\alpha 1} T_{-\beta 1} \cong (\hat{A}^{-1} A)(\Delta P)^{-1} \quad (6.13)$$

and

$$A \cdot T_{-\alpha 1} T_{-\beta 1} = \hat{A} T_{-\hat{\alpha}1} T_{-\hat{\beta}1} (\Delta A)(\Delta P)^{-1} \quad (6.14)$$

$$\Delta A = T_{\Delta\phi 0} T_{\Delta\theta 0} T_{\psi 0} = \begin{bmatrix} 1 & \Delta\psi_0 & -\Delta\theta_0 \\ -\Delta\psi_0 & 1 & \Delta\phi_0 \\ \Delta\theta_0 & -\Delta\phi_0 & 1 \end{bmatrix} \quad (6.15)$$

Substitution of (6.14) and (6.13) into (6.12) gives

$$V_R = \frac{C}{\rho^3} \Delta A (\Delta P)^{-1} S \cdot \Delta P \cdot V_o = \frac{C}{\rho^3} R \cdot V_o \quad (6.16)$$

where

$$R = \Delta A (\Delta P)^{-1} S \Delta P$$

$$= \begin{bmatrix} 1 & \frac{3}{2}\Delta\alpha_0 - \frac{1}{2}\Delta\psi_0 & -\frac{3}{2}\Delta\beta_0 + \frac{1}{2}\Delta\theta_0 \\ \frac{3}{2}\Delta\alpha_0 - \Delta\psi_0 & -\frac{1}{2} & -\frac{1}{2}\Delta\phi_0 \\ -\frac{3}{2}\Delta\beta_0 + \Delta\theta_0 & \frac{1}{2}\Delta\phi_0 & -\frac{1}{2} \end{bmatrix} \quad (6.17)$$

The fundamental excitation pattern sequentially excites each coil, which is equivalent to the following three unit vectors:

$$V_0(S_1) = \begin{bmatrix} 1 \\ 0 \\ 0 \end{bmatrix} \quad V_0(S_2) = \begin{bmatrix} 0 \\ 1 \\ 0 \end{bmatrix} \quad V_0(S_3) = \begin{bmatrix} 0 \\ 0 \\ 1 \end{bmatrix} \quad (6.18)$$

Using that excitation pattern and (6.17), the angular errors are resolved:

$$
\begin{aligned}
\Delta\alpha_0 &= \tfrac{2}{3}(\hat{\rho}^3/C)[2V_{Rx}(S_2) - V_{Ry}(S_1)] \\
\Delta\beta_0 &= \tfrac{2}{3}(\hat{\rho}^3/C)[V_{Rx}(S_1) - 2V_{Rx}(S_3)] \\
\Delta\psi_0 &= 2(\hat{\rho}^3/C)[V_{Rx}(S_2) - V_{Ry}(S_1)] \\
\Delta\theta_0 &= 2(\hat{\rho}^3/C)[V_{Rx}(S_1) - V_{Rx}(S_3)] \\
\Delta\phi_0 &= 2(\hat{\rho}^3/C)V_{Rz}(S_2) = -2(\hat{\rho}^3/C)V_{Ry}(S_3) \\
\hat{\rho} &= \sqrt[3]{C/V_{Rx}(S_1)} = \sqrt[3]{-C/2V_{Ry}(S_2)} = \sqrt[3]{-C/2V_{Rx}(S_3)}
\end{aligned}
\tag{6.19}
$$

The final stage is to convert the angle errors to the angular time increments. The position angles are converted by equating (6.10) and (6.11):

$$
\begin{aligned}
\Delta\alpha_1 &= \Delta\alpha_0/\cos\hat{\beta}_1 \\
\Delta\beta_1 &= \Delta\beta_0
\end{aligned}
\tag{6.20}
$$

The attitude angle are converted using (6.13) and (6.14), resulting in

$$
\begin{bmatrix} \Delta\phi_1\cos\hat{\theta}_1 \\ \Delta\theta_1 \\ \Delta\psi_1 - \Delta\phi_1\sin\hat{\theta}_1 \end{bmatrix}
$$

$$
= \begin{bmatrix} \cos\hat{\beta}_1\cos(\hat{\psi}_1 - \hat{\alpha}_1) & \sin(\hat{\psi}_1 - \hat{\alpha}_1) & -\cos(\hat{\psi}_1 - \hat{\alpha}_1)\sin\hat{\beta}_1 \\ \cos(\hat{\psi}_1 - \hat{\alpha}_1)\sin\hat{\beta}_1 & \cos(\hat{\psi}_1 - \hat{\alpha}_1) & \sin(\hat{\psi}_1 - \hat{\alpha}_1)\sin\hat{\beta}_1 \\ \sin\hat{\beta}_1 & 0 & \cos\hat{\beta}_1 \end{bmatrix} \cdot \begin{bmatrix} \Delta\phi_0 \\ \Delta\theta_0 \\ \Delta\psi_0 \end{bmatrix}
\tag{6.21}
$$

The complete computation scheme is summarized in Figure 6.5.

6.3.1 Implementation and Interference Sources

The electromagnetic system of helmet-position and orientation measurements is implemented in several ways. The simplest and most common one is the

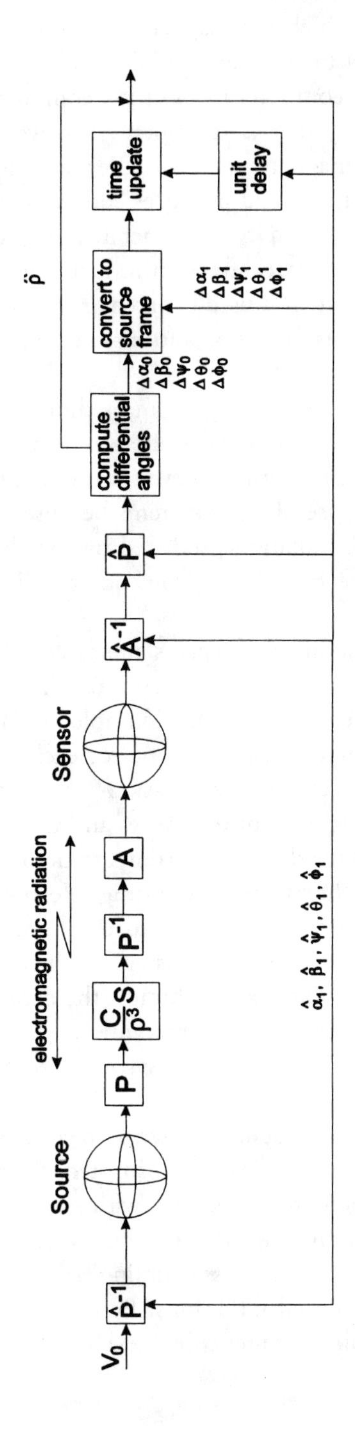

Figure 6.5 The computation scheme of the electromagnetic head-position measurement system.

ac implementation [9], in which the magnetic radiation is modulated with a carrier frequency of 7 to 14 kHz. The received signals are demodulated, sampled, and transferred to the computer for position computations.

The electromagnetic system is extremely sensitive to the presence of nearby metal objects. Ferromagnetic materials alter or distort the shape of the magnetic field. The ac magnetic field produces eddies, or circulating, currents in nearby conductive metals, which in turn generate a secondary magnetic field, which distorts the primary magnetic field, introducing errors in the measurements of the system. If the system is used in an environment with a fixed-metal structure, the magnetic field is measured and mapped and used for correcting the system readings.

The calibration procedure is a lengthy and tedious procedure and has to be repeated every time the environment is altered, because of the removal or addition of metal objects. Furthermore, even with calibration, special care must be taken when in the use of these systems because they are sensitive to all kinds of electromagnetic radiation produced by switching power supplies or the magnetic field of the HMD and from other radiating sources in the cabin [10].

The sensitivity to close metal objects is alleviated to some extend when a different implementation of the system based on dc emitters [11] is used instead of the common ac emitters. In the dc implementation, the magnetic field is generated by a sequence of pulses, one for each measurement. Since the eddy currents are generated only by variations of the magnetic field, they are created only by the rising edge of the pulses and decay soon after. So, by waiting a very short period for the eddy currents to decay, the measurements are obtained unaffected by the metal surrounding. However, the dc system is sensitive to ferromagnetic metals as well as to the local magnetic field. To overcome that, the system measures the constant magnetic field before transmitting the electromagnetic pulses and subtracts the constant field from the readings. In general, dc systems can operate with a generic mapping of the magnetic field in the cockpit, whereas ac systems require unique cockpit mapping.

One difficulty with electromagnetic systems, in contrast to electro-optical systems, is the low-level analog signal transmitted from the helmet and processed by the system. The signal is received with noise that affects the measurement accuracy. Elimination of noise is possible with suitable filtering, but that introduces additional lag. Also, cross coupling between the trackers' magnetic field and the head-mounted CRT must be eliminated by proper MU-metal shielding and by keeping magnetic field levels low.

6.4 Electro-Optical Methods

6.4.1 Rotating Infrared Beams

In the rotating IR beam approach, adopted by Honeywell and used with the IHADSS system in the Apache helicopter [12], two sensor surveying units (SSUs) are hard-mounted to the aircraft on each side of the user. The SSUs generate pairs of thin, collimated, fan-shaped IR light beams rotating at a constant angular velocity [13]. Two small IR detectors are mounted on the side of the helmet. The arrangement of the rotating SSU and the helmet-mounted detectors is shown in Figure 6.6.

The IR beams trigger signal pulses whenever the beams sweep over the helmet detectors. Another set of detectors installed within the SSU establish rotational location reference. Four angles are measured using the rotating beams, measurement accomplished by measuring the time or number of pulses between the triggering of the reference detectors and each of the two helmet

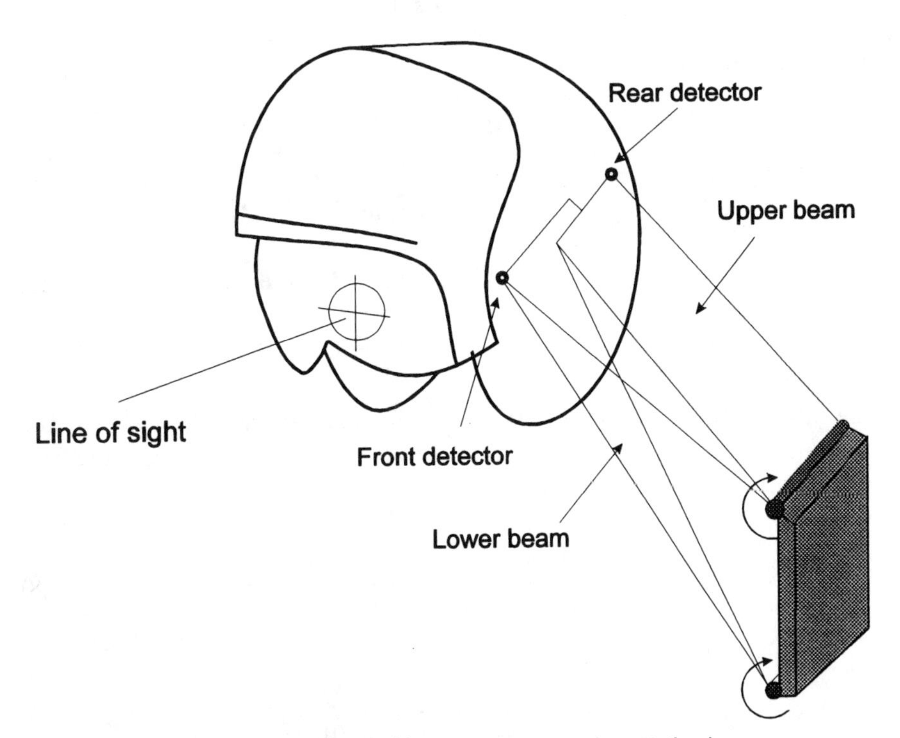

Figure 6.6 The geometry of helmet-position measurement using rotating beams.

detectors as triggered by each beam. The system relies on the constant angular velocity of the rotating beam:

$$\omega = \frac{d\theta_i}{dt} = const \tag{6.22}$$

Thus, all the angles are measured by integration of the angular velocity in the interval between the triggering of the reference detector and the helmet-mounted respective detector:

$$\theta_i = \int_{\theta_0}^{\theta} d\theta_i = \omega \int_{t_0}^{t} dt = \omega(t - t_0) \tag{6.23}$$

That constitutes four angles, which are defined in Figure 6.7: θ_1, θ_2, θ_3, and θ_4. All angles are referenced to the SSU internal reference.

The system calculates the elevation angle (ϑ) and the azimuth angle (ψ).

The light box shown in Figure 6.8 is defined by the known distance σ between the two detectors. The height of the light box at any instance is given by

$$z_0 = -\sigma \sin \vartheta \tag{6.24}$$

and the width of the light box is

$$y_0 = \sigma \cos \vartheta \sin \psi \tag{6.25}$$

Using the four measured angles, θ_1, θ_2, θ_3, and θ_4, and the geometry in Figure 6.7, the following quantities are computed:

$$y_0 = y_1 - y_2 = l_1 \cos \theta_1 - l_2 \cos \theta_2 \tag{6.26}$$

$$z_0 = z_1 - z_2 = l_2 \sin \theta_2 - l_1 \cos \theta_1 \tag{6.27}$$

The distance between the axes of the rotating light sources, K, is known and constant. l_1 and l_2 are computed from the constant distance between the light sources and using the measured angles:

$$l_1 = \frac{K \sin(\pi/2 + \theta_3)}{\sin(\theta_1 - \theta_3)} = \frac{K \cos\theta_3}{\sin(\theta_1 - \theta_3)} \tag{6.28}$$

$$l_2 = \frac{K \sin(\pi/2 + \theta_4)}{\sin(\theta_2 - \theta_4)} = \frac{K \cos\theta_4}{\sin(\theta_2 - \theta_4)} \tag{6.29}$$

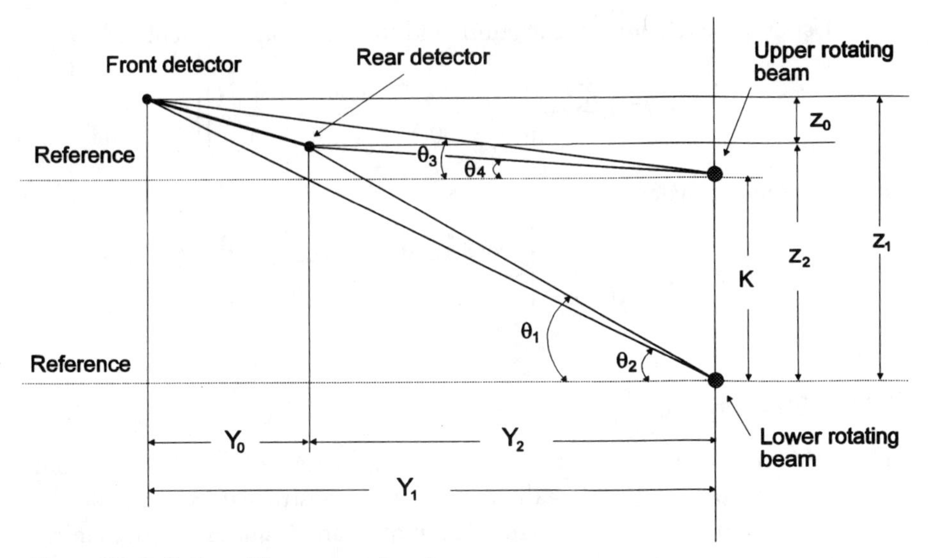

Figure 6.7 Definition of the measured angles.

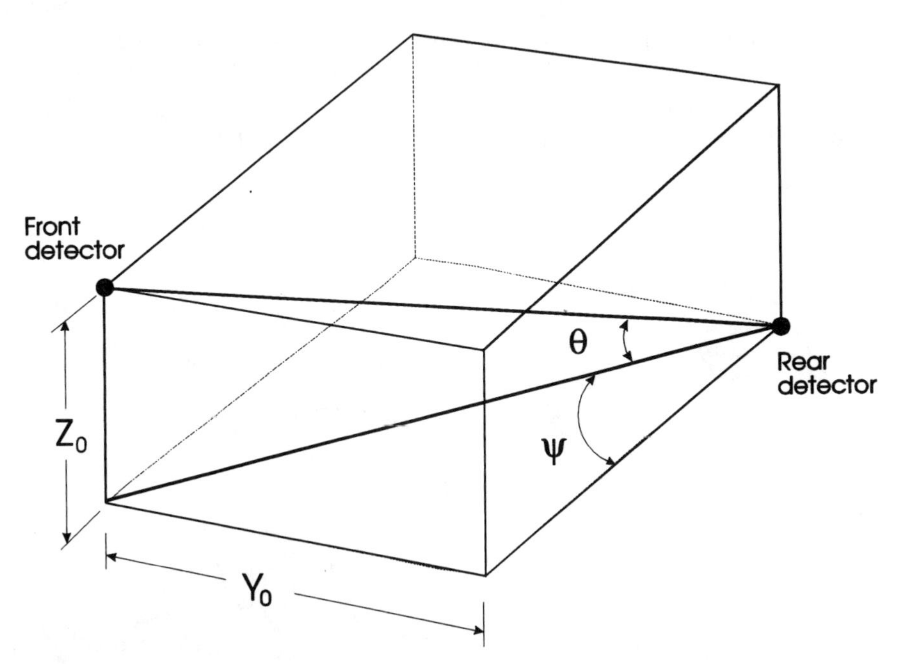

Figure 6.8 Definition of the light box.

Using those relations, the azimuth and the elevation are calculated by

$$\vartheta = \sin^{-1}\left\{\frac{K}{\sigma}\left[\frac{\sin\theta_1\cos\theta_3}{\sin(\theta_1 - \theta_3)} - \frac{\sin\theta_2\cos\theta_4}{\sin(\theta_2 - \theta_4)}\right]\right\} \tag{6.30}$$

and the azimuth angle by

$$\psi = \sin^{-1}\left\{\frac{K}{\sigma\cos\vartheta}\left[\frac{\cos\theta_1\cos\theta_3}{\sin(\theta_1 - \theta_3)} - \frac{\cos\theta_2\cos\theta_4}{\sin(\theta_2 - \theta_4)}\right]\right\} \tag{6.31}$$

6.4.2 LED Arrays

The LED arrays concept centers on remote measurement of three vectors, from which the complete 6-DOF of the helmet are reconstructed by triangulation [14]. This method relies on imaging techniques and requires no mechanical moving elements, as does the rotating-beam method.

The LEDs whose locations on the helmet are accurately known are sensed by a camera that typically is either charge-coupled device (CCD)-based or a position-sensitive device (PSD). The basic mechanization is illustrated in Figure 6.9.

PSDs are simple and rugged detectors and have been used since the early 1960s in optical inspection and measuring instruments, with more recent application in auto-focus cameras. PSDs use a three-layer silicon substrate: N layer, insulator and P layer, and electrodes are connected to all four sides. Incident light generates a charge that is driven through the P layer and collected

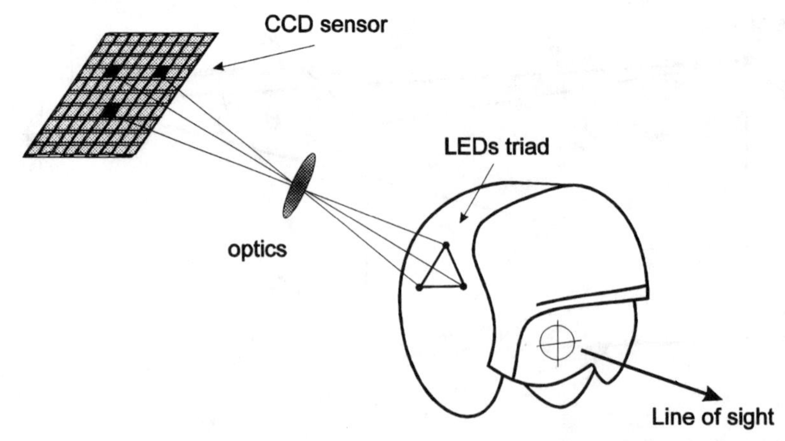

Figure 6.9 Electro-optical system of helmet-position measurement using an LED array.

by the electrodes. The resistivity of the P layer is uniform, so the current collected is inversely proportional to the distance between the incident-light impact point and the electrode. The distance of the spot on the PSD is determined by the ratio of the difference between the current on the two opposite electrodes to the sum of the currents; thus, the absolute value of the current is not important, nor is the light intensity. Moreover, resolution is limited only by noise and analog-to-digital (A/D) quantization.

The spot can be focused or diffuse but must be symmetrical. Sunlight must be prevented from saturating the PSD, which is done, first, with a narrow band filter matched to the LEDs and, second, by measuring the sunlight during a period when all the diodes are off.

The imaging of the LEDs is shown in Figure 6.9. The origin of the coordinate system is placed at the center of the lens, with the x axis along the optical axis of the camera, and the image lies in the y-z plane of the camera. Referring to Figure 6.9, l_i is the vector to the *i*-th LED, p_i is the image of that LED in the camera plane, $\vec{x}$ is a unit vector along the optical axis, f is the focal length of the lens, and m_i are the imaging scale factors:

$$l_i = m_i f \vec{x} - m_i \vec{p}_i \tag{6.32}$$

The distances d_{ij} between the LEDs on the helmet are known and given by

$$(\vec{l}_i - \vec{l}_j)^2 = d_{ij}^2 \tag{6.33}$$

introducing the following normalized variables:

$$\begin{aligned} \vec{\pi}_i &= \vec{p}_i / f \\ \rho_i &= m_i \sqrt{1 + \vec{\pi}_i^2} \\ \gamma_{ij} &= 1 - \frac{1 + \vec{\pi}_i \cdot \vec{\pi}_j}{\sqrt{(1 + \vec{\pi}_i^2)(1 + \vec{\pi}_j^2)}} \end{aligned} \tag{6.34}$$

Using (6.32) through (6.34), the three imaging equations are obtained:

$$\begin{aligned} (\rho_1 - \rho_2)^2 + 2\gamma_{12}\rho_1\rho_2 &= (d_{12} / f)^2 \\ (\rho_2 - \rho_3)^2 + 2\gamma_{23}\rho_2\rho_3 &= (d_{23} / f)^2 \\ (\rho_3 - \rho_1)^2 + 2\gamma_{13}\rho_1\rho_3 &= (d_{13} / f)^2 \end{aligned} \tag{6.35}$$

The equations are solved analytically by the introduction of the new variables:

$$u = \rho_2 - (1 - \gamma_{12})\rho_1$$
$$v = \rho_3 - (1 - \gamma_{13})\rho_1 \tag{6.36}$$

and the parameters $\delta_1 = (d_{12} / f)^2$, $\delta_2 = (d_{23} / f)^2$, and $\delta_3 = (d_{13} / f)^2$:

$$\gamma_{12}(2 - \gamma_{12})\rho_1^2 + u^2 = \delta_1$$
$$\gamma_{13}(2 - \gamma_{13})\rho_1^2 + v^2 = \delta_3 \tag{6.37}$$
$$c_1\rho_1^2 + 2c_4uv + c_5 = c_2u\rho_1 + c_3v\rho_1$$

where $c_1, \ldots, c_5$ are constants. Squaring the third equation in (6.37) and substituting the two first equations yields

$$a_1\rho_1^4 + a_2\rho_1^2 + a_3 = (a_4\rho_1^2 + a_5)u \cdot v \tag{6.38}$$

Squaring (6.38) and using again the relations from (6.37) results in a single fourth-order polynomial equation of $\rho \equiv \rho_1^2$:

$$\rho^4 + s_3\rho^3 + s_2\rho^2 + s_1\rho + s_0 = 0 \tag{6.39}$$

The coefficients $s_0, \ldots, s_3$ are expressed in terms of the coefficients of (6.37) and (6.38). Closed-form solution of a fourth-order equation is obtained by reducing the equation to a third-order equation. Rewrite (6.39) as

$$(\rho^2 + B_1\rho + B_0)^2 - (C_1\rho + C_0)^2 = 0 \tag{6.40}$$

where

$$s_3 = 2B_1$$
$$s_2 = 2B_0 + B_1^2 - C_1^2$$
$$s_1 = 2B_1B_0 - 2C_1C_0 \tag{6.41}$$
$$s_0 = B_0^2 - C_0^2$$

Substitution of the first equation in (6.41) and rearranging gives

$$C_1^2 = 2B_0 + s_3^2 / 4 - a_2$$

$$2C_1C_0 = s_3B_0 - s_1 \tag{6.42}$$

$$C_0^2 = B_0^2 - s_0$$

Equating the second equation squared with the product of the first and third equations in (6.42) gives, finally, a third-order equation that has a closed-loop solution:

$$\eta^3 - s_2\eta^2 + (s_1s_3 - 4s_0)\eta - (s_1 + s_0s_3^2 - 4s_0s_2) = 0 \tag{6.43}$$

with $\eta = 2B_0$. Once the roots of the polynomial equation are computed, the $\vec{l}_i$ vectors to the LED triad are computed by back substitution in (6.32) through (6.42).

Having reconstructed the positions of the LED triad relative to the sensing camera, it remains to compute the position and the orientation of the helmet. To do that, a local coordinate system associated with the triangle whose sides are $\vec{l}_1 - \vec{l}_3$, $\vec{l}_1 - \vec{l}_2$, and $\vec{l}_2 - \vec{l}_3$ generated by the LEDs is defined in the way that the triangle lies in the local x-y plane:

$$\vec{v} = \vec{l}_1 - \vec{l}_3$$

$$\vec{\upsilon} = \vec{l}_2 - \vec{l}_3 \tag{6.44}$$

Using those two vectors, the local coordinate system is defined as

$$\vec{x}^T = \vec{v} \,/\, (\vec{v} \cdot \vec{v})^{1/2}$$

$$\vec{y}^T = [\vec{\upsilon} - \vec{v}\,(\vec{\upsilon} \cdot \vec{v})] / (\vec{\upsilon} \cdot \vec{\upsilon})^{1/2} \tag{6.45}$$

$$\vec{z}^T = \vec{x}^T \cdot \vec{y}^T$$

This triangle is related to the helmet coordinate system by a fixed transformation D^{HT}, which is a known transformation because the locations of the LEDs on the helmet are known, so the helmet axes are computed by applying this transformation:

$$\vec{x}^H = D^{HT}\vec{x}^T$$

$$\vec{y}^H = D^{HT}\vec{y}^T \tag{6.46}$$

$$\vec{z}^H = D^{HT}\vec{z}^T$$

Finally, the helmet coordinates are used to extract the Euler angles: azimuth, elevation and roll, since those are the three unity vectors, (1,0,0), (0,1,0), and (0,0,1), rotated by the three Euler angles. Denoting the three components of each vector by the subscripts *i, j,* and *k* and using the three rotation matrices in Table 6.3 yields

$$\vartheta = -\sin^{-1}(x_k^H)$$

$$\psi = \tan^{-1}(x_j^H / x_i^H) \tag{6.47}$$

$$\varphi = \tan^{-1}(y_k^H / z_k^H)$$

The helmet's center position is computed by

$$\vec{R}^H = R_i^T \vec{x}^T + R_j^T \vec{y}^T + R_k^T \vec{z}^T \tag{6.48}$$

where R_i^T, R_j^T, and R_k^T are the center of the helmet coordinates in the LEDs triad coordinate frame and are known by measurement of their location on the helmet.

The solution of the imaging equation, (6.39), admits, in principle, four solutions, of which only one is correct. Four real solutions are obtained only in the head-on case, in which the normal to the triangle plane is parallel to the optical axis of the camera. In all other cases, only two real solutions are obtained having opposite directions of their local z axes, which are easily distinguishable. The head-on case or proximity to head-on should be avoided. That is done by ensuring that more than one LED symbol is visible to the camera during every instance of operation. Because the helmet has a spherelike shape, no adjacent two symbols lie in the same plane. Because the measurement process is continuous, as soon as the system predicts that in the next time sequence the symbol will approach a head-on condition, it automatically switches to the next symbol. Once the other symbol is utilized, the two-solution case is obtained, and measurement accuracy is restored.

6.4.2.1 Image Processing and Accuracy Enhancement

The system is implemented by using a standard CCD camera and a frame grabber that digitizes the video image of the camera and stores the image as gray levels in the frame memory.

The measurement process involves computing the LED locations on the CCD plane, solving the imaging equations and resolving the helmet position and orientation. The coordinates of the LEDs are determined with subpixel accuracy by employing a centroid algorithm. The image of the LED is diffused over several pixels, so the centroid computation weighs the light energy over a large area on the CCD plane and accumulates sufficient information to compute the center of gravity of the LED with subpixel accuracy. Using that method, an accuracy of up to 1/50 pixel is attainable in practical conditions. Figure 6.10 illustrates the LED light distribution by the continuous curve and the discrete gray levels as sensed by the CCD camera. The centroid algorithm essentially reconstructs the continuous curve from the discrete data. The center of gravity coordinates are computed by

$$p_{iy} = \frac{\sum_{l} \zeta_l \chi_l}{\sum_{l} \chi_l} \qquad p_{iz} = \frac{\sum_{k} \zeta_k \chi_k}{\sum_{k} \chi_k} \tag{6.49}$$

where p_{iy} and p_{iz} are the y and z coordinates, respectively, of the i-th LED in the image plane, and χ_l and χ_k are the gray levels of the l and k pixels, and ζ_l and ζ_k their respective locations in the CCD plane.

6.4.3 V-Slit Cameras

A variation of the LED method is the early proposed Marconi Avionics V-slit camera, which uses a linear CCD array as opposed to the matrix CCD used in the preceding method [15]. The second dimension of the array is obtained by using the V-slit arrangement.

The camera is illustrated in Figure 6.11, and its principle of operation, showing the image of the light source on the CCD plane, is depicted in Figure 6.12.

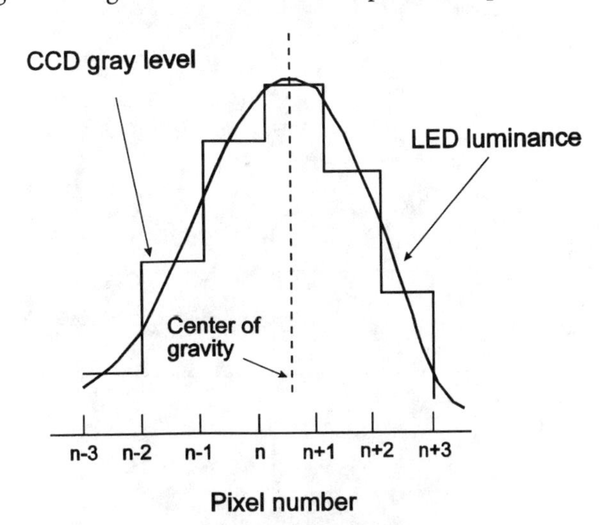

Figure 6.10 Cross-section plot of the LED luminance distribution across the camera focal plane and the image CCD gray levels of the neighboring pixels.

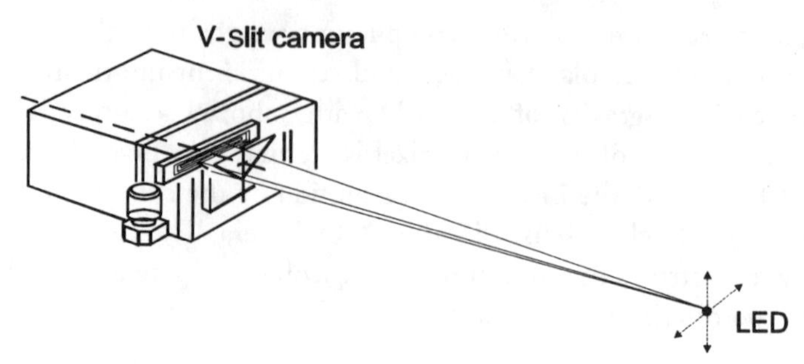

Figure 6.11 The V-slit camera.

Light from the helmet-mounted LED passes through a narrow 0.2-mm V-slit, forming a V-shaped image across the linear CCD, which consists of 1,728 elements. The camera has no optics; thus, it retains the shape of the image regardless of the light source deviation from the prime axis of the camera.

When the LED moves up and down, the images on the CCD move apart or come closer, thus changing the separation between the two images. When the LED moves to the left or the right, the images move to the right or the left, but the spacing remains the same. The LEDs are pulsed sequentially, so at each instance only one LED position is measured. The advantage of the V-slit cam-

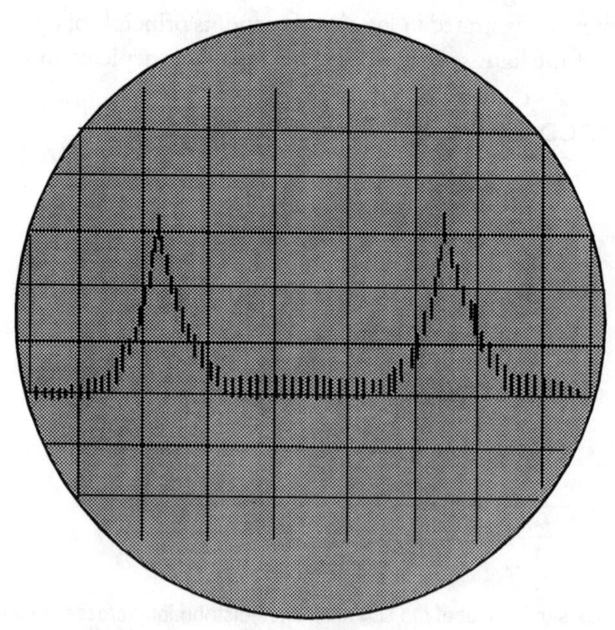

Figure 6.12 The LED image on the linear CCD array.

era over the matrix-array CCD camera is the higher obtainable sampling rate using a linear CCD for comparable resolution image.

6.5 Acoustic or Ultrasonic Sensors

Acoustic head-position measurement devices have been implemented using two basic approaches: time-of-flight measurement and phase coherence measurement [6]. Both methods measure distances between emitters and receivers and implement triangulation principles to compute the head position and orientation. Typically, the transmitters are mounted on the helmet, whereas the receivers are located around the pilot's head in the cockpit [16]. The ultrasonic radiation is in the range of 20 to 40 kHz. Implementation of ultrasonic systems in a fighter or helicopter environment is problematic because the intense external acoustic noise easily penetrates the low-absorptive structure of the cockpit and is picked up by the system's detectors.

6.5.1 Time-of-Flight Trackers

The typical distance between the emitter and the receiver is 10 to 50 cm. The sonic speed ranges from 340 m/s at sea level to 295 m/s in the stratosphere. That implies that the travel time of an acoustic pulse is on the order of 1 ms.

Sonic speed is influenced by environmental temperature and pressure fluctuations. That dictates constant measuring of the sonic speed. To measure local sonic speed, a reference transmitter and receiver are placed in the cockpit and separated by a constant and known distance.

6.5.2 Phase-Coherent Trackers

A phase-coherent tracker measures the distance between a transmitter and a receiver by comparing the phase shift between the transmitted and the received signals. It continuously tracks the phase changes. Because the wavelength of the sonic wave is short, on the order of several millimeters, the tracker also must track the number of wavelengths between the two measurements. However, because the measurement is continuous, sufficiently fast electronics can be realized to avoid wavelength losses. The system is calibrated during the initial boresight procedure in which any number-of-wavelength ambiguities are removed. To ease the detection process, each helmet-mounted transmitter transmits a slightly different frequency signal [17].

6.6 Head Tracking Using Inertial Sensors

Helmet trackers that use inertial sensors have the advantage of being autonomous. They do not require any cockpit-mounted transmitters or receivers and thus do not restrict measurements to a confined motion box. That is advantageous in applications such as for tanks, whose crew members are required to change position from inside the turret to outside.

Another advantage of inertial head-tracking methods is the capability of modern inertial sensors to measure high angular rate rotations. Solid-state gyroscopes and miniature fiber-optic gyroscopes (FOGs), for example, have measurement ranges of 1,000 deg/sec with a scale-factor accuracy of 100 parts per million (PPM) and as such enable accurate measurement of the fastest head motion. Furthermore, their size and weight are steadily decreasing; a complete inertial measurement unit (IMU) weighing less than 0.5 kg is now available. The main drawback of using inertial sensors is their bias or drift, which causes measurement error to grow with time. For that reason, some scheme of calibration or drift compensation must be employed. Sometimes the error growth can be limited by external information from, for instance, the master navigator of the vehicle.

Head-orientation measurement using inertial sensors is mechanized by one of two basic methods: (1) using miniature rate gyros to measure the inertial angular rates of the head in elevation and azimuth and to steer a visually coupled device and (2) using an IMU to measure the 6-DOF head position and orientation relative to the inertial space.

6.6.1 Gyroscopes

With the rate-gyroscope method, the head acts like a joystick: it uses head movements to steer the VCS device, and the feedback to the user is the visual information from the sighting device. As such, the accuracy requirements from the gyros are moderate, since any pointing errors are continuously compensated by the user. The LOS position eventually is measured by reading the angular sensors mounted on the VCS gimbals. Figure 6.13 illustrates the concept of head-motion tracking using rate gyroscopes.

Rate gyros also can be used for reduction of lag and "smoothing" the head position measurements by other devices, such as electromagnetic or electro-optic systems, that operate with insufficient sampling rate. So far, the size and weight of gyros have prevented their use, but with future technologies of miniature, lightweight, micromachined rate gyros [18], that obstacle should be removed.

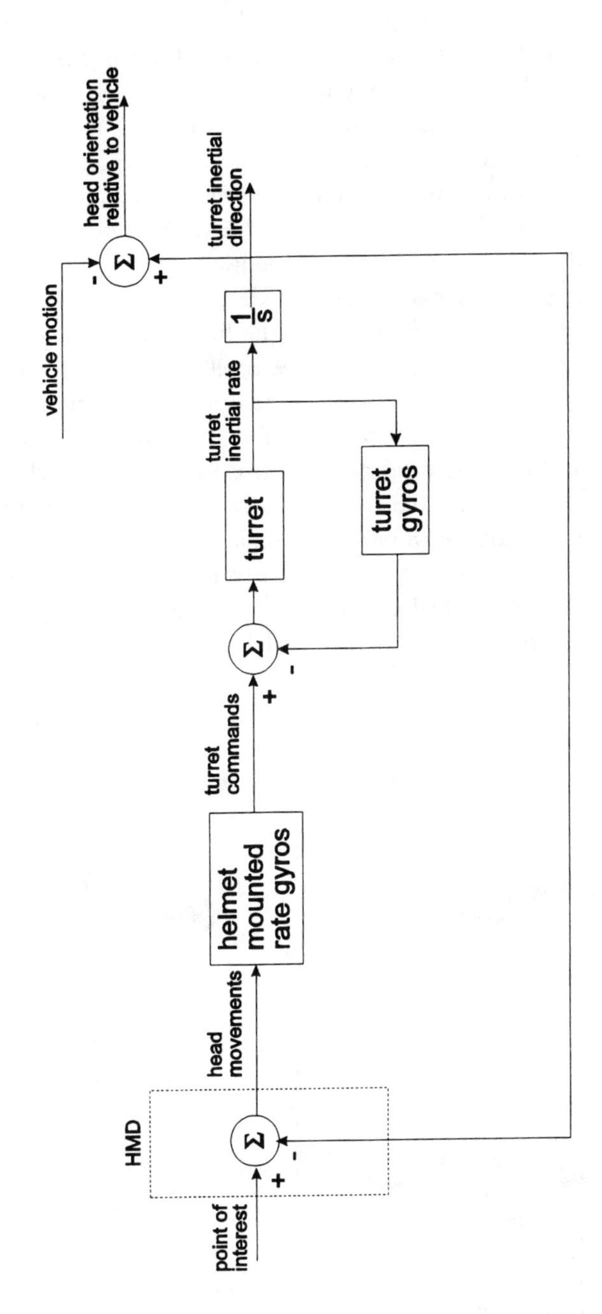

Figure 6.13 Head-motion tracking using rate gyroscopes.

6.6.2 Inertial Measurement Units

The second method utilizes a strapdown inertial attitude and heading reference unit (AHRU) concept. A miniature IMU, composed of three miniature gyroscopes and three accelerometers, is mounted on the helmet. The gyroscopes of the IMU measure the angular increments of the head motion while the accelerometers measure the increments of the linear velocities of the head. As the inertial sensors measure angular and linear velocity increments relative to the inertial space rather than relative to the vehicle, the motion of the vehicle itself is subtracted from the computed head motion.

A functional block diagram of an AHRU is shown in Figure 6.14. The inertial angular rates and specific forces of the head are sensed by the IMU, which consists of three gyroscopes and three accelerometers. The sensor readings are compensated for bias and scale-factor errors. The inertial rate data is corrected for the effects of Earth rotation and the vehicle motion over the Earth's surface to obtain the head rates relative to the local level coordinate frame. Those rates are utilized to derive the direction cosine matrix (DCM) and the associated helmet attitude and azimuth angles.

The inertial body axis accelerations are transformed to the local level frame, are compensated for the local gravity acceleration and Coriolis acceleration, and are integrated to obtain the local level velocities. The velocity is

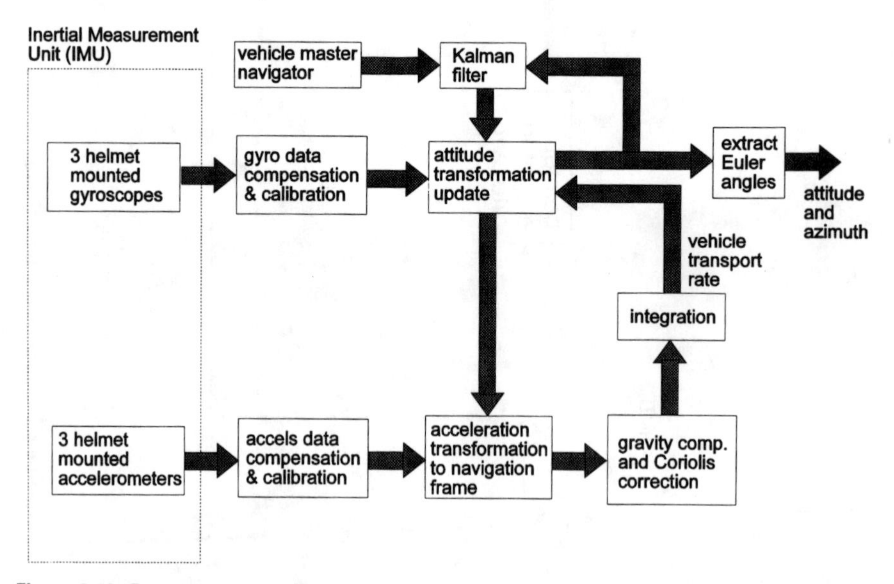

Figure 6.14 Functional block diagram of helmet-pose measurement using strapdown AHRU concept.

divided by the local radius of the Earth to obtain the angular transport rates for compensation of the inertial angular rates.

The attitude with respect to the navigation frame of coordinates is determined from the DCM, C_H^N, which is updated by

$$\dot{C}_H^N = C_H^N[\omega_{IH}^H \times] - [\omega_{IN}^N \times] C_H^N \tag{6.50}$$

and

$$\omega_{IH}^N = \omega_{IE}^N + \omega_{EN}^N \tag{6.51}$$

where

the index N = the navigation coordinate frame;

H = the helmet coordinate frame;

I = the inertial coordinate frames;

E = the Earth coordinate frame;

ω_{IH}^H = the angular rate of the helmet, in the helmet coordinate frame, as measured by the gyroscopes;

ω_{IN}^N = the angular rate of the navigation coordinate frame, relative to the inertial space, expressed in the navigation coordinate frame;

ω_{IE}^N = the Earth rotation rate in the navigation coordinate frame;

ω_{EN}^N = the transport rate, or rotational rate of the navigation coordinate frame with respect to the Earth, expressed in the navigation coordinate frame.

The rotation matrix $[\omega \times]$ is defined as

$$[\omega \times] = \begin{bmatrix} 0 & -\omega_z & \omega_y \\ \omega_z & 0 & -\omega_x \\ -\omega_y & \omega_x & 0 \end{bmatrix} \tag{6.52}$$

The helmet velocity in the navigation coordinate frame is updated by

$$\dot{V}^N = C_H^N f^H - (2\omega_{IE}^N + \omega_{EN}^N) \cdot V^N + g^N \tag{6.53}$$

where

f^H = helmet axes specific forces as measured by the accelerometers;

g^N = the gravity vector in the navigation coordinate frame.

The navigation coordinate frame is related to the Earth coordinate frame by the transformation matrix, C_N^E, which is computed by

$$\dot{C}_N^E = C_N^E[\omega_{EN}^N \times] \tag{6.54}$$

The typical navigation coordinate frame is the so-called wander azimuth coordinate frame, which is oriented at an azimuth angle from the local north-east-down frame. For that system, the transport rate is

$$\omega_{EN}^N = [\rho_x, \rho_y, 0] \tag{6.55}$$

with

$$\begin{aligned} \rho_x &= V_y\left(\frac{\cos^2\alpha}{R_n} + \frac{\sin^2\alpha}{R_m}\right) + V_x \cos\alpha \sin\alpha\left(\frac{1}{R_n} - \frac{1}{R_m}\right) \\ \rho_y &= -V_x\left(\frac{\cos^2\alpha}{R_m} + \frac{\sin^2\alpha}{R_n}\right) + V_y \cos\alpha \sin\alpha\left(\frac{1}{R_m} - \frac{1}{R_n}\right) \end{aligned} \tag{6.56}$$

R_m and R_n are the meridional and principal normal radii of the Earth at the specific latitude.

The helmet attitude and azimuth angle are finally computed from the DCM (6.50). The DCM matrix is related to the Euler angles by

$$C_H^N = \begin{bmatrix} \cos\theta \cos\psi & \sin\phi \sin\theta \cos\psi - \cos\phi \sin\psi & \cos\phi \sin\theta \cos\psi + \sin\phi \sin\psi \\ \cos\theta \sin\psi & \sin\phi \sin\theta \sin\psi + \cos\phi \cos\psi & \cos\phi \sin\theta \sin\psi - \sin\phi \cos\psi \\ -\sin\theta & \sin\phi \cos\theta & \cos\phi \cos\theta \end{bmatrix} \tag{6.57}$$

The Euler angles are readily extracted from (6.57) by

$$\begin{aligned} \phi &= \arctan\left[\frac{C_H^N(3,2)}{C_H^N(3,3)}\right] \\ \theta &= \arcsin[-C_H^N(3,1)] \\ \psi &= \arctan\left[\frac{C_H^N(2,1)}{C_H^N(1,1)}\right] \end{aligned} \tag{6.58}$$

The Euler angles from (6.58) measure the head orientation and azimuth relative to the navigation frame of reference. They easily can be converted to the vehicle coordinate frame by using the vehicle-orientation measurements obtained by the vehicle master navigator, if measurements relative to the vehicle are required.

6.7 Dynamic Response Improvements

All head-pose measurement systems introduce a delay or lag for several reasons:

- Regardless of the technology used, the sensor has an inherent response time.
- All systems sample the data of the sensors and use some computation scheme to resolve the head pose. Even with modern computers' high computing speed, computation time is not negligible because of the large number of coordinate transformations and trigonometric function calculations that are needed.
- Filters added to the schemes to reduce the noise of the sensors further increase the lag of the system.

The lag of helmet-position trackers disturbs the user because it impairs the correspondence of the displayed information on the HMD with the outside world, and it impairs the tracking performance of the user of HMDs [19]. Furthermore, when used in conjunction with VCSs, it limits the agility of those devices.

The obvious solution to reduce the lags of the systems is to increase the sampling rate and have a faster computation cycle. However, other methods have been devised that are based on predictors or the addition of auxiliary sensors, which assist in predicting the head motion.

6.7.1 Lag Compensation Using Predictors

A head-position predictor extrapolates the current head position τ milliseconds ahead by computing the head position given the current velocity. The velocity is computed from the previous time history of the head motion, assuming moderate changes in the velocity. Frequently, the predictor is implemented by using a Kalman filter. Consider the simple one-dimensional case in which a state vector composed of the head position and velocity is given as

$$x_t = [\vartheta \quad \dot{\vartheta}] \tag{6.59}$$

A state-space discrete model for the head kinematics is

$$x_{t+1} = \begin{bmatrix} 1 & T \\ 0 & 1 \end{bmatrix} x_t + \begin{bmatrix} T^2/2 \\ T \end{bmatrix} \nu_t = \phi x_t + \Gamma \nu_t \quad (6.60)$$

where ν_t represents the randomness of the head velocity and is modeled for simplicity as a Gaussian white noise with zero mean and variance of σ_1^2. The position of the head is measured also with noise n_t and is given by

$$y_t = [1 \quad 0] x_t + n_t = H x_t + n_t \quad (6.61)$$

The optimal Kalman filter that minimizes the variance of the prediction error, $E[(x_{t+1} - \hat{x}_{t+1})(x_{t+1} - \hat{x}_{t+1})^T]$, is

$$\hat{x}_{t+1} = \Phi \hat{x}_t + K_t (y_t - H \hat{x}_t)$$

$$K_t = \Phi P_t H^T [H P_t H^T + R]^{-1} \quad (6.62)$$

$$P_{t+1} = [\Phi - K_t H] P_t [\Phi - K_t H]^T + Q + K_t R K_t^T$$

where $R = \Gamma\Gamma^T \sigma_1^2$ and $Q = \Gamma\Gamma^T \sigma_2^2$.

The measurement noise also is considered to be Gaussian white noise with zero mean and variance equal to σ_2^2.

6.7.2 Lag Compensation Using Auxiliary Acceleration Measurements

A predictor compensates for lags essentially by differentiating the head-position motion to obtain head velocity or acceleration. That differentiation usually is noisy and lags the head motion. A similar concept can be implemented using additional accelerometers mounted on the helmet to measure head accelerations directly [20]. The accelerometers measure the head motion without lag or with only very small lag, thus potentially increasing the prediction accuracy. Miniature accelerometers with very low weight and cost are available. Using pairs of accelerometers mounted about the helmet-rotation center provide angular acceleration measurements of head motion. In a similar fashion, miniature rate gyros instead of the accelerometers can be used to obtain angular rate measurements.

The computational scheme employs a complementary filter, as shown in Figure 6.15. The inherent characteristic of the filter is to blend the quick response of the accelerometers with the accuracy of the helmet-position tracker. The noise and the bias of the accelerometers essentially are eliminated, while the position lag due to the delay of the head tracker is reduced.

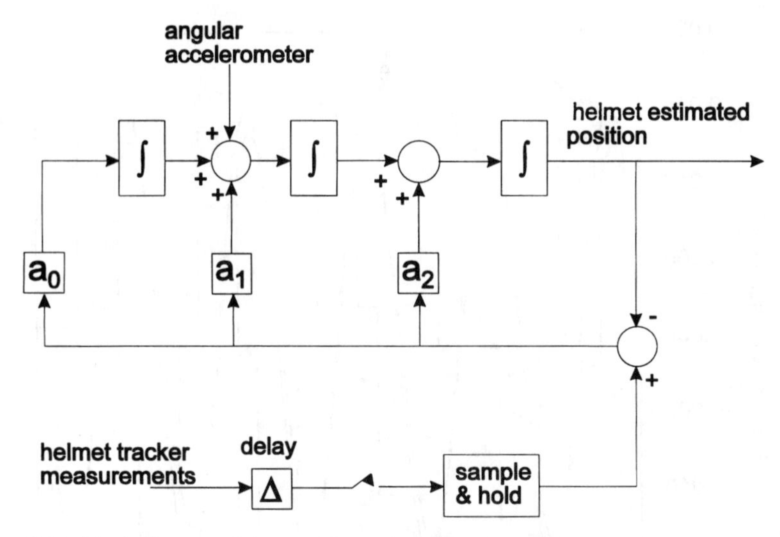

Figure 6.15 Block diagram of the complementary filter.

Figure 6.16 illustrates by simulation the effect of the complementary filter. The head motions are simulated by a sine wave motion of an amplitude of 15 mrad and a frequency of 5 Hz. Shown are the sampled lagged response of the head tracker and the output of the complementary filter and the true head motion. It can be seen that the complementary filter tracks the head motion without a delay and with a peak error of less than 10%.

6.8 Eye Tracking

Helmet trackers measure the head orientation, or the LOS defined by the reticle of the display (see Figure 6.1). That dictates that users align their LOS with the reticle. In reality, during head tracking, the eyes' motion exceeds the head motion, and in many target-searching conditions an angular offset between eye orientation and head orientation is maintained [21].

Figure 6.17 shows typical patterns of head and eye motions during target search. Faster and more accurate measurement of LOS pointing potentially is achievable by incorporating an eye tracker with the helmet tracker. Also, during air-to-air combat under high-g conditions, it sometimes is not possible for pilots to move their heads, whereas they can still track with their eyes. Thus, eye trackers would enable continuous target tracking, whereas head tracker would fail.

A simple and straightforward way to measure eye movements is by electro-occulography. The method of electro-occulography is based on the physio-

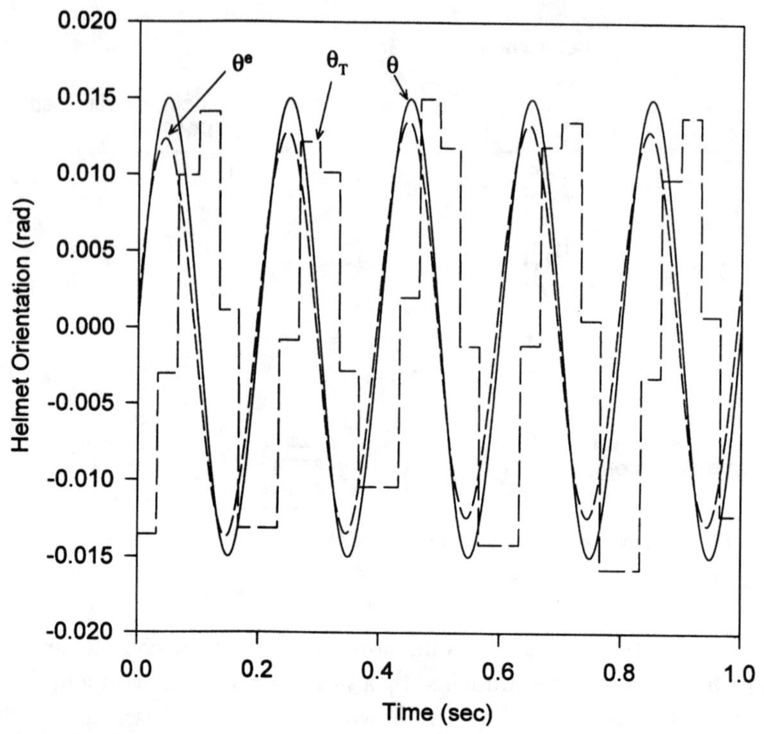

Figure 6.16 Steady-state response of the complementary filter. Shown are the true head motion, the head motion estimated by the filter, and the delayed and sampled head motions as obtained from the head tracker.

logical property that the eye has a dc potential between the cornea and the retina on the order of 1 mV. The field from that eye dipole varies almost linearly, with horizontal rotations up to about 30 degrees on either side of the center [22]. By using silver–silver chloride skin electrodes to measure the eye's electric potential, the eye rotation is measured. Other methods use the corneal reflection of eye illumination as detected by a position detector mounted in front of the eye.

Although eye trackers can measure very small eye movement, their accuracy generally is poor. That is because they are incapable of distinguishing between eye translations and rotations. Very small eye translations significantly impair accuracy. For instance, a translation of 0.1 mm is equivalent to a rotation of approximately 1 degree. Accurate eye trackers are used that utilize the Purkinje images of the eye that are obtained by light reflections of an illuminated eye form the eye's optical elements [23,24]. The double-Purkinje-image methods use the first and fourth images reflected from the front of the cornea

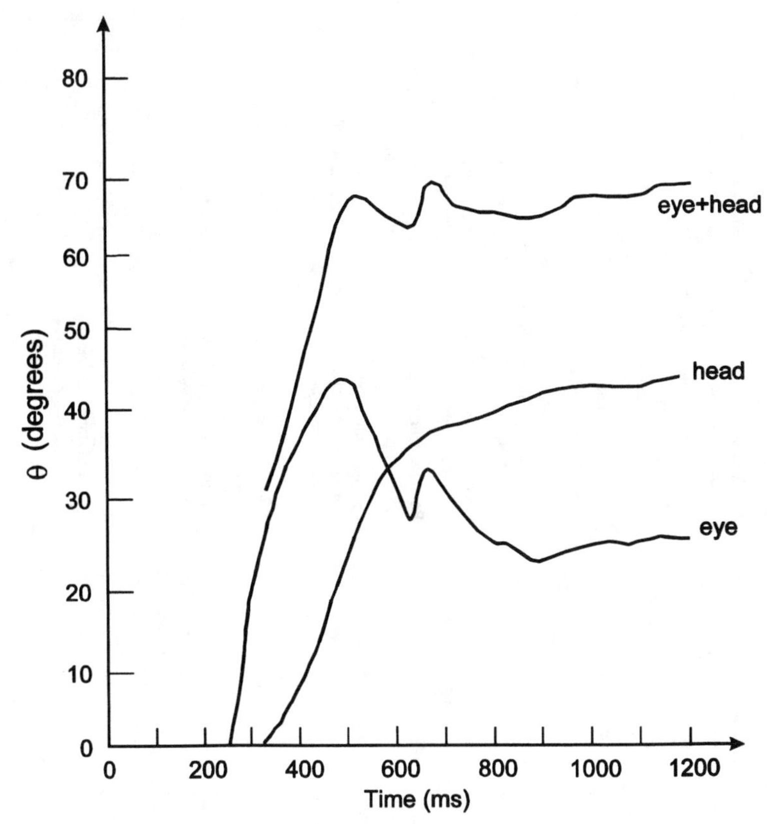

Figure 6.17 Typical head and eye motions during target search. The LOS is determined by the sum of the head rotation and eye gaze.

and from the rear surface of the lens, respectively. The two images are formed at about the same plane with similar intensity. The two other images, the second and third Purkinje images, are either coincident with the first or formed at a much farther plane than the other images and thus not utilized. Owing to the different distance between the two used images, they can be used to determine the angular rotation of the eye without being affected by translations that act to shift both images at the same direction and magnitude.

With the double-Purkinje-image, eye-tracking accuracy is on the order of 1 arc-min. Yet most eye trackers are large, heavy, and bulky and require special attachment to the helmet. For those reasons, their usefulness in operational HMDs is still limited.

Eye-tracking systems operating by illumination of the eye employ a noncolimated beam of invisible near IR light (typically 850 nm). For eye safety, the illumination level must be kept small.

Table 6.4
Summary of Head Pose Measurement Systems

Category	Technology	Realization	Strengths	Weaknesses	Interference Sources
Direct measurement	Mechanical bars and angular sensors: resolvers or shaft encoders	Mechanical structure	Accuracy, immunity to interference	Bulky; added weight increases helmet inertia	None
Remote sensing	Electromagnetic	Single emitter in cockpit, single receiver on helmet	Low added helmet weight; simple mechanization	ac systems require complex cockpit mapping	Very sensitive to presence of metals
	Electro-optic	Rotating IR beams	Proven technology	Mechanical elements reduce reliability	IR energy source reflections
	Electro-optic	LED array	Accurate, very low added helmet weight	Requires many symbols for full coverage	IR energy source reflections
	Electro-optic	V-slit camera	Potentially high sampling rate	Requires four cameras for full coverage	Ambient light, IR energy source reflections
	Ultrasonic	Time-of-flight measurement	Low added helmet weight	Requires continuous measurement of sonic speed; partial blockage	Ultrasonic noise sources, multipath signals
	Ultrasonic	Phase coherence	Low added helmet weight	Requires continuous measurement of sonic speed; partial blockage	Ultrasonic noise sources, multipath signals
Self-contained	Inertial	Rate gyros	Independent of cabin-mounted source unit; easily integrated with VCS	Can be used only with visual feedback; adds weight	Gyro bias
	Inertial	IMU	Autonomous system, independent of vehicle; unlimited motion box; capable of high angular rate	Requires external measurements to relate to vehicle coordinates and to bound errors and drifts	Gyro bias and noise

6.9 System Comparison

Each of the implementations or technologies introduced in this chapter has some strengths or advantages over other systems but also some inherent deficiencies [8]. Almost all the systems potentially can meet the accuracy required from a head tracker. Table 6.4 summarizes the main characteristics of the various head-pose measurement systems.

6.10 Summary

Head-pose measurement systems are implemented using various technologies. The most common ones are electromagnetic, electro-optical, ultrasonic, mechanical attachment, and most recently, inertial-sensor systems. Almost all systems can provide measurement accuracies of about 2 mrad near the nominal LOS direction where boresighting usually is performed. The accuracy is reduced to about 10 to 15 mrad at large angles away from the nominal LOS direction. Also, the measurement rate of most systems is comparable to the HMD video rate of 50 to 60 Hz. Some methods have no inherent restriction to increase the measurement rate to higher values.

The appropriate method depends on the specific application, the range of head motion, the required motion box, and the potential sources of interference in the particular environment. When considering suitability of a measurement system, attention must be paid to the attainable accuracy of the system, the measurement lag it introduces, the added weight to the helmet, and the feasibility of installing the system in the cabin of the vehicle. For example, a magnetic tracker is fairly easy to install, while an optical system with multiple cameras requires a minimum distance to the helmet and angular coverage.

References

[1] Clubine, W. R., "Modeling the Helmet-Mounted Sight," M.Sc. Thesis, Air Force Institute of Technology, Wright-Patterson AFB, Ohio, 1982.

[2] List, U. H., "Nonlinear Prediction of Head Movements for Helmet-Mounted Display," Operation Training Div., AFHRL-TP-83-45, Williams AFB, AZ, December 1983.

[3] Wells, M. J., and M. W. Haas, "Head Movements During Simulated Air-to-Air Engagements," in *Helmet-Mounted Displays II*, SPIE, Vol. 1290, 1990, pp. 246–257.

[4] Jacobs, R. S., T. J. Triggs, and J. W. Aldrich, "Helmet Mounted Display/Sight System Study," AFFDL-TR-70-83, April 1970.

[5] CAE Electronics Ltd., "Wide-Field-of-View, Helmet-Mounted Infinity Display System Development," AFHRL-TR-84-27, December 1984.

[6] Meyer, K., H. L. Applewhite, and F. A. Biocca, "A Survey of Position Trackers," *Presence*, Vol. 1, No. 2, Spring 1992, pp. 173–200.

[7] Raab, F. H., et al., "Magnetic Position and Orientation Tracking System," *IEEE Trans. on Aerospace Electronic Systems*, Vol. AES-15, No. 5, September 1979, pp. 709–718.

[8] Ferrin, F. J., "Survey of Helmet Tracking Technologies," in *Large-Screen-Projection, Avionic, and Helmet-Mounted Displays*, SPIE, Vol. 1456, 1991, pp. 86–94.

[9] Polhemus Navigation Sciences Division, McDonnell Douglas Electronics Company, "3Space Tracker Manual," Colchester: Vermont, 1985.

[10] Bohm, H. D. V., H. Schreyer, and R. Schranner, "Helmet Mounted Sight and Display Testing," in *Large-Screen-Projection, Avionic, and Helmet-Mounted Displays*, SPIE, Vol. 1456, 1991, pp. 95–123.

[11] Ascension Technology Corporation, "Questions and Answers About the Bird," 1990.

[12] Berry, J., et al., *PNVS Handbook*, Fort Rucker, AL: Directorate of Training and Doctrine, 1984.

[13] Honeywell Avionics Division, "Introduction to Honeywell Helmet Sight Systems," Minneapolis, April 1977.

[14] Toker, G., and M. Velger, "Measurement of Pose by Computer Vision and Applications to Control," *Proc. CompEuro*, Bologna, May 13–16, 1991.

[15] Stephenson, D., and D. Fosberry, "A Helmet-Mounted Sight System for Fighter Aircraft," *GEC J. Science and Technology*, Vol. 46, No. 1, 1980, pp. 33–38.

[16] Axt, W. E., "Evaluation of a Pilot's Line-of-Sight Using Ultrasonic Measurements and a Helmet Mounted Display," *National Aerospace and Electronics Conference*, Dayton, OH, May 18–22, 1987, pp. 921–927.

[17] Sutherland, I. V., "A Head-Mounted Three Dimensional Display," *AFIPS Conference Proc., Fall Joint Computer Conf.*, Vol. 33, 1968, pp. 757–764.

[18] Connelly, J. H., and G. N. Brand, "Advances in Micromechanical Systems for Guidance, Navigation and Control," presented at the *AIAA Guidance, Navigation, and Control Conf.*, New Orleans, August 11–13, 1997.

[19] So, R. H. Y., and M. J. Griffin, "Compensation Lags in Head-Coupled Displays Using Head Position Prediction and Image Deflection," *AIAA J. Aircraft*, Vol. 29, No. 6, November–December 1992, pp. 1064–1068.

[20] Merhav, S. J., and M. Velger, "Compensating Sampling Errors in the Stabilization of Helmet Mounted Displays Using Auxiliary Acceleration Measurements," *AIAA J. Guidance, Control, and Dynamics*, Vol. 14, No. 5, September–October 1991, pp. 1067–1069.

[21] Robinson, G. H., B. W. Koth, and J. P. Ringenbach, "Dynamics of the Eye and Head During an Element of Visual Search," *Ergonomics*, Vol. 19, No. 6, 1976, pp. 691–709.

[22] Fleming, D. G., et al., "Adaptive Properties of the Eye-Tracking System as Revealed by Moving-Head and Open-Loop Studies," *Annals New York Academy of Sciences*, 1969, pp. 825–850.

[23] Cornsweet, T. N., and H. D. Crane, "Accurate Two-Dimensional Eye Tracker Using First and Fourth Purkinje Images," *J. Optical Society of America*, Vol. 63, No. 8, August 1973, pp. 921–928.

[24] Crane, H. D., and C. M. Steele, "Accurate Three-Dimensional Eyetracker," *Applied Optics*, Vol. 17, No. 5, March 1978, pp. 691–705.

7

Display Symbology and Information

Display symbology and information are the essence of the HMD. Of all the attributes of HMDs, the issues of how much and what information and symbology should be presented on the display and the display organization and format are the most indeterminate and insufficiently defined subjects.

In the early HMDs, the straightforward approach was to adopt the symbology and information that were presented to the user in the display that the HMD replaced. In that approach, for application of an HMS, the symbology imitated the fixed gunsight and for applications in which the HMD was intended to be used as a primary flight reference (PFR), the symbology was copied from the traditional HUD. Still, in most current HMDs (excluding displays that are used as a sight only), the symbology is largely an outgrowth of the existing HUD symbology, with the addition of some specialized symbols and the modification of others.

Despite the wide use and experience gained from the use of HUDs, there has been little standardization in the format and symbology of HUDs. The same holds true for HMDs [1].

The HMD offers many advantages because of its capability to present constant and continuous information and symbology to the pilot regardless to the pilot's direction of gaze. However, two factors distinguish the HMD from all other conventional displays: first, the HMD symbology always appears within the pilot's field of view, and, second, the display changes its orientation relative to both the aircraft and the outside world. As a result, the same advantages create the unique problem of the HMD of what and how to present that important information without creating confusion and interfering with the pilot's view of the external world.

It is widely acknowledged that the information presented on the HMD must be the data that is essential for the specific task [2–6]. Considering that the display has a restricted available display area and that every piece of information presented on the display requires some mental processing and attention, anything not vital to the immediate comprehension of the display may only distract the viewer's attention, obscure the outside world view, and cause confusion.

7.1 Information and Symbology Requirements

The purpose of the symbology of the HMD is to provide the pilot with valuable information, indications, and cues to reduce response time to information flow by relieving the pilot from the need to look at the aircraft's fixed displays, to reduce workload, and to increase situational awareness.

The natural tendency of the display designer is to present simultaneously as much information as possible. Obviously, the available display area is limited, and each piece of information produces clutter, which overloads the user who must process a large amount of information. To reduce the amount of presented information and symbology, most displays have several display modes among which the pilot can switch manually or automatically according to the mission or task for which the display is being used at a particular instance. In addition, for each mode the user can press a button to switch to a declutter mode, in which less essential symbology is removed.

The symbology must be presented to the user in a clear, unambiguous, and readable format. Therefore, it is clearly advantageous to present the symbology in stroke mode. Numerous experimental works have examined the requirements of the symbology [7]. On the basis of on those studies, it is accepted that for good readability of the display, the symbol height should be at least 16 to 20 arc-min, the optimal symbol width-to-height ratio 1:5, the stroke line width 1/6 to 1/10 the symbol height (thus, larger than about 2 arc-min), and the minimum symbol spacing 10 to 15 arc-min. As a rule of thumb, the symbology typically extends to about 70% of the display's FOV to the left, to the right, and above the display center and extends to over 80% downward. That practice minimizes the clutter at the top of the display, which must be left clear in the air-to-air mode of fighters.

7.2 Principles of Symbology and Information Presentation

7.2.1 Display Information Levels

The information presented on the display is categorized according to the level of information shown. Information is perceived and requires different ways of interpretation or understanding according to how it is presented. Some variables require mental processing and interpretation if they are presented directly.

When the same variables are presented differently, they are perceived and understood naturally. Consequently, certain variables are shown simultaneously in two different ways, for example, as meter-type information and as symbols. Figure 7.1 shows the Honeywell IHADSS symbology. The IHADSS HMD is the first widely used operational system and still is in service in the Apache helicopter. It combines analog-type scales, numerical information, alphanumeric symbols, and status and command symbology overlaid on a video imagery from a thermal night vision sensor.

The seven different levels are as follows.

- *Status information* is the current state of the controlled element or the state of a certain variable in the environment, such as the aircraft altitude or speed, what missile is ready for launch, or what mode of oper-

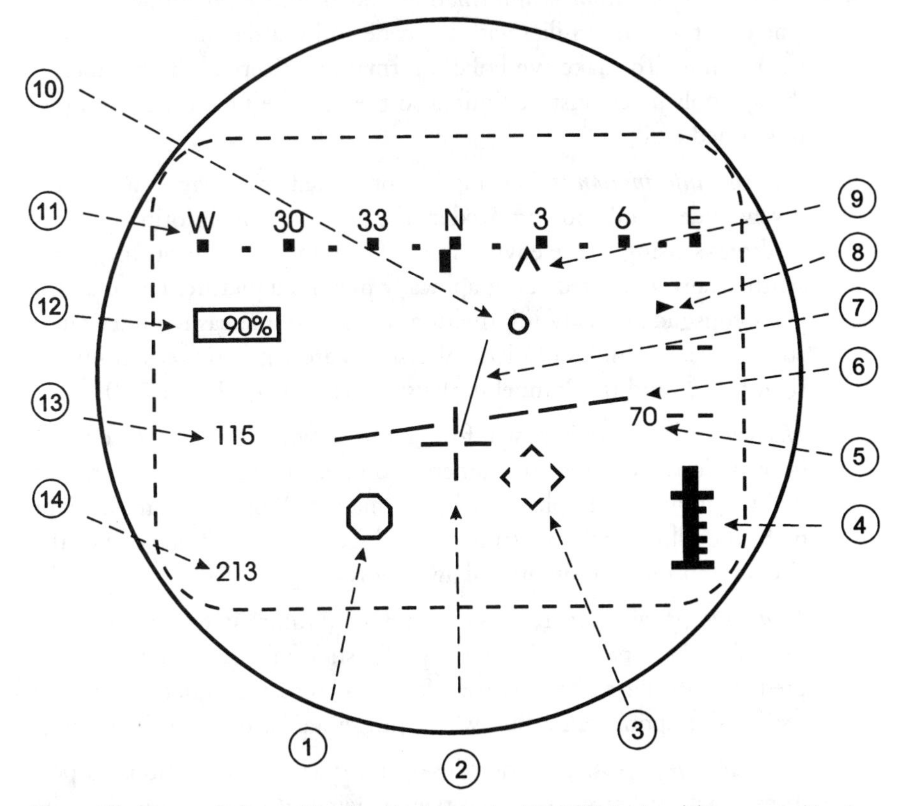

Figure 7.1 The symbology of the Apache helicopter IHADSS display: (1) hover-position box, (2) sensor's LOS reticle, (3) aircraft LOS, (4) analog altitude, (5) digital altitude, (6) horizon line, (7) velocity vector, (8) vertical-speed indicator, (9) Doppler steer indicator, (10) acceleration cue, (11) heading tape, (12) digital torque, (13) digital indicated speed, and (14) digital ground speed.

ation is selected. The status information commonly is presented by a numerical value or a symbol. Usually, the information on the status display is the kind of information the user is required to view not lengthily but infrequently and with short glimpses only. Therefore, the status information must be easy to interpret and free from any likely ambiguities.

- *Indicators and meters* display information that shows the continuously changing state of the vehicle, such as the speed bar or the rate of climb bar. Typical indicators are presented as dials or a thermometer-type bar. Among the main advantages of such a presentation is the ability of the user to extract rate-of-change information and quickly get the approximate value of the presented variable.
- *Symbolic information* is information coded into alphanumeric form. The user must be familiar with the code and translate it to understand its meaning. To make symbolic information effective, the number of the symbols used must be limited so they can be memorized and distinguished easily.
- *Pictorial information* is information presented in a way that is intuitively understood and perceived by the operator. The information may be direct sensing of the environment, for example, a sensor image or a synthetically generated and graphically presented picture. Pictorial displays consolidate many information items into one easily understandable image. Examples of pictorial displays are synthetic terrain images (Figure 7.2) and the "tunnel in the sky" symbology (Figure 7.3).
- *Historic information* is past information shown together with current information to enhance the understanding of the recent state. Graphic flight path is an example of such information. From the graphic flight path, the pilot is able to estimate the direction and the velocity of the aircraft and increase situational awareness.
- *Predictive information* presents the computed future state of the variable to assist the operator in making decision of the current action. Predictive displays often are used for vehicles that are characterized by a sluggish response and, thus, whose future state is not easily perceived.
- *Command information* guides the operator as to what actions to perform to achieve immediate objectives. Placing an aiming symbol on the target, for example, ensures that the weapon will hit the target. Command display relieves the pilot from the need to calculate what action must be taken. The pilot needs only to follow a simple instruction. In that manner, the response time is reduced. Command displays

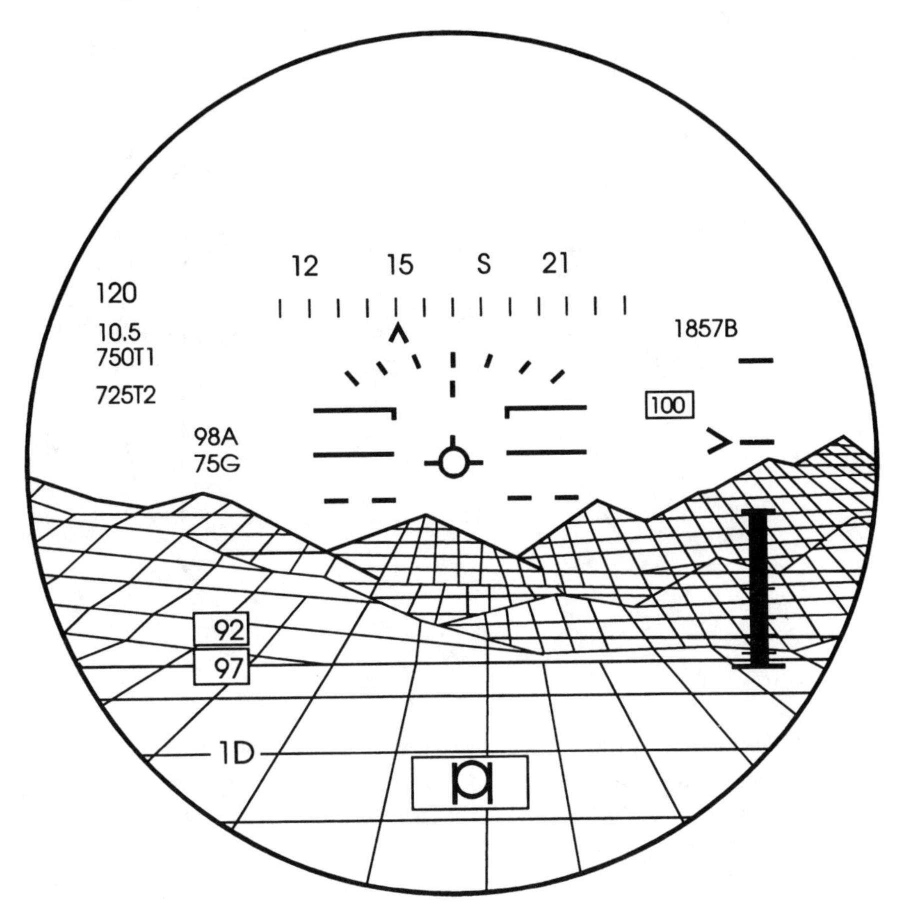

Figure 7.2 Pictorial display is used to enhance situational awareness in poor-visibility conditions. The terrain mesh is generated by using a stored digital map and measurement of the altitude above the ground.

are used whenever a quick response is required or in stressful conditions, when an erroneous action cannot be easily corrected by the pilot and may be fatal to aircraft survival.

7.2.2 Analog Versus Digital Format

Information that the user does not need to know exactly but does need to perceive qualitatively within a range or relative to the extreme values is better and more easily interpreted when it is displayed in an analog format. Analog presentation ordinarily is portrayed as a dial, a scale, or a thermometer-type expanding or contracting bar. Information that is constantly changing also is better pre-

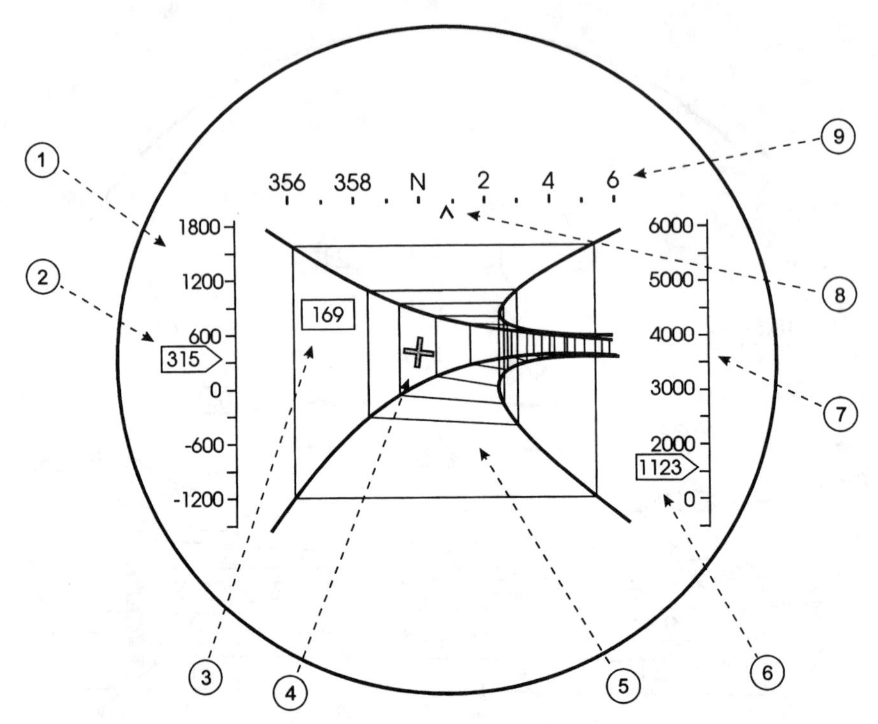

Figure 7.3 Pictorial display with a tunnel symbology: (1) vertical-speed scale, (2) vertical-speed value, (3) aircraft speed, (4) predictor, (5) tunnel, (6) numerical altitude, (7) altitude scale, (8) heading symbol, and (9) heading scale. The cross represents the future position of the aircraft. The command symbology guides the pilot into the tunnel by placing the predictor symbol into the square rib of the tunnel.

sented in an analog format. When presented as numerical values, the same information, which normally may change quickly, requires a longer time to resolve the current value. The analog presentation of a dynamically changing variable enables the operator to extract not only the current value but also its rate of change, thus allowing prediction of its future value. Analog scales can be perceived sensibly out of the center of the FOV, and the information can be assessed rapidly in the normal fast scan around the primary and backup instruments.

Information that is either static or must be known exactly is better perceived when presented by digital values.

Generally, analog information occupies more display area than digital information, which usually needs a single digit or a pair of digits. However, too many parameters that are presented numerically may increase confusion and require more time for the user to process.

7.2.3 Central and Peripheral Presentation

Apart from the type of information presented on the HMD, its format and organization are important issues. Almost always, the criterion adopted to determine which data is shown in the central region of the display and what is shifted to the periphery is simple: items that must be monitored constantly are placed at the center of the display. Less frequently used symbols, such as numerical indications of speed or height, that are not used frequently are moved away from the center toward the periphery of the display. As a result, symbology that is used for the control of the aircraft and information needed for weapons aiming and ground avoidance are placed in the central area of the display and contend for the pilot's attention in the central area of vision [8].

Placing a large amount of information in the central area of the display results in overcrowding of the display and may cause confusion between adjacent symbols and obscure the external world [6].

Information that can be monitored out of the corner of the eye, such as the vertical speed bar or the heading bar, is placed at the periphery. Evidently, the fewer items that require direct foveal view, the less the pilot is required to divide attention among those items. In the design of a display symbology, each item must be considered carefully if it can be presented as a symbol suitable for presentation at the periphery of the display and thus save valuable central display area. On the other hand, a balance between central and peripheral symbology must be ensured; otherwise, too much symbology may clutter the display edges and limit the effective field of view of the external world, increasing the risk of missing a view of potential hazards and threats.

7.2.4 Pictorial Presentation

In many cases, the external view must be aided by more comprehensive visual cues. For that purpose, pictorial display symbology can be used. Common examples are conditions of low or poor visibility and extremely hazardous flight environments. In NOE flights, for example, when there is a danger of hitting obstacles, enhancements of the terrain structure are beneficial. That can be achieved by adding to the display a graphic presentation of the surroundings with markings of the significant potential obstacles [9].

Current technology enables pilots of fixed-wing aircraft to perform low-altitude flights in low-visibility day and night conditions using night vision systems, such as the low-altitude navigation and targeting infrared night (LANTIRN) system, used in the F-16 fighter. The quality of the IR image in certain weather conditions, however, sometimes is degraded, and the aircraft pilotage as well as the pilot's situational awareness are impaired. To enhance the pilot's awareness of the terrain, it has been suggested that the image quality be enhanced by a synthetic terrain image that exploits digital terrain-elevation data and the navigational information obtained from the aircraft or pod navigation system. See Figure 7.2 for a display symbology that enhances the terrain structure of the immediate vicinity.

A further step can be achieved by adding an aiding display that shows a guiding path the pilot is required to follow. One example for such an aiding display is the tunnel display, which is useful in helping the pilot perform complex maneuvers [10,11]. Figure 7.3 showed a display format that includes a tunnel-in-the-sky symbology. The tunnel guides the pilot into, for example, the landing path to achieve an accurate flight toward the landing runaway.

The tunnel in the sky is an example of a three-dimensional perspective display symbology that combines status, command, and predictive information into a single pictorial display. The tunnel describes the desired flight path the pilot is required to follow. It shows a predictor symbol of the aircraft, portraying its future orientation and path for the current pilot action. To follow the tunnel path, the pilot must place the predictor symbol in the square rib of the tunnel.

7.2.5 Symbol Coding

Symbol coding increases the amount of information given to the user while minimizing the display area. Coding enables quick detection, processing, and interpretation of the information. To make coding effective, it must be distinguishable, clear, and limited to a small group of codes.

Coding can be introduced by the use of several methods. Commonly, coding is achieved by using different shapes, alphanumeric symbols, colors, and flashing symbols. Color coding has been found to be the single most effective type of coding available, superior to all other codings [12]. However, color coding is rarely used in HMDs, simply because most displays are monochromatic and current displays cannot provide sufficient brightness in color. Flashing symbols, because they attract attention, are used mainly as warning indications [6].

7.3 Primary Flight Reference Symbology and Displays

PFR symbology displays the essential information required for the pilot to fly the aircraft [2]. The PFR information includes the air speed, aircraft altitude and attitude, heading, aircraft climb or descend, and the magnitude and direction (left or right) of the turn.

In most aircraft, the HUD is used as the PFR, although some aircraft still use a dedicated classic mechanical indicator for each parameter. Many of the HMD symbols are adopted from the HUD.

A typical PFR symbology normally used in HMDs is shown in Figure 7.4. The symbology is divided into the central symbology and the peripheral symbology. The descriptions of the symbols used in the HMD are listed in Table 7.1.

The main elements of the primary flight reference are the artificial horizon line, the pitch ladder, the azimuth or heading scale, the vertical-speed scale, the altitude bar, numeric altitude, numeric air speed, and the slide or slip ball.

7.3.1 Artificial Horizon Line

A horizon line is presented relative to a fixed symbol of the aircraft. The aircraft symbol is always shown in the center of the display.

7.3.2 Pitch Ladder

The pitch ladder is a scale that shows the pitch angle of the aircraft. It is composed of a series of horizontal lines with vertical tick marks and a numerical indication at either end, with increments of 5 degrees between successive lines.

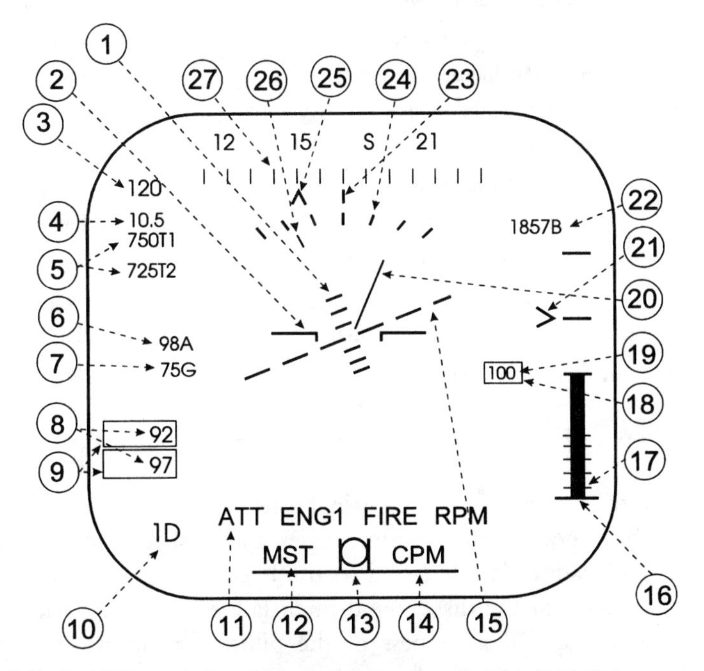

Figure 7.4 Typical PFR symbology. Item descriptions are listed in Table 7.1.

Table 7.1
HMD Symbology Shown in Figure 7.4

Number	Symbol Description	Units
1.	Pitch ladder	degrees
2.	Sensor LOS fixed symbol	—
3.	Bearing to waypoint	degrees
4.	Distance to waypoint	kilometers
5.	Engine temperature	degrees
6.	Indicated air speed	knots
7.	Ground speed	knots
8.	Torque	percentage
9.	Torque limit indication	—
10.	Display mode (D indicates declutter mode)	number
11.	Alphanumeric warning codes	alphanumeric
12.	Alphanumeric messages	alphanumeric
13.	Sideslip ball	—
14.	Fail message	alphanumeric
15.	Horizon line	degrees
16.	Vertical-speed scale	feet per minute
17.	Radar altitude analog bar	feet
18.	Minimum altitude warning	frame symbol
19.	Numeric radar altitude	feet
20.	Velocity vector	knots
21.	Rate-of-climb pointer	feet per minute
22.	Numeric barometric altitude	feet
23.	Aircraft-heading fixed lubber line	—
24.	Roll-angle scale	degrees
25.	Bearing-to-waypoint pointer	degrees
26.	Roll-angle pointer	degrees
27.	Compass or azimuth scale	degrees

Very often, positive pitch angles are distinguished from negative angles by the use solid symbol lines for the positive angles and dashed lines for negative angles. The pitch ladder is referenced to the fixed aircraft symbol, which is located at the center of the display and pitch ladder. The pitch ladder is an important cue for spatial awareness of the pilot, particularly during large maneuvers. To assist in recovery from extreme attitude conditions, most displays use specific symbols to designate the zenith and nadir poles, which are visible only at extreme angles [13]. In some displays, the pitch ladder uses "bendy bars," in which the pitch lines are canted to indicate the direction of the horizon when it is no longer in view [14]. Their deficiency is that they make it more difficult to estimate accurately the aircraft roll attitude. Figure 7.5 illustrates several formats commonly used for the pitch ladder.

Different symbology for presenting aircraft attitude, designed specifically for HMDs for use in large-attitude conditions are the arc-segmented attitude reference (ASAR), commonly termed the "orange peel," and the theta symbology [5,15,16]. Both are shown in Figure 7.6, along with conventional attitude symbology.

The ASAR symbology presents a circle surrounding a fixed inverted T, the climb-dive symbol of the aircraft. The upper part of the circle, which represents the area above the horizon, is invisible; the visible portion of the circle represents the area below the horizon. During level flight, half the circle is visible; it becomes less than a half circle during climb. When the aircraft dives, more of the circle becomes visible. Roll angles are indicated by tick marks displayed at the ends of the visible circle.

The theta attitude symbol portrays a three-dimensional globelike hemisphere, which is free to rotate about the three axes, with its longitudinal lines indicating the heading in 45-degree increments. The upper portion of the ball is drawn with solid lines and the lower with broken lines. Meridional tick marks indicate the pitch angles.

Within the ball, there is an inverted-T climb-dive symbol to indicate the aircraft reference. Cardinal headings are marked by the letters N, S, E, and W.

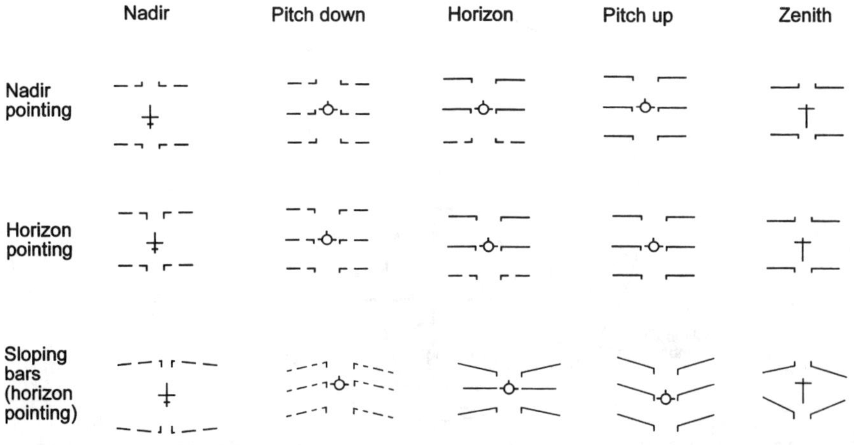

Figure 7.5 Pitch-ladder display formats as shown at several pitch attitudes. The vertical edge tags indicate nadir or horizon pointing.

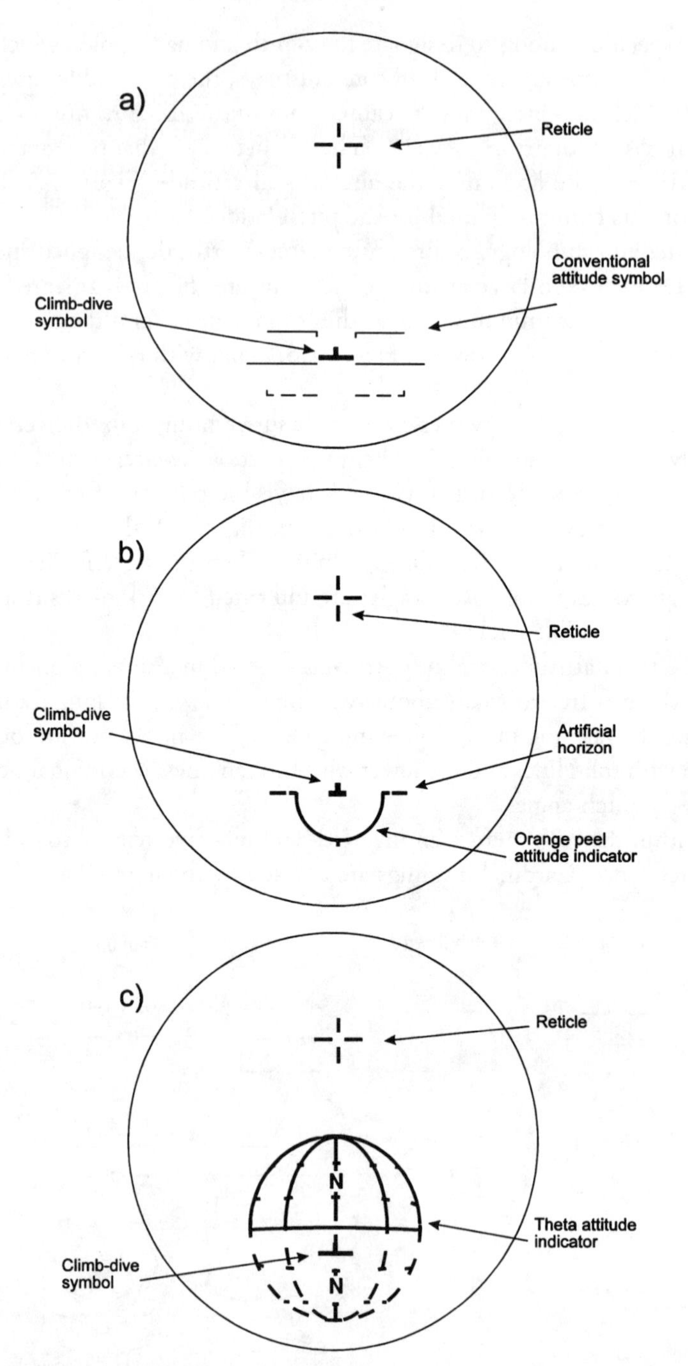

Figure 7.6 Three attitude symbols: (a) conventional pitch ladder, (b) ASAR symbol, and (c) theta globe-attitude symbol.

7.3.3 Azimuth or Heading Scale

The azimuth or heading scale consists of a movable heading scale and a fixed lubber line. It is positioned horizontally across the top of the display and displays 180 degrees heading, with the cardinal headings marked by the letters N, S, E, and W, and numerical headings for each 30-degree increment in units of tens of degrees (e.g., N, 03, 06, E).

7.3.4 Vertical-Speed Scale

The vertical-speed scale, a vertical bar placed at the right edge of the display, shows the rate of climb or descent of the aircraft. Some formats use a dial instead of the vertical bar.

7.3.5 Altitude Bar

The altitude bar is a vertical bar usually placed at the right side of the display, near the vertical speed bar. It shows by a thermometer-type indicator the altitude of the aircraft.

7.3.6 Numeric Altitude

The numeric altitude displays numerically the current altitude of the aircraft.

7.3.7 Numeric Air Speed

The numeric air speed shows numerically the current aircraft speed. Normally, both numeric altitude and numeric air speed are placed at the periphery of the display.

7.3.8 Slide or Slip Ball

The slide or slip ball is placed at the bottom of the display and shows the side forces acting on the aircraft. When the ball is within the two vertical dashes, it indicates straight flight or a coordinated turn.

A representative HMD symbology is the IHADSS symbology (see Figure 7.1). To reduce clutter, not all the symbols are shown simultaneously but are selected according to the display mode. The HMD has four operational modes in which different combinations of the symbology are presented: cruise mode, transition mode, hover mode, and bob-up mode.

7.4 Symbology Frame of Reference

The symbology of the HMD can be presented in one of two distinct ways: inside-out or outside-in [17]. The inside-out presentation is an egocentric frame of reference, also termed "pilot's view." In this frame of reference, a fixed aircraft symbol represents the aircraft attitude, while the horizon symbol rolls and pitches with the aircraft motion. In the outside-in presentation, also termed "God's view," the aircraft attitude symbol is presented with the orientation as would be seen by a stationary viewer behind and above the aircraft. Here, the aircraft symbol is no longer fixed in the display frame of reference but rather is rolling and pitching. Figure 7.7 depicts the two types of symbologies.

Inside-out symbology is considered more appropriate for flight control tasks; outside-in is more appropriate for navigation tasks. Some display arrangements combine both types of presentations. For example, in the perspective tunnel-in-the-sky display (see Figure 7.3), a three-dimensional predictor symbol in an outside-in configuration is used instead of the plain cross symbol, together with the conventional inside-out attitude symbology. The three-dimensional aircraft symbol is useful for perceiving future aircraft attitude, thus better assisting execution of control commands.

Perhaps the most appropriate use of outside-in symbology is in the landing display mode. By showing a three-dimensional symbol of the aircraft with a graphic image of the landing strip, the pilot gets a view of the position and orientation of the aircraft relative to the landing strip, and the landing task becomes easier (Figure 7.8).

7.5 Conformal Versus Nonconformal Symbology

The traditional HUD presents attitude information that conforms with the outside world, in the sense that the artificial-horizon symbol coincides with the real horizon and inclines to the body axes of the aircraft. In the HMD, the additional attitude between the head and the aircraft axes introduces the dilemma of how to present the information:

- Conforming to the outside world, often referred to as world or roll-stabilized symbology, which changes with the head orientation;
- Relative to the aircraft axes, referred to as aircraft stabilized symbology, as in the HUD.

Improper implementation of the symbology orientation can lead the pilot to execute incorrect control actions and aggravate spatial disorientation [18,19].

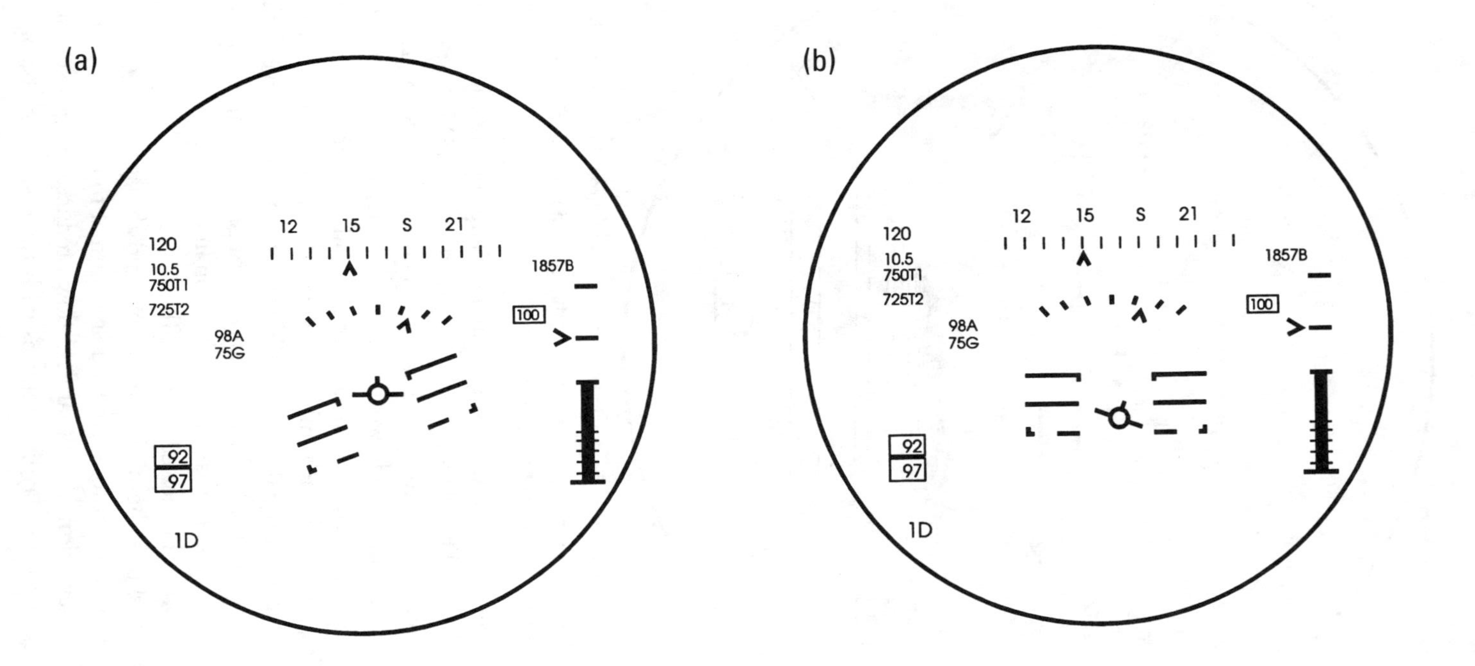

Figure 7.7 Symbology frame of reference: (a) inside-out symbology and (b) outside-in symbology.

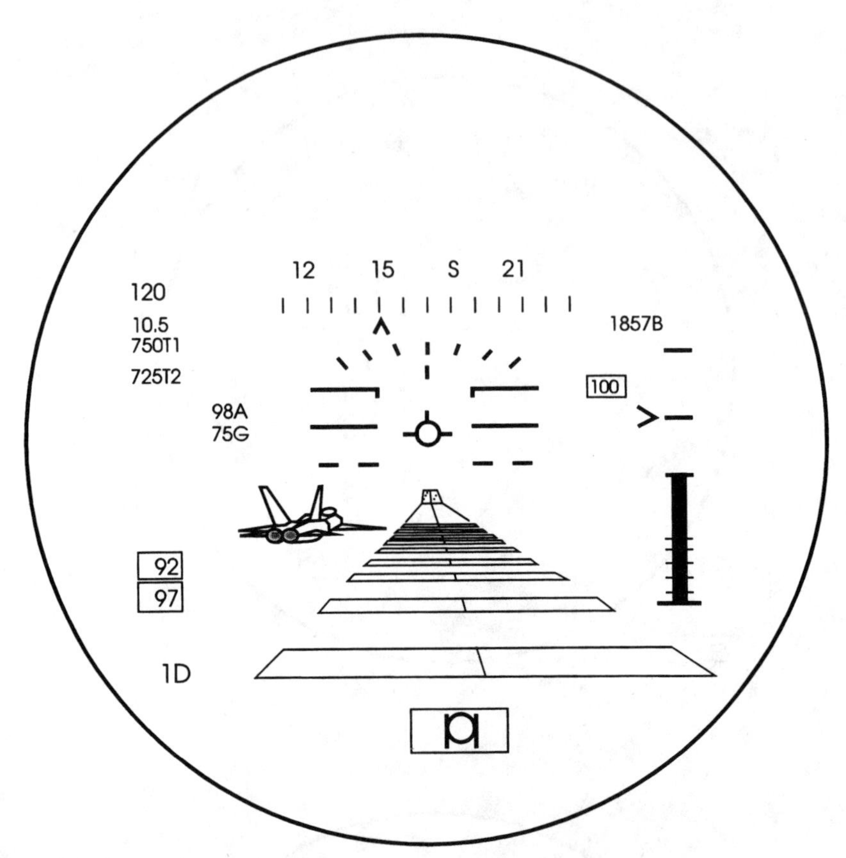

Figure 7.8 Perspective pictorial display in landing mode. The three-dimensional aircraft symbol shows the future position and attitude of the aircraft relative to the runway for the current control commands.

That kind of mismatch between the symbology and the outside world is evident in some VCSs. This class of systems presents on the HMD a raster image that comes from an electro-optical sensor, such as the PNVS used by the Apache helicopter. The outside-world view of the pilot is the image of the sensor that is panned and elevated at large angles and co-aligned with the pilot's LOS. On the other hand, the symbology is referenced, or stabilized, to the aircraft axes, thus fixed in the display coordinates. When pilots move their heads to an off-axis position, the flight symbology no longer conforms with the image presented by the sensor. In that case, the horizon line and the velocity vector no longer coincide with the viewed image [18].

Figure 7.9 shows the optical field flow and the symbology for forward flight in two head orientations: forward and normal to the flight path. In the forward head orientation, the visual field flow produces image expansion. For the aircraft-stabilized display, it conforms with the display symbology. When

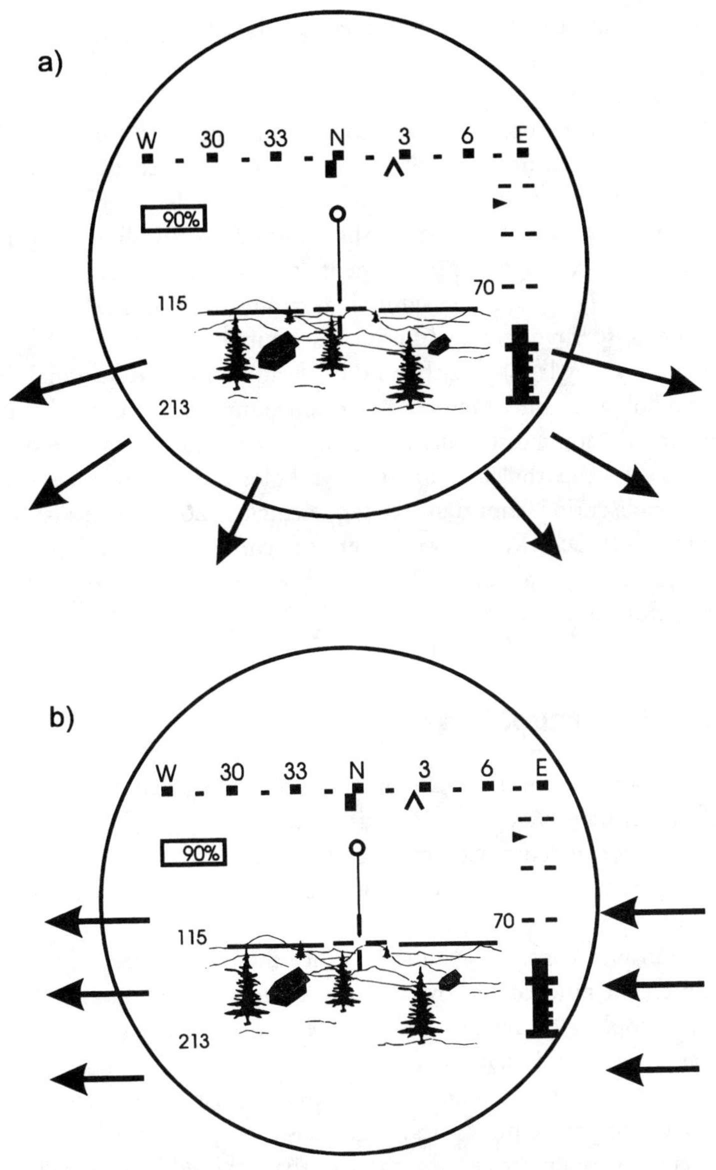

Figure 7.9 Visual field flow and symbology as seen for forward flight and (a) forward head position and (b) side head position. The arrows indicate the direction of the image flow.

the head is rotated to the side, the image flow is to the side of the display, while the symbology remains as in the forward head position. Experience has shown that this awkward combination of two unaligned coordinate systems makes it difficult for the pilot to integrate the visual sensor information and the symbolic spatial information. Consequently, the pilot's spatial awareness is impaired, and frequent and extensive training on the part of pilot is required to maintain proficiency.

Conformal symbology is not easy to implement. It must combine measurements of the attitude of both the aircraft and the head; thus, it is affected by measurement errors of the helmet-tracker system. That effect is manifested in two ways: first, the symbology is shown not perfectly aligned with the outside world, and second, the measurement errors are essentially a temporal noise that continuously agitates the symbology around the outside scene. To some extent, the noise is reduced by filtering the symbology motion, but by doing so the symbology may lag slightly behind the head motions. Additional time delay to the symbology is due to the head-position computations of the helmet tracking system. Another disadvantage of using a conformal symbology on the HMD results from the fact that the symbology always overlies the outside scene, thus obscuring important scenery elements [20]. In contrast to the case of viewing a HUD, HMD viewers cannot circumvent the symbology by shifting their heads to get a view of the scene, because the symbology remains constantly in front of their eyes.

7.6 Display Control Laws

Display control laws refer to the equations and scaling that determine the position of the central dynamically changing symbology [4,21,22]. The main symbols that are controlled dynamically are the pitch ladder and the roll indicator, the velocity vector, the acceleration cue symbol, and the helmet-mounted sight-aiming reticle.

The display control dynamics fall into three main categories: the scaling or compression ratio of the dynamically changing symbology, the effects of time and sampling delays on the response of the symbology, and the control laws that govern the dynamics of symbols.

Scaling is used to enhance or reduce sensitivity of the dynamics of the symbology. For precision control, high sensitivity is required, while for large maneuvers or smoother response, the sensitivity should be reduced.

Time delay of the symbology presentation due to computation cycle, tracker lag, or graphics refresh rate may impair the stability of the combined human-machine system, or force may slow down the pilot's response to changes in the aircraft state.

An example of a display control law is the velocity vector and acceleration cue used in the IHADSS display shown in Figure 7.10. The velocity vector is a line extending from the LOS cross symbol as a vector sum [23]. The length of the velocity vector indicates the aircraft speed, and its inclination indicates the flight direction relative to the ground. Hence, forward velocity is shown by a line in the up direction, and the vector growth to the side indicates aircraft translations to that side. The sensitivity of the vector growth depends on the display mode: in the hover mode, it is 10 times more sensitive than in the transition mode; in cruise mode it is not shown at all. The acceleration cue, marked as a circle, is referenced to the tip of the velocity vector. When the aircraft inertial acceleration is zero, the acceleration cue is located at the tip of the velocity vector. Both the velocity vector and the acceleration symbol are shown in Figure 7.10(a).

The equations governing the position of the tip of the velocity vector are (see Figure 7.10)

$$
\begin{aligned}
V_x &= K_{sc}\dot{x} \\
V_y &= K_{sc}\dot{y}
\end{aligned}
\tag{7.1}
$$

The equations governing the acceleration cue are

$$
\begin{aligned}
A_x &= K_{sc}(\ddot{x} + \dot{x}) \\
A_x &= K_{sc}(\ddot{y} + \dot{y})
\end{aligned}
\tag{7.2}
$$

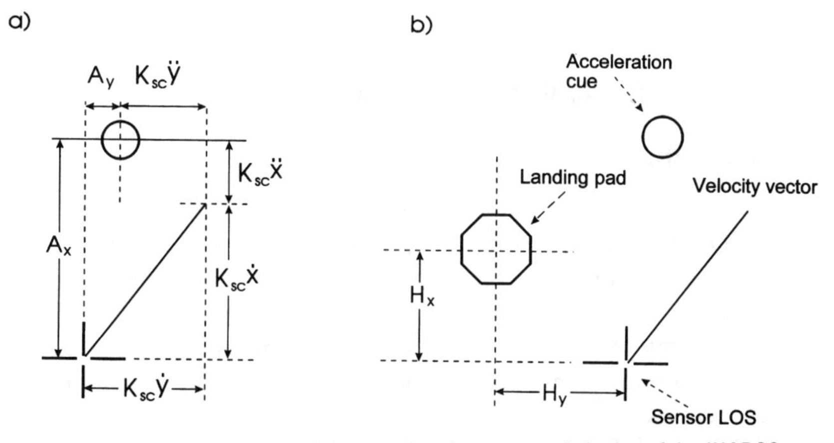

Figure 7.10 The velocity vector and the acceleration cue symbologies of the IHADSS display: (a) control laws that govern the dynamics of the velocity vector and the acceleration (circle) symbol and (b) the control law for hovering above a desired point. Placing the acceleration cue on the pad symbol generates aircraft commands that fly the aircraft to the desired hover point.

In (7.1) and (7.2), $\dot{x}$ and $\dot{y}$ are the aircraft forward and side velocity components, $\ddot{x}$ and $\ddot{y}$ are the aircraft accelerations, and K_{sc} is the symbol scaling. This symbology constitutes a status display showing the velocity and acceleration of the aircraft. One of the major uses of this symbology is to assist the pilot during hovering of the helicopter. To do that, a hover pad symbol is added to the display, transforming the display to a command display. The position of the hover pad on the display is determined by the distance between the aircraft, which is represented by the cross, and the desired hover point. The equations of the hover pad symbol to the cross are

$$
\begin{aligned}
H_x &= K_{hv}(x_c - x) \\
H_y &= K_{hv}(y_c - y)
\end{aligned}
\tag{7.3}
$$

where K_{hv} is the hover-symbol scaling. The pilot is required to fly the aircraft so the acceleration cue coincides with the hover pad; thus, $A_x = H_x$ and $A_y = H_y$. It can be shown [21] that by following that control law, the aircraft will approach the landing pad with an exponential path.

The display control laws greatly affect the handling qualities of the aircraft, that is, the ease of controlling the aircraft, the accuracy, and the ability to meet the desired responses of the vehicle. The display laws also determine the tendency of the pilot to execute erroneous commands. Because the display control laws interact with both the pilot and the aircraft, they must be formulated using both aircraft and pilot control dynamics.

7.7 Virtual HUD and Virtual Cockpit

7.7.1 Virtual HUD

Some functions or information usually presented on the HUD and transferred to the HMD is required only when the pilot's LOS is along the aircraft's longitudinal axis [24]. That includes main piloting information and aiming weapons that are rigidly fixed to the aircraft. In that case, the aircraft's HUD is the best display for presenting the information and symbology. However, the presence of the HMD permits elimination of the HUD, thus saving the volume and weight it occupies. For such purposes, the concept of the virtual HUD was devised. The virtual HUD automatically presents the normal HUD symbology each time the pilot's gaze is in the direction where the HUD is normally placed, that is, along the longitudinal axis of the aircraft. The main advantage of the virtual HUD concept is that the information presented in this mode of operation is not required to be presented in off-boresight conditions, thus reducing the clutter on the HMD. Also, the virtual HUD is less prone to pilot errors and disorientation risks because it forces the user to perform actions associated with this mode of operation only when the pilot's LOS is along the aircraft axes.

The virtual HUD requires, certainly, an accurate helmet-position sensor. The accuracy of the head tracker must be comparable to the conventional HUD, which is about 1 to 2 mrad. Also, measures must be taken to reduce symbology agitation caused position-measurement sensor errors and continual head motions.

7.7.2 Virtual Cockpit

The virtual cockpit concept relates to situations in which the view of the outside-world scene is either poor or absent. In such conditions, the HMD can be used to present the information that substitutes for the view of the real world [25–27]. Typical examples for such a requirement are the futuristic hypersonic aircraft, which is envisioned with a windowless cockpit; aircraft that are planned to operate in exceptionally hazardous missions and in which the crew will be enclosed in a protective capsule; tanks and other armored vehicles in nuclear/biological/chemical-regime operations, in which the tank hatches are sealed for protection; spacecraft; submarines; and various types of remotely controlled vehicles. The source of information may come from a variety of sensors: televisions and other electro-optical sensors, scanning radars, or even prestored terrain data matched with the onboard navigation system [28].

Beyond the external-world view, in the virtual cockpit concept some traditionally cockpit-mounted displays are replaced by the HMD and, in a fashion similar to that of the virtual HUD concept, are projected on the HMD when the user's gaze is directed at the point where the original display would have been. The virtual cockpit concept drives the need for color HMD to allow simulation of the outside world and existing head-down color displays.

7.8 Display Automation

The tendency toward crowded display symbology, on the one hand, and the availability of numerous sources of information and sensors in the advanced aircraft (e.g., accurate inertial navigation systems, radars, altimeters, threat-warning systems, digital terrain maps, and FLIRs), on the other hand, combined with powerful computers and sophisticated software techniques, have pushed the natural approach of automating display symbology. With automation, the display mode can be selected automatically by software controls according to the current situation.

Air-to-air weapon-aiming symbology, for example, is necessary but can be shown only when a threat has been detected. Landing symbology can be displayed only if some conditions of height above ground and speed are met. Another example might be if the pilot does not respond as expected to a warning indication within a certain time period. Then the display symbology may switch to a warning mode, in which most symbology is removed and the warning indication appears in the center of the display.

The compression ratio of some symbols may be changed at certain flight conditions. For example, in helicopters flying at the NOE, attitude indication needs to be more sensitive at lower altitudes than at higher altitudes. That can be achieved by changing the compression ratio of the attitude symbol. Automatic decluttering also may be useful, for example, to remove the altitude bar at high-altitude conditions, leaving only the numerical altitude indication.

7.9 Summary

The HMD symbology is designed to provide the user vital information on the current situation of the vehicle and cues that otherwise are difficult to get solely from the visual scene. The information presented on the HMD varies from simple sight-type symbology through more comprehensive PFR data similar to the symbology of the conventional HUD, up to the sophisticated pictorial display symbology that substitutes or complements the outside-world scene.

The powerful computers and graphics processors currently available, together with the advanced imagery and navigation sensors, enable display designers to incorporate within the display many informational items and cues. The symbology may be in the form of analog or digital information, as well as sophisticated graphics that translate the sensor's information into readable and intuitively interpretable and naturally perceivable cues. This type of symbology eventually helps relieve the pilot from the need to process the instrument data and outside-world information.

Finally, because it is difficult to predict the appropriateness and usefulness of the display symbology, each item and every display format have to be tested extensively before being implemented in the final display configuration. The task of testing the display format must be performed with attention and caution. Experienced operational pilots tend to react conservatively to new display formats and are inclined to accept symbology that resembles traditional displays. That hinders the introduction of novel display formats and concepts. Therefore, evaluations of new designs should be carried out with wide user populations, and the pilots' acceptance and opinion should be interpreted carefully. Furthermore, when experienced pilots test a display symbology and format, they should be given sufficient time and training to experience the new display, to reduce the effects of old habits and familiarity with conventional displays.

References

[1] Haworth, L. A., and R. L. Newman, "Test Techniques for Evaluating Flight Displays," NASA Technical Memorandum 103947, February 1993.

[2] Bailey, R. E., "HUD Lessons Learned for HMD Development," in *Helmet- and Head-Mounted Displays and Symbology Design Requirements*, SPIE, Vol. 2218, 1994, pp. 216–225.

[3] Garman, P. J., and J. A. Trang, "In Your Face! The Pilot's/Tester's Perspective on HMD Symbology," in *Helmet- and Head-Mounted Displays and Symbology Design Requirements*, SPIE, Vol. 2218, 1994, pp. 274–280.

[4] Newman, R. L., and L. A. Haworth, "Helmet-Mounted Display Requirements: Just Another HUD or a Different Animal Altogether?" in *Helmet- and Head-Mounted Displays and Symbology Design Requirements*, SPIE, Vol. 2218, pp. 226–237, 1994.

[5] Geiselman, E. E., and R. K. Osgood, "Toward an Empirically Based Helmet-Mounted Display Symbology Set," *Proc. Human Factors and Ergonomics Society 37th Annual Meeting*, 1993, pp. 93–97.

[6] Haidn, H., and G. Odendahl, "Symbology Requirements in Head-Up and Head-Down Displays for Helicopters in NOE-Flight," in *Display Systems*, SPIE, Vol. 1988, 1993, pp. 108–114.

[7] Meister, D., "Human Engineering Data Base for Design and Selection of Cathode Ray Tube and Other Display Systems," Report WPRDC-TR-84-51, Navy Personnel Research and Development Center, San Diego, 1984.

[8] Doyle, A. J. R., "The Eye as a Velocity Transducer: An Independent Information Channel?" in *Helmet- and Head-Mounted Displays and Symbology Design Requirements*, SPIE, Vol. 2218, 1994, pp. 339–350.

[9] Rate, C., et al., "Subjective Results of a Simulator Evaluation Using Synthetic Terrain Imagery Presented on a Helmet Mounted Display," in *Helmet- and Head-Mounted Displays and Symbology Design Requirements*, SPIE, Vol. 2218, 1994, pp. 306–315.

[10] Grunwald, A. J., J. B. Robertson, and J. J. Hatfield, "Experimental Evaluation of a Perspective Tunnel Display for Three-Dimensional Helicopter Approaches," *AIAA J. of Guidance and Control*, Vol. 4, No. 6, November–December, 1981, pp. 623–631.

[11] Theunissen, E., "Factors Influencing the Design of Perspective Flight Path Displays for Guidance and Navigation," *Displays*, Vol. 15, No. 4, 1994, pp. 241–254.

[12] Stokes, A. F., and C. D. Wickens, "Aviation Displays," in E. L. Wiener and D. G. Nagel, eds., *Human Factors in Aviation*, San Diego: Academic Press, 1988, pp. 347–431.

[13] Taylor, R. M., "Aircraft Attitude Awareness From Visual Displays," *Displays*, Vol. 9, No. 2, 1988, pp. 65–75.

[14] Taylor, R. M., "Some Effects of Display Format Variables on the Perception of Aircraft Spatial Orientation," in *Human Factors Considerations in High Performance Aircraft*, AGARD-CP-371, 1984, pp. 14.1–14.14.

[15] Boehmer, S. C., "X-31 Helmet Mounted Visual and Audio Display (HMAD) System," in *Helmet- and Head-Mounted Displays and Symbology Design Requirements*, SPIE, Vol. 2218, 1994, pp. 150–160.

[16] Geiselman, E. E., and R. K. Osgood, "Utility of Off-Boresight Helmet-Mounted Symbology During a High Angle AirboTarget Acquisition Task," in *Helmet- and Head-Mounted Displays and Symbology Design Requirements*, SPIE, Vol. 2218, 1994, pp. 328–338.

[17] Stokes, A. F., C. D. Wickens, and K. Kite, *Display Technology Human Factors Concepts*, Warrendale, PA: Society of Automotive Engineers, 1990.

[18] Haworth, L. A., and R. E. Seery, "Helmet Mounted Display Symbology Integration Research," *Proc. 48th Annual Forum of the American Helicopter Society*, Washington, DC, 1992, pp. 197–213.

[19] Jones, D. R., T. S. Abbott, and J. R. Burley II, "Evaluation of Conformal and Body-Axis Attitude Information for Spatial Awareness," in *Helmet-Mounted Displays III*, SPIE, Vol. 1695, 1992, pp. 146–153.

[20] Long, J., and C. D. Wickens, "Conformal Versus Non-Conformal Symbology and the Head-Up Display," in *Helmet- and Head-Mounted Displays and Symbology Design Requirements*, SPIE, Vol. 2218, 1994, pp. 361–368.

[21] Hess, R. A., and P. J. Gorder, "Design and Evaluation of a Cockpit Display for Hovering Flight," *AIAA J. Guidance, Control, and Dynamics*, Vol. 13, No. 3, May–June 1990, pp. 450–457.

[22] Eshow, M. M., and J. A. Schroeder, "Improvements in Hover Display Dynamics for a Combat Helicopter," *Proc. 48th Annual Forum of the American Helicopter Society*, Washington, DC, 1992, pp. 793–808.

[23] Berry, J., et al., *PNVS Handbook*, Fort Rucker, AL: Directorate of Training and Doctrine, 1984.

[24] Danneberg, E., et al., "Investigation of Interactions Between Helmet-Mounted Sight/Display, Sensor Platform and Human Pilot," European Space Agency Technical Translation, ESA-TT-746, June 1983.

[25] Ineson, J., "Imagery for a Virtual Cockpit," *Displays*, Vol. 12, No. 3/4, 1991, pp. 129–140.

[26] Kaye, M. G., et al., "Evaluation of Virtual Cockpit During Simulated Missions," in *Helmet-Mounted Displays II*, SPIE, Vol. 1290, 1990, pp. 236–245.

[27] Adam, E. C., "Tactical Cockpits—The Coming Revolution," *IEEE Systems Mag.*, Vol. 9, No. 3, March 1994, pp. 20–26.

[28] Swenson, H. N., et al., "Design and Flight Evaluation of Visually-Coupled Symbology for Integrated Navigation and Near-Terrain Flight Guidance," in *Helmet- and Head-Mounted Displays and Symbology Design Requirements*, SPIE, Vol. 2218, 1994, pp. 161–172.

8

Biodynamic Effects and Image Stabilization

Biodynamic interference results from the exposure of the human operator to vibrations and accelerations. Vehicle vibrations are transmitted to the operator's head, causing it to vibrate in motions that are correlated with the vehicle accelerations. Due to the mechanically flexible nature of the human body, it tends to amplify vibrations at a certain frequency range and to attenuate lower and higher frequencies.

When the head or body moves, there are characteristic associated eye movements evoked by the various sensory systems that affect the perception of the image presented on the HMD.

The effect of head vibrations is manifested by increased pointing or aiming accuracies when the sight is used and by impairment of the legibility of information displayed on an HMD due to impairment of visual acuity.

Visual acuity degradation in the presence of vehicle vibrations results from the reflexive eye movements induced by stimulation of the vestibular system of the inner ear. The normal function of the vestibulo-ocular reflex (VOR) is to induce eye movement opposite to that of the head, thus maintaining a stationary point of regard and enabling Earth-fixed objects to be viewed during normal locomotor activities.

This chapter describes the type of eye movements, the vestibular system, and how it reacts to head vibrations, leading to a discussion of the perception of operating an HMD under vibrations and concluding with an analysis of HMD image stabilization.

8.1 The Control of Eye Movements

Human eye movements are required to orient and maintain position of the high-acuity foveal section of the eye on the object of interest. The eye movements are of several types and are governed by different mechanisms that respond to different types of stimuli [1]. Some eye movements are voluntary, and some are reflexive, or involuntary. The main eye movements are saccadic, pursuit, and nystagmus movements.

8.1.1 Saccadic Eye Movements

Saccadic eye movements, or saccades, are rapid eye movements that are used to redirect the eyes from one target to another in the shortest amount of time [2]. Saccadic eye movements enable the observer to view an image that extends more than the small field of view of the high acuity of the eye, which is restricted to only about 2 to 4 degrees. The saccades are voluntary and usually occur as a sequence of discrete high velocity, up to about 700 to 1,000 deg/sec. During the saccade, the visual function is suppressed. The saccade typically positions the target within 0.3 degrees of the fovea centralis and does not occur for target displacements smaller than the same angle. The saccadic motion is initiated after a latency period, depending on the target displacement. The latency, that is, the time between change of the target's position and commencement of saccadic eye movement, can be several hundred milliseconds; pilot practice can reduce latency. Latency is less for upward motion of the eye than down.

If the target is being tracked by the observer, the target pursuit is quick and accurate. Even if the target disappears, the eyes will still track to the estimated position.

8.1.2 Physiological Nystagmus

Physiological nystagmus are small (less than 0.3 degrees), involuntary eye movements that continuously occur between saccades. The physiological nystagmus are characterized by small flicks (termed microsaccades), slow drift, and high-frequency tremor. The physiological nystagmus has been proved to be necessary to maintain vision; when an object was stabilized on the fovea, it caused image fading.

8.1.3 Optokinetic Nystagmus

Optokinetic nystagmus occurs when the visual environment rotates about the stationary observer. The eyes involuntarily follow a passing object for a while and then make a rapid movement in the opposite direction to acquire a new object. The sequence of movements repeats itself and results with a sawtooth-like motion pattern of the eyes. The optokinetic nystagmus is often called "train nystagmus" because of the association of looking at the passing scene through a train window. Relative movement of the scene with respect to the pilot, for example, in a spinning aircraft, can cause vertigo due to inappropriate eye movement. The effect can continue and slowly decay after the stimulus is removed. The optokinetic nystagmus is composed of two phases of movements: a slow phase of a rapid, smooth-tracking movement interrupted by fast-phase or saccadic-type movements to reset the eye direction to a new point in the scene so that tracking can resume.

The optokinetic nystagmus is suppressed only if a fixation point is provided; otherwise, it is very difficult to suppress.

8.1.4 Pursuit Eye Movements

Pursuit eye movements are smooth and have the role of stabilizing retinal images during the viewing of a moving object, by matching the angular velocity of the eye to that of the object [3]. Pursuit eye movements rotate the eye at a velocity that approximates that of the target, up to 25 to 30 deg/sec. In a nonstressful situation, tracking is straightforward, but under stress the movement may be interspersed with saccades and even head movements.

8.2 The Vestibular System

The vestibular organ consists of two sensory systems: the semicircular canals, which are sensitive to angular accelerations, and the otoliths, which are sensitive to specific forces and linear accelerations. The vestibular organ is located in the inner ear.

8.2.1 The Semicircular Canals

The semicircular canals are three canals located in approximately orthogonal planes (Figure 8.1). The vertical canals line in planes that make angles of 45 degrees with the lateral axis, and the horizontal canals are tilted by 30 degrees to the horizontal plane. Each canal has a coplanar canal in the opposite ear.

Each canal forms two-thirds of a circle, closed by a collective base, the utricle. The diameter of the circle is about 6 mm; the diameter of the cross section of the canal is about 0.2 mm. The canals are filled with fluid called

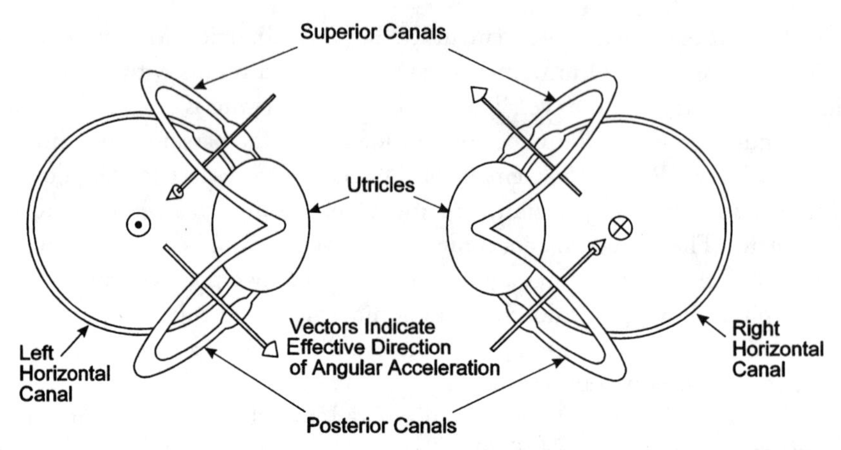

Figure 8.1 The structure and orientation of the semicircular canals (*From:* [4], *After:* [5]).

endolymph. The canals are widened near the utricle. That widening, called the ampulla, contains the cupula; its base is called the crista. The cupula acts as a flapperlike valve. The cupula and the crista together seal the ampulla. A simplified drawing of the cupula within the ampulla is depicted in Figure 8.2.

In principle, the semicircular canal acts like a torsion pendulum. When subjected to angular acceleration around the normal to the plane of the canal, the endolymph flows through the canal. In doing so, it deflects the cupula from its resting position. In the crista, there are a large number of sensory hair cells. From these hair cells are the cilia, which are fiberlike extensions that project into the gelatinous cupula. Bending of the cilia affects the discharge rate of neurons in the ampullary branch of the vestibular nerve. Thus, for greater angular acceleration, the more the cilia bend and the higher the neuron discharge rate.

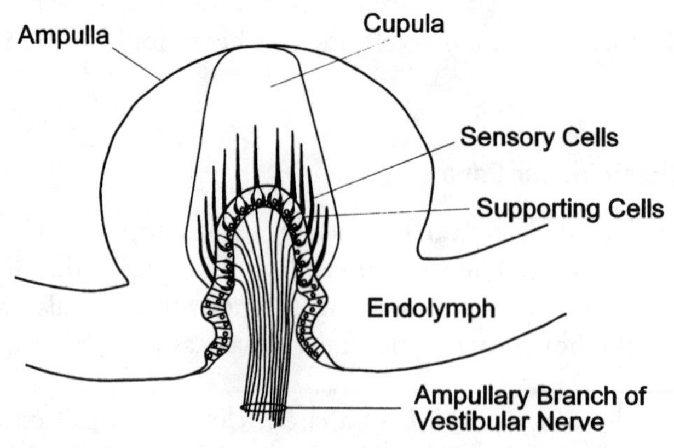

Figure 8.2 Schematic drawing of the ampulla sectioned in the plane of the semicircular canal (*From:* [4], *After:* [5]).

8.2.2 The Otolith System

In each inner ear are two otolith systems: one in the utricle and one in the saccule, which is a beam-shaped downward extension of the utricle. Both the utricle and the saccule are filled with endolymph. The otolith is constructed of the macula, which is a slightly curved supporting base and contains sensory hair cells and supporting cells. On the macula rests a gelatinous membrane filled with a large number of calcium carbonite crystals, termed the statoconia. A schematic drawing of a cross section of an otolith and its macula is shown in Figure 8.3.

Specific forces parallel to the plane of the maculae cause the stoconia to shear relative to the macula. That causes the cilia of the hair cells to bend and change the firing rate of the afferent neurons. The average plane of the utricular macula is roughly parallel to the horizontal canal, while the saccular macula is approximately parallel to the vertical axis. The polarization of the hair cells in the utricular and saccular maculae gradually changes from place to place, and the polarization changes 180 degrees on a dividing line, the striola.

8.2.3 The Vestibulo-Ocular Reflex

The normal function of the VOR is to induce eye movements that are opposite to those of the head. The VOR has evolved as a consequence to the need to preserve maximum visual acuity of objects fixed in space during activities that are vital for human survival, such as walking, jumping, and running. During normal locomotor activities, the head experiences accelerations in the frequency range of 2 to 5 Hz.

The VOR response is characterized by two eye-movement components: slow and smooth eye movements and fast, saccadic-type movements, which act in an anticompensatory manner to reset the eye in its orbit [6,7].

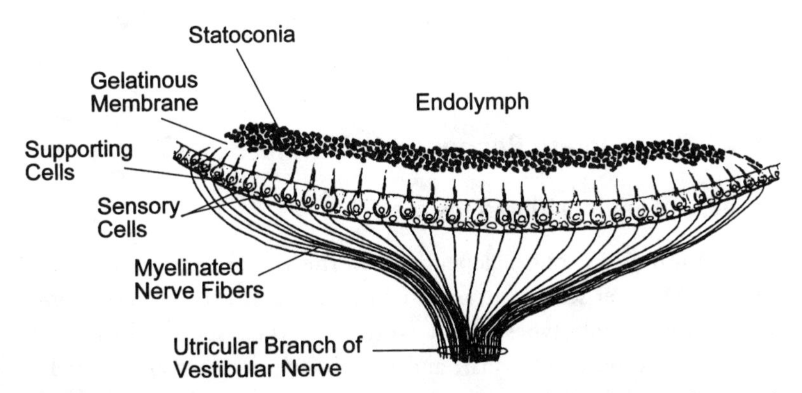

Figure 8.3 Cross section of an otolith and its macula. The otolith slides over the macula while restrained by supporting hairs and sensory cells (*From:* [4], *After:* [5]).

In situations in which the observer has to view objects or images that are not fixed in space but are attached to a moving vehicle or to the head, as in the case of viewing an HMD, the VOR becomes inappropriate. To maintain visual acuity in those situations, it must be suppressed [8].

8.2.4 The Pursuit Reflex

The function of the pursuit reflex is to track moving targets with the eyes alone. During vibration, the HMD vibrates with the head; hence, the image also vibrates relative to the viewing eye. To preserve acuity and prevent blurring, the eye tracks the vibrating image. However, the pursuit reflex stops being effective when either the velocity of the target is too great (i.e., greater than 40 to 60 deg/sec) or the frequency of the target motion exceeds 1 to 3 Hz. The pursuit reflex is the principal mechanism capable of suppressing the vestibular reflex during viewing of an HMD during vibrations and head movements [8]. That close interaction between the vestibular system and the visual system is supported by the findings that breakdown of the pursuit reflex occurs in the same frequency range as the breakdown of the vestibulo-ocular suppression [9]. Furthermore, comparison of the recorded eye movements of subjects attempting to read numbers presented on HMDs revealed that during the reading tests very small eye movements relative to the display occurred, indicating suppression of the VOR. When the subjects closed their eyes and hence took away the stimulus for the pursuit reflex, large eye movements appeared, indicating normal function of the VOR.

An interesting fact is that visual acuity is relatively poor for a moving object, even when the eyes appear to be successfully pursuing it [10]. That is because of the inhibition of vision that is coupled with the voluntary eye movements. That clearly indicates that it is more appropriate to avoid involuntary image movements than to force the viewer to compensate the image motion by eye pursuit movements.

8.3 Head Biodynamic Response to Whole-Body Vibrations

8.3.1 Nature of Vehicle Vibrations

Vehicle vibrations are strongly dependent on the vehicle type, ride conditions, and HMD/HMS-user seating arrangements. In aircraft, the vibration level and spectra vary for aircraft types, flight conditions (i.e., speed, altitude, height above terrain, weather conditions), and the tasks (e.g., close target tracking or aerial refueling). Aircraft vibrations are multi-axial, although mostly pronounced in the vertical axis and, to a lesser extent, in the lateral axis.

Vehicle vibrations are random in nature, but often deterministic vibrations also exist. The latter are caused by, for example, rotor rotations in helicopters. Vehicle vibrations typically are specified by the power spectral density (PSD) function, which describes the acceleration intensity distribution at each frequency. Accordingly, white noise is described by a flat function over the frequency range, while single-frequency vibration is described by a spike in the PSD function.

The PSD function does not always describe the realistic vibration behavior of the vehicle. Most time histories of acceleration recordings show frequent quiet periods interspersed with large amplitude bumps. Because the PSD is computed from recordings of long runs and involves data averaging, in the course of the computation of the PSD curves, the large-amplitude but short-duration accelerations are averaged and thus masked. On the other hand, the short periods of high vibrations occur often during the instances at which clear readability of the display is required most. Typical examples for such occurrences of short-interval, high-intensity vibrations are during dogfight maneuvers or high-speed, low-altitude penetration and ground attack flights. Consequently, the examination of the PSD curves alone may lead to misjudgment of the levels of vibrations that the user may encounter. Hence, the possibility of reduced display visibility may not be apparent in advance.

The most often cited data on aircraft vibrations that relates to biodynamic interference is that recorded by Jarrett [11]. That reference provides data from a Canberra and a Phantom F-4 aircraft during low-altitude, high-speed flight. Jarrett's data is shown in Figure 8.4. The root mean square (RMS) values of the accelerations were found to be 0.25g in heave and 0.1g in sway for the Canberra and 0.25g in heave and 0.13g in sway for the Phantom. Although the RMS values of the Phantom accelerations are higher, the Canberra is considered to be a rougher aircraft, probably because at higher frequencies (higher than 2 Hz), the accelerations of the Canberra are greater than those of the Phantom. In that frequency range, the Canberra's heave accelerations are two to three times greater than those recorded in the Phantom.

Some limited information about the levels of accelerations in a more advanced aircraft, the F-15, is given in Sisk and Matheny [12] and Scherz and Tucker [13]. Sisk and Matheny reported that the accelerations during close target tracking were found to be in the levels of 0.1g RMS, before buffet intensity rise, and 0.3g RMS at maximum values of aircraft normal-force coefficient. Their analyses indicate that a significant vibration power lies in the range from 3 to 4 Hz. Scherz and Tucker [13] provide data on F-15 accelerations during high-speed flight at 200 ft above the terrain. Their data, shown in Figure 8.5, indicates predominant acceleration levels in the range of 3 to 10 Hz.

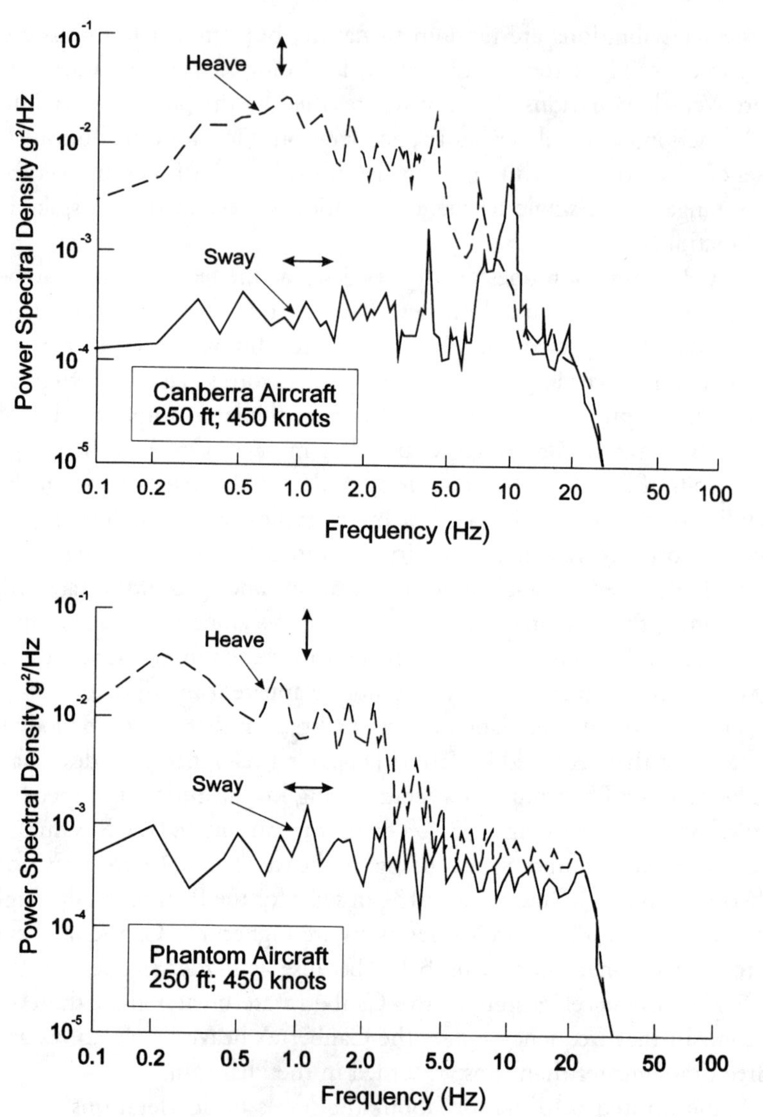

Figure 8.4 PSD function curves for the seat accelerations of a Canberra aircraft and a Phantom F-4 aircraft, both flown 250 ft above ground level and at 450 knots (*After:* [10]). The data shown was used to drive a vibration rig so it was bandpass filtered between 0.5 and 25 Hz.

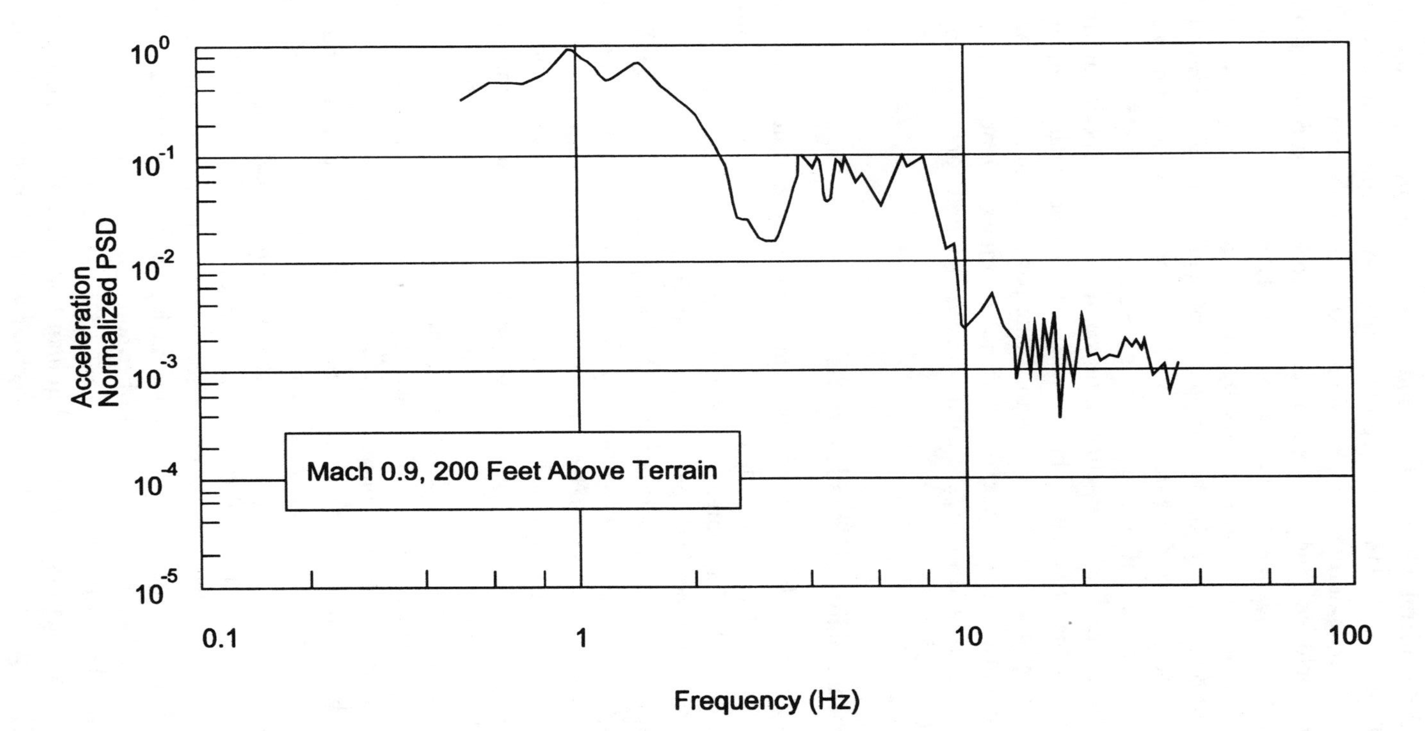

Figure 8.5 Power spectral density function for the pilot station acceleration of an F-15 aircraft flown at 5,000 ft, 200 ft above terrain, and at Mach 0.9 (*After:* [13]).

Head and aircraft vibrations in realistic aircraft maneuvers were measured by Barzilay and Velger [14]. The measurements were performed in an operational F-15 aircraft during routine training exercises of a combat pilot. The maneuvers included dogfight exercises, high-speed low-altitude flights, medium-altitude flights, and aircraft take-off and landing.

The F-15 aircraft is characterized by a prominent vibration mode at the frequency of 8 to 8.5 Hz. In that frequency, the angular accelerations of the head in pitch were 25 to 37 rad/sec^2. Those levels of accelerations imply head vibrations on the order of 10 to 15 mrad at the frequency of 8 Hz.

In contrast to fixed-wing aircraft, helicopter vibration spectra are more deterministic in nature, that is, harmonic sine waves at specific frequencies characterized by peaks at the main rotor blade passage frequency (rotor frequency multiplied by the number of blades) and multiples thereof, the tail rotor frequency, gearbox, and so on. The cabin vibration levels of current helicopters are about 0.1g [15].

8.3.2 Transmissibility of Vibrations to the Head

The human body can be visualized as a complex mechanical system composed of a large group of elements having masses, supported by a relatively stiff structure (the skeleton), and connected via visco-elastic soft tissues. The human body changes its geometric shape according to its posture.

From the mechanical point of view, the human body can be regarded as an elastic system composed of a series of masses (the various organs) interconnected by springs and dampers (flesh, muscles, tendons, and skin tissues) [16]. As such, when the body is exposed to mechanical forces or motions, it does not react as a solid body but deforms and undergoes changes in shape similarly to any other flexible elastic structure. The head responses to the aircraft motions in a complex motion in all six degrees of freedom, that is, in three linear displacements and three angular rotations [17,18]. The transmission of seat vibration to the head is a function of many parameters, including posture, muscle tension, body size and mass distribution, head position and attitude, and helmet mass, center-of-gravity, and attachments. Figure 8.6 depicts a biomechanical model of a sitting pilot. The model shows the main body elements that influence the transmissibility of vibrations to the head.

The motion of the body includes resonances that result from the spring-mass combinations. Some organs respond in higher-frequency modes while others respond in lower frequencies. Furthermore, some organs produce larger displacements than others.

Many investigators have studied human body vibration transmissibility [19–22]. In those studies, the subjects were vibrated on a vibration table

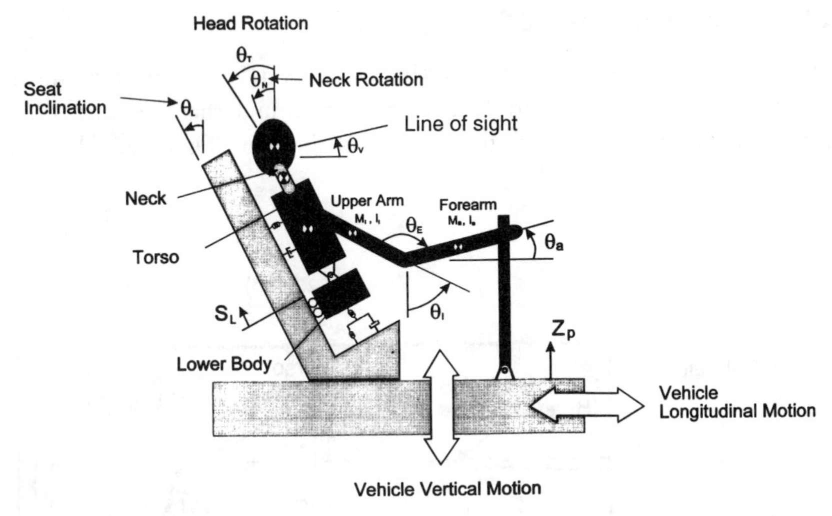

Figure 8.6 Biomechanical model of a sitting pilot.

or in a moving flight simulator and the head vibrations measured by linear or angular accelerometers. To ensure reliable and correct readings and to avoid artifacts due to installation effects (e.g., accelerometer slippage or unintended movements), the measuring devices are usually rigidly mounted on a bite bar that is gripped in the teeth like a pipestem. The bite bar is an individualized dental impression formed from a thermosetting plastic. Slow-motion movies help in the identification of various natural vibration modes of the body and what organs are resonating. The ratio of the head accelerations to the moving-base accelerations gives the transmissibility of the human body to vibrations.

Most human body transmissibility studies have found that the human body has several resonance peaks in the frequency range of 1 to 40 Hz. The first resonance typically is obtained at 1.5 to 2 Hz and is manifested by head bobbing. In that frequency, the torso starts rocking in a surge, or fore-aft motion. This dynamic mode is sensitive to the interface of the seat back to the torso and the tension of the restraint straps and is excited by fore-aft vibrations. The head response is followed by a dip, or decrease in the head motion, in the frequency of about 3 Hz. When the frequency is increased to the next resonance, typically about 4 to 5 Hz, the abdomen and consequently the head start to vibrate so violently that it becomes hard for the subject to read cockpit instruments. Further increase of the frequency again reduces head vibrations. In the frequencies higher than 10 Hz, two prominent resonances are again encountered: around 20 Hz and around 40 Hz.

8.3.3 Effects of Posture and Seating Conditions

Vibration transmissibility to the head highly depends on the posture and the seating configuration. Jex and Magdaleno [23] found that the seat tilt, for example, affects the head-bobbing response of the pilot. Figure 8.7 shows the

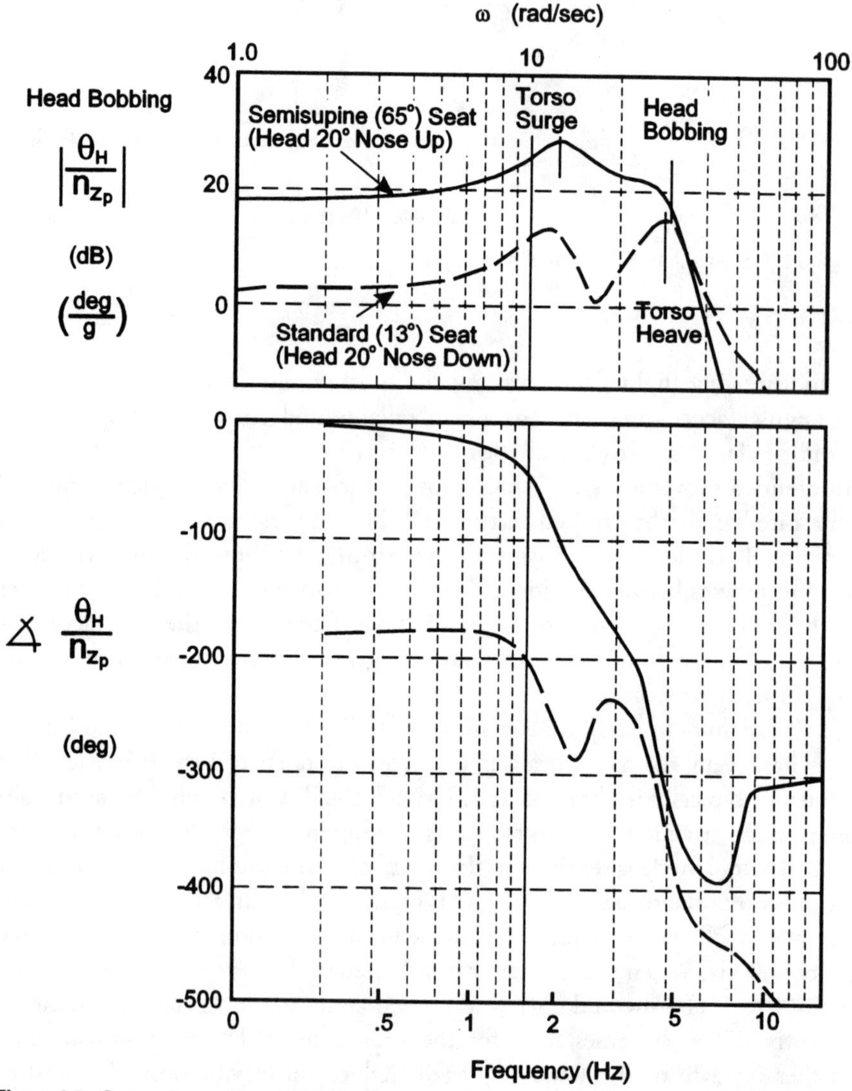

Figure 8.7 Comparison of head-bobbing response for a standard (13-degree) seat and a semisupine (65-degree) seat (*From:* [23]).

head-bobbing response for standard and semisupine seats. With the 65-degree semisupine inclination seat, typical for a high-acceleration fighter aircraft cockpit, the investigators predicted head movements higher by a factor of 4 in comparison with the conventional 13-degree inclination seat.

The transmissibility of vibrations to the head also is affected if the subject is restrained against the seat back and uses a soft or hard seat base. Reutlinger and Van der Vaart [21] found that cushions of the type used in commercial aircraft have a notable influence on the vibration transmissibility to the head in the sense that they amplify the head vibrations in the frequency range of 1 to 7 Hz and attenuate it at higher frequencies.

The orientation of the head also has a great effect on the transmissibility of vibrations. Subjects looking up at large angles above the horizontal plane increase the seat-to-head transmissibility considerably compared to downward positions, mainly due to the stiffening of the neck muscles [24].

8.3.4 Helmet Slippage

One major side effect of vehicle vibrations on the user of HMDs and HMSs is the slippage of the helmet due to head vibrations and voluntary and involuntary accelerations. Helmet slippage results in the display moving relative to the eyes, making it look blurred. The main cause for helmet slippage—more accurately, helmet rotation around the skull—is not due to inadequate fitness or inappropriate structure of the helmet but rather to scalp tangential compliance [11].

Jarrett [11] measured helmet displacement on the head relative to the eye while subjects were stationary and exposed to the Canberra and Phantom vibrations. With the subjects viewing a stationary target during the vibrations, helmet displacements up to 2 mm vertically and 1.3 mm horizontally were measured. On the basis of on the helmet dimensions provided by Jarrett and assuming that the center of rotation of the helmet is in the center-of-the-head volume, those helmet displacements are equivalent to about 14 and 9 mrad, respectively. The consequence of those findings is that during vibrations the HMD images move with the helmet relative to the eye.

When the subjects executed voluntary head movements to simulate a task of visual search, the helmet displacements increased to about 7 mm vertically and 12 mm horizontally. Although those findings seem intolerable, during the voluntary head movements visual acuity is degraded anyway because of the central inhibition of vision [10]. Therefore, essentially, it must be ensured only that no hysteresis exists; once displaced, the HMD will return to (or close to) its former position relative to the eye.

8.4 Aiming With an HMS During Vibrations

A direct effect of vibration on the use of HMDs and HMSs is the increase in aiming errors. The vibrations are transmitted directly to the head, producing vibration-induced head motions, termed biodynamic feedthrough. If under stationary conditions a person can aim at a target with the head with accuracy of about 0.1 degrees, during vibrations the aiming error often may exceed 1 to 1.5 degrees, depending on the vibration intensity and spectra [25]. Wells and Griffin [26] found that aiming performance mostly deteriorated during vibrations at the frequencies of 3.2 to 5 Hz. At that frequency, the time on target decreases from 95% to about 23%, as shown in Figure 8.8(a). Below about the frequency of 1 Hz, the subjects were able to compensate for a notable amount of head vibrations, but as the vibration frequency exceeded 1 Hz, a marked decline in aiming performance was evident. In another experiment, they found that aiming error increases almost linearly with vibration intensity. The time on target, on the other hand, decreases nonlinearly with the vibration intensity, as shown in Figure 8.8(b).

Tatham [27] used the Canberra vibration data to evaluate experimentally the tracking error in azimuth and elevation. He found an RMS tracking error of about 2 degrees for the elevation and 1 degree for the azimuth. His experiments also supported the findings of Wells and Griffin, stating that the tracking error varies linearly with the vibration level.

8.5 Visual Acuity

8.5.1 Viewing Vibrating Objects

When viewing an object that is vibrating at frequencies lower than 1 Hz, the eyes' smooth pursuit movements match the object movements, thus retaining good visual acuity. The eye movements are reflexive or compulsive and are referred to as a psycho-optical reflex [28].

When the oscillation frequency of the object is increased to 1 to 5 Hz, a significant degradation of visual performance, as evident by a marked increase in text and number reading errors, occurs [28,29]. At frequencies up to about 3 Hz, the eyes still attempt to track the object motion but with smaller amplitudes, and they lag behind the object. When the eye movements exceed the angular velocity of the pursuit system, which is about 30 to 60 deg/sec, the smooth pursuit movements are interrupted by saccades, which attempt to compensate for the limitation of the pursuit reflex by evoking high-speed movements to track the object.

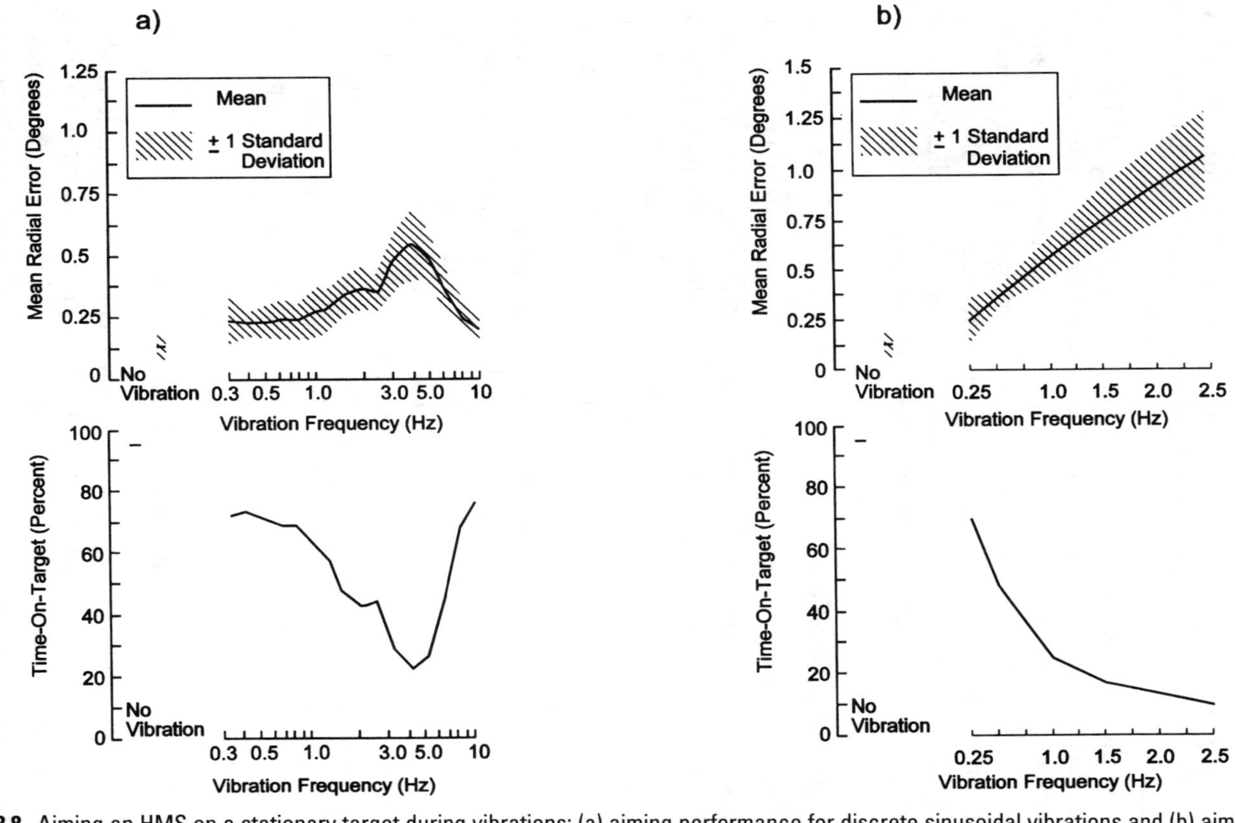

Figure 8.8 Aiming an HMS on a stationary target during vibrations: (a) aiming performance for discrete sinusoidal vibrations and (b) aiming performance during random vibrations at various amplitudes (*After:* [26]).

In sinusoidal object vibrations, nodal image points are formed. The nodal points are obtained at the extremities of the motion where the object velocity is zero and the displacement is a maximum. The eyes fixate on the nodal points and as a result are capable of perceiving the image [30]. Most studies have found that in the range of vibration frequency (i.e., between 3 and 5 Hz) the reading performance is degraded most significantly.

As the frequency of the object vibrations is further increased, the reading errors decrease with the increase of the vibration amplitude, in contrast to expectations. That may be explained by the fact that at the higher frequencies, as the pursuit reflex is no longer effective, the observers concentrate on fixating on the nodal images. Therefore, for smaller amplitudes, the object's images overlap, causing confusion and thus reading errors. As the amplitude increases, the images become separate and distinct, so they can be perceived readily.

8.5.2 Observer Vibration

Tracking a vibrating object is performed by the pursuit reflex. However, the effectiveness of the pursuit reflex declines rapidly as the frequency of the object increases beyond about 1 Hz [31]. Because of the failure of the pursuit reflex to track the object, the performance of reading a vibrating image is impaired. On the other hand, when the subject is vibrating and the object is stationary, the VOR maintains stabilization of the eye; hence, no relative motion between the eyes and the viewed object occurs.

The difference between reading performance in the case of vibration of the object while the viewer is stationary and the case of viewer vibration and stationary object is given by Benson and Barnes [6] and is shown in Figure 8.9. The reading performance is expressed as the percentage of errors made by the viewers and by the number of digits read during a period of 10 sec. While for the case of object vibrations the reading performance decreased rapidly as the oscillation frequency exceeded 0.5 to 1 Hz, for the subject-vibration case, no significant performance degradation over the static case is evident up to frequencies of 5 to 6 Hz.

8.5.3 Visual Perception of Displays Under Vibration

Lee and King [32] measured eye-to-head movements by having subjects adjust the gain and the phase of target displacement until it appeared stationary in their LOS. Both subjects and targets oscillated at the same single frequency at a time, in the frequencies range of 3 to 70 Hz. The findings show marked eye movements relative to the head, indicating that a significant image blur is expected during the viewing of a display attached to the head.

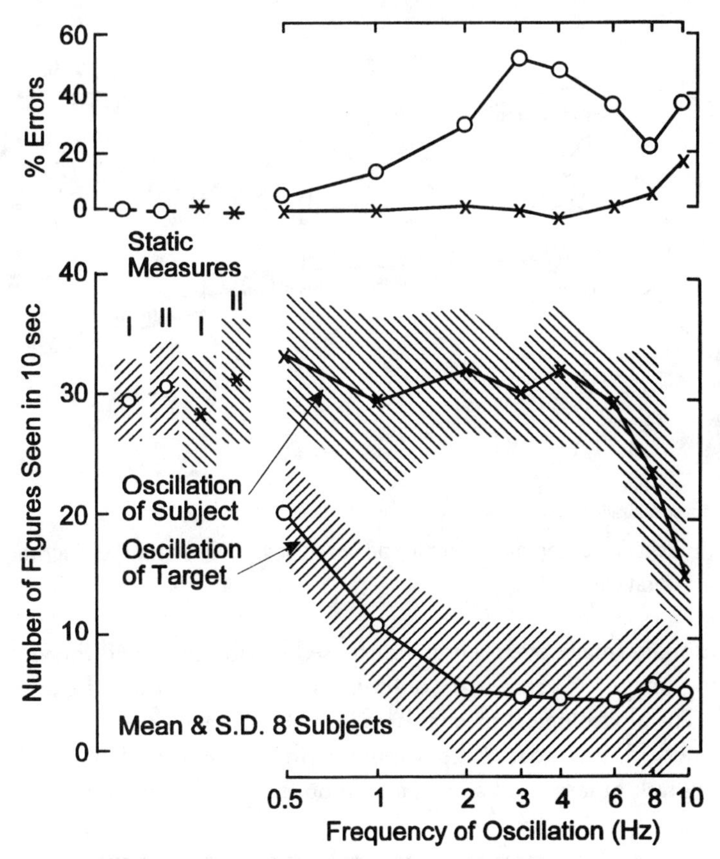

Figure 8.9 Comparison of reading performance during observer oscillations and object oscillations (*From:* [6]).

In using HMDs, viewers are required to view images that are moving with their heads while their eyes are stabilized in the space (Figure 8.10). The clear difference between the reading performance of a stationary observer viewing an oscillating target and the reading performance of an observer viewing a stationary target during whole-body vibrations implies that the perception of an HMD differs from the perception of a panel-mounted display. If for the panel-mounted display, owing to the stabilizing effect of the VOR, the degradation of reading performance in the lower frequency range is minor, for the HMD the viewer is required to view an oscillating image that vibrates in coordination with the head. Thus, similar deterioration of legibility of the display as for the case of the oscillating target may be expected.

The difference between the perception of a panel-mounted display or HUD and that of an HMD is the fundamental difference between the of view-

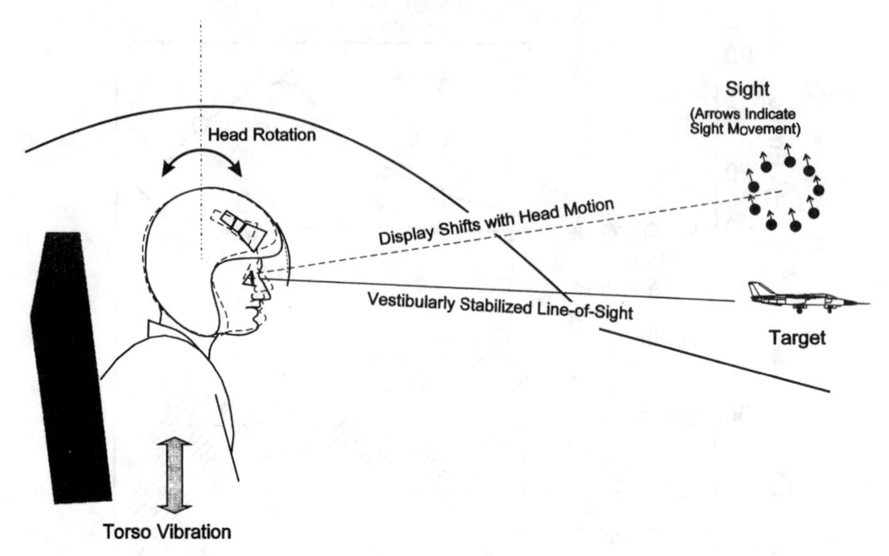

Figure 8.10 Schematic diagram illustrating an observer viewing an HMD during whole-body vibrations.

ing a stationary image by an observer exposed to vibrations and the viewing of a vibrating image by a stationary observer. The two cases differ because of the two different physiological mechanisms involved in each case.

Furness [33] compared the reading performance of an HMD with similar reading task of large and small panel mounted display. Figure 8.11 shows a

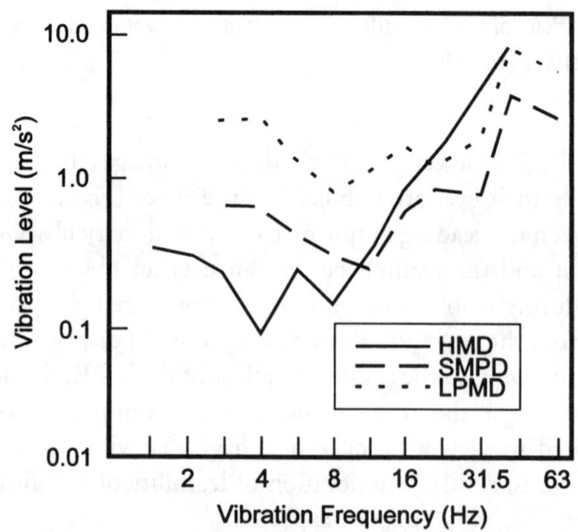

Figure 8.11 Contours of 10% error for an HMD, a small panel-mounted display (SPMD), and a large panel-mounted display (LPMD) (*After:* [33]).

comparison between the reading performance in terms of the vibration level for which at each frequency 10% reading errors are obtained for the three types of displays. It shows that for the lower frequencies up to about 8 Hz the vibration level for which 10% errors are obtained is significantly lower than the level for which the same error percentage occurs for the panel mounted displays.

Banbery et al. [34] have demonstrated that a significant deterioration of effective resolution of a panel-mounted display occurs while the subject is exposed to vibrations. The most deterioration was at the frequencies of 4 and 6 Hz. Collimating the display resulted with a significant improvement in the effective resolution of the display, clearly indicating the superiority of HUDs that use collimating optics over regular displays that lack such optics.

Laycock [35] measured the reading errors of numerals presented on an HMD under conditions of vibrations of the Canberra aircraft (Figure 8.12). He found that for numerals 18 arc-min in height, reading errors increase from about 4% in the stationary case up to 45% in the full Canberra vibration level.

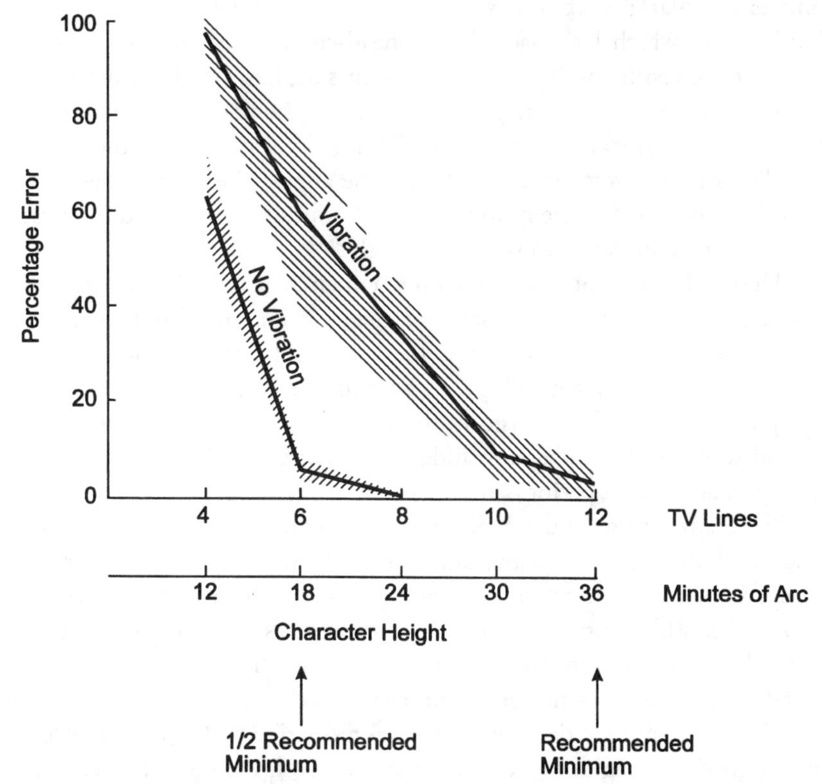

Figure 8.12 Reading-error percentage as function of numeral size under simulated Canberra aircraft vibration conditions (*After:* [35]).

The reading error is reduced with the enlargement of the numerals. For numerals 36 arc-min in height, the error percentage was restored to the static case.

8.5.4 Mathematical Model of Eye-Control System in HMD Viewing During Vibrations

The eye-control system when viewing an HMD during vibrations is illustrated schematically in Figure 8.13. The model, which was originally formulated by Magdaleno and Jex [36], includes two main elements: the pursuit or fixation reflex, which describes the viewer's efforts to track the target by evoking compensatory eye movements, and the VOR, which describes vibration-induced eye movements attempting to stabilize the LOS in the inertial space. The model describes a human subject on a moving platform subjected to vertical vibrations (G_z). The model is adequate only for describing the effects of vehicular vibrations during the viewing of either a stationary target or a target moving in a smooth motion in a small field of view. It does not account for some of the prominent ocular nonlinearities evident in those situations, such as target acquisition, in which large saccadic eye-head-coordinated motions take place [37] or for the vestibular nystagmus that occurs during large head motions [8].

The vehicle vibrations transmitted through the body cause head angular rotation ($\Delta\theta_H$) and translation (ΔZ_H). The HMD is attached to the head and thus follows the movements of the head. The angular head movements excite the VOR, which evokes eye movements (θ_V) opposite to the head movements so as to compensate for head rotation.

The VOR incorporates a model the semicircular canals. The semicircular canals are best described as a double-lagged torsion pendulum type model [38–40]. In the mid-frequency range, the semicircular canals, however, behave like an angular velocity sensor [39]. To account for that, a pseudo-integrator is employed in the model to transform the angular accelerations to angular rates. The lead term in the model is included to equalize the VOR loop for appropriate eye stabilization performance.

Because the image of the display moves with the head and the eyes compensate for head movements, a relative displacement between the image and the viewing eye occurs. To view a stationary image, eye movements (θ_F) are induced by the fixation reflex, which corrects for the image movements. The fixation reflex is represented by a model with a lead term representing the operator's capability to respond to both target position error and the rate of change of the error. The model also includes a pure time delay of the central processing and a pseudo-integrator, enabling good tracking of low-frequency targets. Both the pursuit reflex and the VOR commands activate the eye-control system via the extrinsic ocular muscles, which in turn steer the eyes relative to the head to the desired eye point of regard.

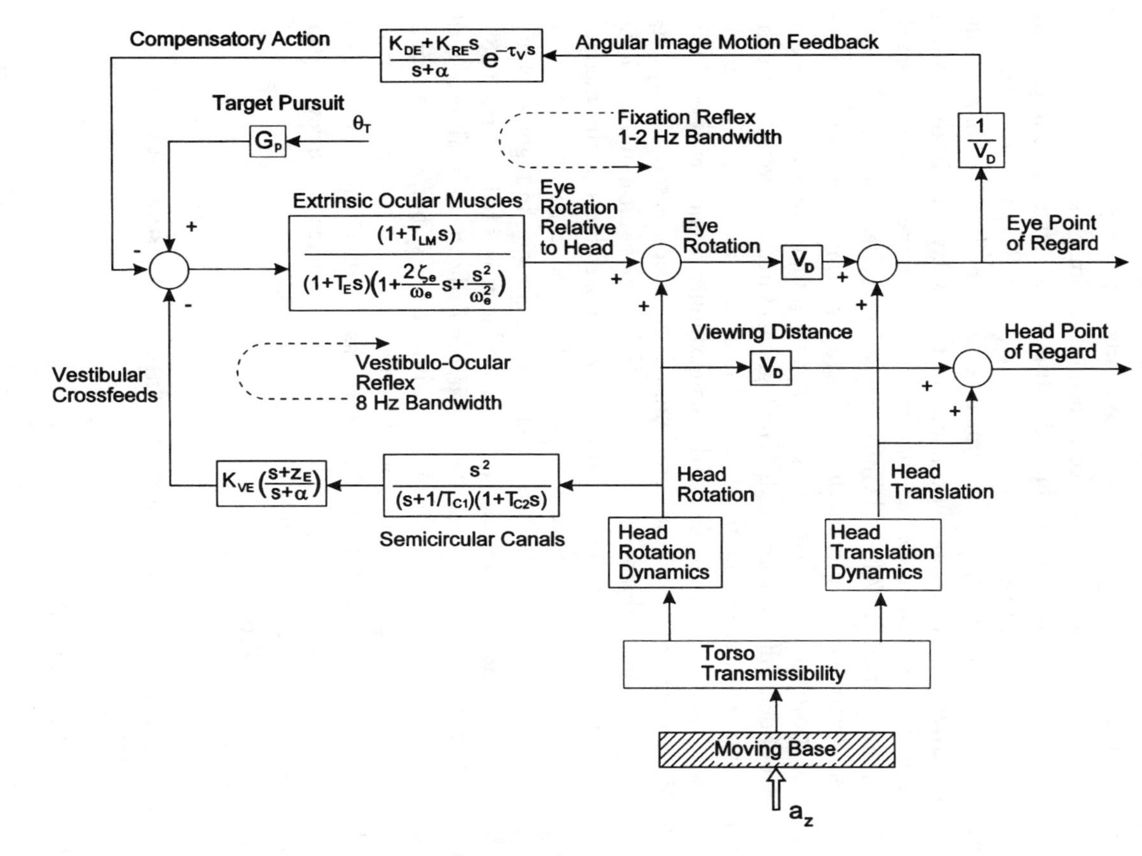

Figure 8.13 Block diagram of the eye-control system for a human viewing an HMD on a vibrating platform.

8.6 HMD Image Stabilization

The natural approach for retaining the visual acuity in HMDs is to minimize the relative motion between the viewed image and the eye by shifting the image on the display in the same direction and magnitude as the induced reflexive eye movement (i.e., stabilize the image in the space), similarly to how it is done by the VOR.

Wells and Griffin [41] proposed a method for image stabilization. They measured the rotational acceleration at the display and, after twice integrating the angular acceleration signal, fed the signal to the CRT so as to deflect the image in antiphase to the display movement. That method resulted in a considerable improvement in reading performance of the display; that is, the number of numerals read correctly increased and the time to read the numerals lessened. However, the method of Wells and Griffin has two main deficiencies. The use of angular acceleration as a source for the stabilization scheme necessitates highpass filtering of the input signal to remove dc components that otherwise might produce unbounded shifting commands. The highpass filter, however, distorts the shifting commands; as a result, the image stabilization is inaccurate. The second deficiency results from the inability of the highpass filter to completely eliminate the low-frequency voluntary head commands. These residuals still may be large compared to the head vibrations and consequently may result in unintentional large image motions. That limits considerably the usefulness of the method, since most applications involve users' voluntary head movements.

Wells and Griffin tested their stabilization technique in flight tests in a helicopter [42]. The mean reading error of six subjects was reduced from 18% without the stabilization system to 4% with it. Similarly, the mean time to read a set of numerals decreased from 40 sec without the stabilization to about 25 sec with stabilization. On the ground, while stationary, the subjects' mean reading error was 0.4% and the reading time was 20 sec. During the tests, the subjects did not attempt to make voluntary head movements. However, inadvertent voluntary head movements produced some low-frequency image drift to the edges of the display.

8.7 Reduction of Biodynamic Interference by Adaptive Filtering

8.7.1 Adaptive Noise Canceling

The deficiencies of the simple filtering of the biodynamic interference by the use of a combination of integrations and highpass filtering can be overcome by a method of adaptive noise canceling. Adaptive noise canceling has been implemented successfully in many applications in communications and bioengineering. In those implementations, it has proved to be efficient in estimating a weak signal corrupted by noise even in cases when the signal-to-noise ratio is very low, provided a signal correlated with the estimated signal is available [43].

Widrow's adaptive noise-canceling technique makes use of the least-mean-squares (LMS) adaptive filter algorithm [44]. In the sequel, another adaptive filtering algorithm [45] is described, and its performance in reducing the effects of biodynamic interference is demonstrated by computer simulations.

8.7.2 The Adaptive Filter Algorithm

The basic requirement of the adaptive filter is to estimate the involuntary head motions with error smaller than 2 mrad RMS and with a sufficiently short convergence time. To be successful, the filter must estimate the biodynamic interference in the presence of large-amplitude voluntary head commands, which may be 10 times larger than the biodynamic interference itself.

The following adaptive filter meets those requirements. The algorithm is a variant of the instrumental variable approximate maximum likelihood (IVAML) method [46].

The total head motion (θ_H) is the sum of the voluntary head command (θ_{HC}) and the involuntary head motion (θ_{HB}), that is, the biodynamic interference:

$$\theta_H(t) = \theta_{HC}(t) + \theta_{HB}(t) \tag{8.1}$$

In (8.1) and in the sequel, t is the discrete time. The voluntary head commands (θ_{HC}) typically are low-frequency, large-amplitude signals, while the involuntary head motions (θ_{HB}) are characterized by higher frequencies and smaller amplitudes. The biodynamic system can be represented by the following input-output model:

$$\theta_{HB}(t) = H(q^{-1})a(t) \tag{8.2}$$

where $a(t)$ is the acceleration of the vehicle. $H(q^{-1})$ is a polynomial of the order N_H and the following form:

$$H(q^{-1}) = h_0 + h_1 q^{-1} + \cdots + h_{N_H} q^{-N_H} \tag{8.3}$$

where q^{-1} is the unit delay operator; that is, $q^{-1}a(t) = a(t-1)$. Substituting (8.2) in (8.3) yields

$$\theta_H(t) = H(q^{-1})a(t) + \theta_{HC}(t) \tag{8.4}$$

Both the total head motion, $\theta_H(t)$, and the vehicle accelerations, $a(t)$, are measurable signals. The formulation of (8.4) is a proper form to estimate the parameters of $H(q^{-1})$ using a standard least-squares technique, with the last term in (8.4) interpreted as a noise in the measurement equation. However, because the voluntary head motions, which in the above formulations are regarded as a measurement noise, are considerably larger than the biodynamic involuntary head motions (the signals that are to be estimated), the convergence rate is very slow. That is overcome by using instrumental variables instead of the measured signals.

The instrumental variables are formulated by expressing the voluntary head command (θ_{HC}) as an autoregressive process:

$$\theta_{HC}(t) = \frac{1}{D(q^{-1})}\eta(t) \tag{8.5}$$

where $\eta(t)$ is a zero-mean Gaussian white noise process. $D(q^{-1})$ is a polynomial of order N_D of the following form:

$$D(q^{-1}) = 1 + d_1 q^{-1} + \cdots + d_{N_D} q^{-N_D} \tag{8.6}$$

Substituting (8.5) in (8.4) and multiplying by $D(q^{-1})$ yields

$$D(q^{-1})\theta_H(t) = H(q^{-1})D(q^{-1})a(t) + \eta(t) \tag{8.7}$$

Define the following data vector:

$$\zeta(t) = [a(t), \ldots, a(t - N_H)]^T \tag{8.8}$$

and the following N_H dimensional instrumental variable vector:

$$\phi(t) = D(q^{-1})\zeta(t) \tag{8.9}$$

Also define the scalar $\bar{\theta}_H$ as

$$\bar{\theta}_H(t) = D(q^{-1})\theta_H(t) \tag{8.10}$$

and the following parameters vector

$$\pi_1 = [h_0, \ldots, h_{N_H}] \tag{8.11}$$

Using the definitions in (8.8) through (8.11), (8.7) becomes

$$\bar{\theta}_H(t) = \pi_1^T\phi(t) + \eta(t) \tag{8.12}$$

Now the noise term in the measurement equation is white, and the convergence of the algorithm is rapid. However, (8.12) cannot be implemented directly, because the parameters of the "whitening filter," $D(q^{-1})$, are unknown. That is circumvented by replacing the filter coefficients with their estimates. To estimate the filter coefficients, $D(q^{-1})$, let us assume that the polynomial $H(q^{-1})$ is known. Then from (8.4) the voluntary head command is

$$\theta_{HC}(t) = \theta_H(t) - H(q^{-1})a(t) \tag{8.13}$$

Because the polynomial $H(q^{-1})$ is unknown, it is replaced by its most recent estimate, $\hat{H}(q^{-1})$, thus forming the following equation:

$$v(t) = \theta_H(t) - \hat{H}(q^{-1})a(t) \tag{8.14}$$

Define the data vector

$$\psi(t) = [-v(t-1), \ldots, -v(t-N_D)]^T \tag{8.15}$$

and a parameters vector

$$\pi_2 = [d_1, \ldots, d_{N_D}]^T \tag{8.16}$$

The estimation process is performed recursively in two steps. In the first step, the parameters vector π_1 is estimated by using the previous estimate of π_2. In the second step, the parameters vector π_2 is estimated utilizing the recent estimate of π_1. In other words, in step 1 the involuntary head motion is estimated, while in step 2 the voluntary head commands are estimated.

The estimation process requires measurements of the helmet angular rotations and of the aircraft accelerations. To achieve an accurate estimation, it must be performed at sufficiently high resolution. Because the head vibrations are mainly in the frequency range of 2 to 10 Hz, the computation must be performed at a rate of at least 100 to 120 Hz. That requires that the estimation scheme be efficient and fast. Therefore, a fast-gain computation scheme [47] is used to implement the filter in high rate.

The complete algorithm is summarized in Table 8.1.

8.7.3 Performance of Adaptive Filters

The adaptive filtering scheme for image stabilization on the HMD is demonstrated by computer simulation. A block diagram of the model of an operator using an HMD/HMS onboard a moving vehicle is shown in Figure 8.14. The scenario described by the model is of a human operator pointing a device at a

Table 8.1
The IVAML Adaptive Filter Algorithm

Define: $\pi_1 \equiv [h_0, \ldots, h_{N_H}]^T; \pi_2 \equiv [d_1, \ldots, d_{N_D}]^T$

Initialize: $\pi_1(0) = 0; \pi_2(0) = 0; A_1(0) = 0; A_2(0) = 0; B_1(0) = 0; B_2(0) = 0$

$R_1^e(0) = \delta; R_2^e(0) = \delta; L_1(1) = 0; L_2(1) = 0$

Update:

Step 1: $D(q^{-1}) = 1 + d_1 q^{-1} + \cdots + d_{N_D} q^{-N_D}$

$H(q^{-1}) = h_0 + h_1 q^{-1} + \cdots + h_{N_H} q^{-N_H}$

$\zeta(t) = [a(t) \ldots a(t - N_H)]^T$

$\phi(t) = D(q^{-1})\zeta(t)$

$\bar{\theta}_H(t) = D(q^{-1})\theta_H(t)$

$e_1(t) = \bar{\theta}_H(t) - \pi_1^T(t-1)\phi(t)$

with $\eta(t) = [\phi(t-1), \ldots, \phi(t - N_H - 1)]^T$; $A(t-1) \leftarrow A_1(t-1)$; $B(t-1) \leftarrow B_1(t-1)$; $x(t) = \phi(t)$; $R^e(t-1) \leftarrow R_1^e(t-1)$; $L(t) \leftarrow L_1(t)$ and using the gain fast-computation scheme, compute $L_1(t+1)$.

Update: $\hat{\pi}_1(t) = \hat{\pi}_1(t-1) + \gamma(t)L_1(t)e_1(t)$

Step 2: $\nu(t) = \theta_H(t) - H(q^{-1})a(t)$

$\psi(t) = [-\nu(t-1), \ldots, -\nu(t - N_D)]^T$

$e_2(t) = \nu(t) - \pi_2^T(t-1)\psi(t)$

with $\eta(t) = [\psi(t-1), \ldots, \psi(t - N_D)]^T$; $A(t-1) \leftarrow A_2(t-1)$; $B(t-1) \leftarrow B_2(t-1)$; $x(t) = \psi(t)$; $R^e(t-1) \leftarrow R_2^e(t-1)$; $L(t) \leftarrow L_2(t)$ and using the gain fast-computation scheme, compute $L_2(t+1)$.

Update: $\hat{\pi}_2(t) = \hat{\pi}_2(t-1) + \gamma(t)L_2(t)e_2(t)$

Gain fast computation:

Given $A(t-1)$, $B(t-1)$, $R^e(t-1)$, $L(t)$, update:

$\epsilon(t) = x(t) + A^T(t-1)\eta(t)$

$A(t) = A(t-1) - L(t)\epsilon(t)$

$\beta(t) = L^T(t)\eta(t)$

$\bar{\epsilon}(t) = [1 - \beta(t)]\epsilon(t)$

$R^e(t) = \lambda R^e(t-1) + \bar{\epsilon}(t)\epsilon^T(t)$

$L^*(t) = \begin{pmatrix} \bar{\epsilon}(t)/R^e(t) \\ L(t) + A(t)\bar{e}(t)/R^e(t) \end{pmatrix}$

Partition $L^*(t)$ as: $L^*(t) = \begin{pmatrix} M(t) \\ \mu(t) \end{pmatrix}$ $\begin{matrix} \text{N elements} \\ \text{last element} \end{matrix}$

$r(t) = x(t-n) + B^T(t-1)\eta(t)$

$B(t) = [B(t-1) - M(t)r(t)] / [1 - \mu(t)r(t)]$

$L(t+1) = M(t) - B(t)\mu(t)$

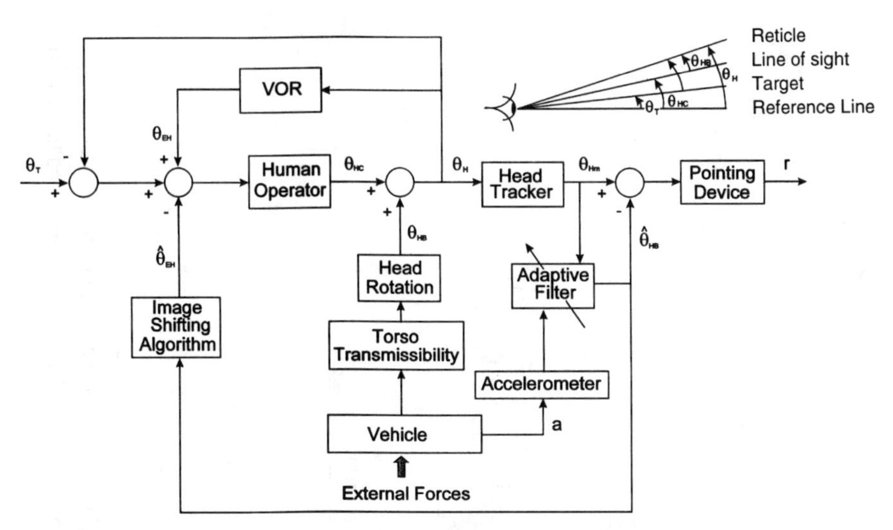

Figure 8.14 System modeling and image stabilization based on the adaptive filtering concept.

moving target using head movements. The head attitude relative to the platform (θ_H) is measured by a head orientation tracking system. The measured head attitude (θ_{Hm}) is used as a command to the pointing device. The head involuntary commands (θ_{HB}) appear as additive noise to the voluntary head commands. The total head motion invokes eye movements (θ_{EH}) relative to the head.

The adaptive filter employs the vehicle accelerations, as measured by accelerometers, and the head-tracker measurements for estimating involuntary head motions. The head-tracker measurements are used, after appropriate transformation, to shift the image on the HMD in the same amplitude but in antiphase to the head motion. Apart from the involuntary head motion, the adaptive filter also provides a smooth estimate of the head voluntary command (r) that is directed to the pointing device.

Figure 8.15 uses a computer simulation to show the performance of the adaptive filtering method. The head involuntary motions are simulated by the biodynamic model of Riedel et al. [48], which describes the rotational head response of a seated pilot to vertical accelerations (see Figure 8.6). The model is formulated as a transfer function of order 11:

$$\frac{\ddot{\theta}_{HB}}{a}(s) = \frac{\kappa(s + 1/T_z)\prod_{1}^{3}(s^2 + 2\zeta_z\omega_z s + \omega_z^2)}{(s + 1/T_p)\prod_{1}^{5}(s^2 + 2\zeta_p\omega_p s + \omega_p^2)} \tag{8.17}$$

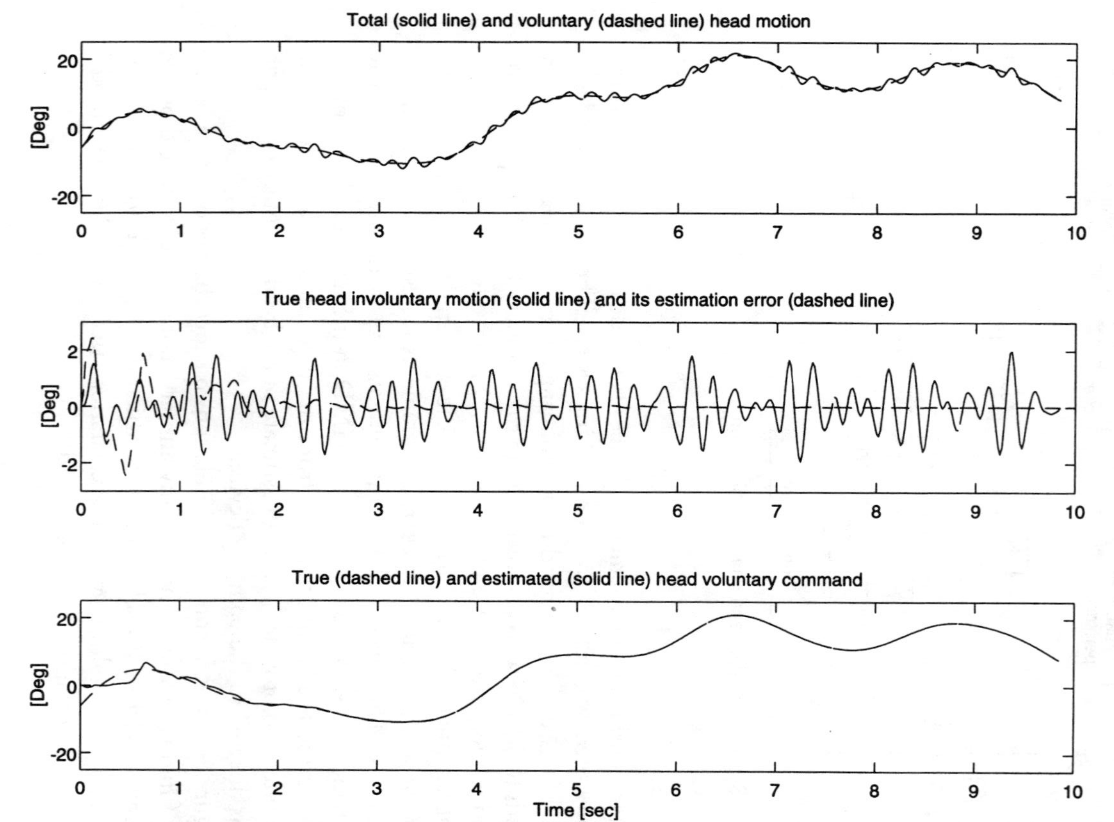

Figure 8.15 Computer simulation of adaptive filtering of biodynamic interference: (top) the total head and voluntary motion (the difference between the two plots is the involuntary head motion); (middle) the involuntary head motion and its estimation error; and (bottom) the voluntary true and estimated head commands.

Table 8.2 summarizes the poles and the zeros of the transfer function.

The total head motion is shown in the top plot in Figure 8.15. It consists of a low-frequency, large-amplitude signal that represents the voluntary head commands, corrupted by a high-frequency and small-amplitude biodynamic interference. The signal representing the voluntary head motion indicates the limited bandwidth of the human operator as well as the large control commands experienced in a typical target tracking task.

In spite of the distinct spectral separation between the two components of the total signal, the obvious approach of highpass-filtering the signal is not effective because of the very high ratio of voluntary signals to involuntary signals. Consequently, that approach would cause significant distortions of the involuntary head motions, mainly because of phase shifts that cannot be avoided using conventional filtering, thus making image stabilization by shifting the image on the display ineffective.

The middle plot in Figure 8.15 exhibits the performance of the adaptive filter by showing the filter's estimation error. The estimation error of the biodynamic interference (i.e., the difference between the true and the estimated involuntary head motions) is drawn, along with the interference itself. It shows that the adaptive filter estimates the involuntary head motions with very high accuracy.

The estimation error represents the residual relative motion between the viewer's eyes and the displayed image. It is shown that the relative motion is essentially zero; thus, a good image stabilization is achieved. Consequently, the voluntary command also is estimated (bottom plot in Figure 8.15) with high accuracy. Furthermore, it is evident that the filter's convergence is rapid.

Table 8.2
Poles and Zeros of the Biodynamic Model

	$1/T$; ζ	ω
Zeros	18.284	
	0.458	10.396
	0.098	16.203
	0.236	51.815
Poles	58.788	
	0.265	10.914
	0.203	12.943
	0.354	26.761
	0.151	29.611
	0.191	57.483

Achieving a satisfactory level of image stabilization requires, however, a very fast head tracker, operating at a frequency of at least 120 Hz, preferably at 240 Hz [49].

8.8 Summary

HMDs introduce a unique situation that the common user does not encounter or experience in any other condition: viewing a display that is attached to the head and that moves with the head when the wearer is exposed to vibrations and accelerations.

The principal physiological mechanism that enables humans on the move to view Earth-fixed objects is the VOR, which invokes eye movements to compensate for head vibrations. However, when during the viewing of an HMD, the vestibular-induced eye movement is undesirable because it produces relative motion between the viewed image on the display and the eye; hence, the VOR becomes inappropriate.

Humans are capable, to some extent, of suppressing the vestibulo-ocular effect. Yet, at the high frequencies often encountered during a ride on vehicles, some level of image stabilization still is required to retain good visual legibility of the display.

Vehicle vibrations can seriously degrade a viewer's ability to perceive information presented on an HMD. That may result with an increase in errors of reading numerical or alphanumeric information, mishap of fine details in images of complex scenery, and an increase in the time required to interpret the displayed information correctly. Consequently, to avoid the effects of vibrations, the HMD designer is forced to enlarge the angular extent of the symbology, essentially limiting the amount of information and the fine details that can be presented. That poses a stringent limit on the resolution and the quality of the HMD.

A possible remedy for those limitations is the use of HMD image stabilization. If implemented properly, image stabilization can reduce significantly the effects of biodynamic interference and enable full exploitation of the capabilities of the display in terms of display information legibility and resolution, amount of displayed information, and image quality.

In most operational situations, the levels of vehicle vibrations are moderate, so the effects of biodynamic interference do not prohibit the use of HMDs, even without any image-stabilization measures. However, those effects occur in certain instances during maneuvers or flights during buffeting or low altitudes. Unfortunately, during such instances, the clear legibility and readability of the display is needed most. Therefore, the display image stabilization is vital, although it is not needed most of the time.

Apart from the effects of vibrations on display legibility, they also degrade the pointing and sighting effectiveness of the HMD/HMS system. When the system is used as a sight, the vibrations of the head are manifested by aiming errors. In VCSs, the vibrations are transmitted as noisy commands to the imaging sensor, causing it to vibrate with the head vibrations. As a result, the image from the sensors may smear further, reducing image quality.

References

[1] Robinson, D. A., "The Oculomotor Control System: A Review," *IEEE Proc.*, Vol. 56, 1968, pp. 1032–1049.

[2] Fuchs, A. F., "The Saccadic System," *The Control of Eye Movements*, New York: Academic Press, 1971, pp. 343–362.

[3] Young, L. R., "Pursuit Eye Tracking Movements," *The Control of Eye Movements*, New York: Academic Press, 1971, pp. 429–443.

[4] Peters, R. A., "Dynamics of the Vestibular System and Their Relation to Motion Perception, Spatial Disorientation and Illusions," NASA CR-1309, 1969.

[5] Hosman, R. J. A. W, and J. C. Van der Vaart, "Vestibular Models and Thresholds of Motion Perception. Results of Tests in a Flight Simulator," Delft University of Technology, Dept. of Aerospace Engineering, LR-265, April 1978.

[6] Benson, A. J., and G. R. Barnes, "Vision During Angular Oscillation: The Dynamic Interaction of Visual and Vestibular Mechanisms," *Aviation, Space, and Environmental Medicine*, Vol. 49, No. 1, 1978, pp. 340–345.

[7] Lau, C .G. Y., et al., "Linear Model for Visual-Vestibular Interaction," *Aviation, Space, and Environmental Medicine*, Vol. 49, No. 7, 1978, pp. 880–885.

[8] Barnes, G. R., A. J. Benson, and A. R. J. Prior, "Visual Vestibular Interaction in the Control of Eye Movement," *Aviation, Space, and Environmental Medicine*, Vol. 49, No. 4, 1978, pp. 557–564.

[9] Barnes, G .R., and G. P. Sommerville, "Visual Target Acquisition and Tracking Performance Using a Helmet-Mounted Sight," *Aviation, Space, and Environmental Medicine*, Vol. 49, No. 4, 1978, pp. 565–572.

[10] Riggs, L. A., "Visual Acuity," in C. H. Graham, ed., *Vision and Visual Perception*, New York: Wiley, 1965.

[11] Jarrett, D. N., "Helmet Slip During Simulated Low-Level High Speed Flight," Royal Aircraft Establishment, Farnborough, England, Tech. Report 78018, 1978.

[12] Sisk, R. S., and N. W. Matheny, "Precision Controllability of the F-15 Airplane," NASA TM-72861, 1979.

[13] Scherz, C. J., and P. B. Tucker, "Flight Test Evaluation of Active Ride Control System for Tactical Aircraft," *Proc. AIAA Guidance Navigation and Control Conf.*, Snowmass, CO, August 1995.

[14] Barzilay, E., and M. Velger, "Measurement of Head Vibrations During Operational F-15 Maneuvers," unpub. tech. report, 1991.

[15] Reichert, G., "Helicopter Vibration Control—A Survey," *Vertica*, Vol. 5, 1981, pp. 1–20.

[16] Von Gierke, H. E., and N. P. Clark, "Effects of Vibration and Buffet on Man," *Aviation Medicine*, 1971, pp. 188–225.

[17] Barnes, G. R., "The Effect of Aircraft Vibration on Vision," AGARD-CP-267, March 1980.

[18] Vogt, L., E. Schwartz, and H. Martens, "Head Movements Induced by Vertical Vibrations," AGARD-CP-267, March 1980.

[19] Garg, D. P., and M. A. Ross, "Vertical Mode Human Body Vibration Transmissibility," *IEEE Trans. on Systems, Man, and Cybernetics*, Vol. SMC-6, No. 2, February 1976, pp. 102–112.

[20] Pradko, F., T. R. Orr, and R. A. Lee, "Human Vibration Analysis," Soc. Automotive Engineers, 1965, pp. 331–339.

[21] Reutlinger, R. A. A, and J. C. Van der Vaart, "Transmission of Sinusoidally Changing Vertical Specific Force to the Heads of Seated Men Measured in a Flight Simulator," Delft University of Technology, Dept. of Aerospace Engineering, LR-319, May 1981.

[22] Jex, H. R., and R. E. Magdaleno, "Progress in Measuring and Modeling the Effects of Low Frequency Vibration on Performance," in *Models and Analogues for the Evaluation of Human Biodynamic Response, Performance and Protection*, AGARD-CP-253, November 1978, pp. A29-1-A29-10.

[23] Jex, H. R., and R. E. Magdaleno, "Biodynamical Models for Vibration Feedthrough to Hands and Head for Semisupine Pilot," *Aviation, Space, and Environmental Medicine*, Vol. 49, No. 1, 1978, pp. 304–316.

[24] Griffin, M. J., et al., "The Biodynamic Response of the Human Body and Its Application to Standards," in *Models and Analogues for the Evaluation of Human Biodynamic Response, Performance and Protection*, AGARD-CP-253, November 1978.

[25] Wells, M. J., and M. J. Griffin, "Flight Trial of a Helmet-Mounted Display Image Stabilization System," *Aviation, Space, and Environmental Medicine*, Vol. 55, 1987, pp. 319–322.

[26] Wells, M. J., and M. J. Griffin, "Tracking With the Head During Whole-Body Vibration," in J. Patrick and K. D. Duncan, eds., *Training, Human Decision Making and Control*, North-Holland: Elsevier Science Publishers, B. V., 1988, pp. 323–333.

[27] Tatham, N. O., "The Effect of Turbulence on Helmet Mounted Sight Accuracies," AGARD-CP-267, March, 1980.

[28] Huddleston, J. H. F., "Tracking Performance on a Visual Display Apparently Vibrating at One to Ten Hertz," *J. Applied Psychology*, Vol. 54, No. 5, 1970, pp. 401–408.

[29] Drazin, D. H., "Factors Affecting Vision During Vibration," *Research*, Vol. 15, 1962, pp. 275–280.

[30] Griffin, M. J., and C. H. Lewis, "A Review of the Effects of Vibration on Visual Acuity and Manual Control, Part I: Visual Acuity," *J. Sound and Vibration*, Vol. 56, No. 3, 1978, pp. 383–413.

[31] Stark, L., "The Control System for Versional Eye Movements," *The Control of Eye Movements*, New York: Academic Press, 1971, pp. 363–428.

[32] Lee, R. A., and A. I. King, "Visual Vibration Response," *J. Applied Physiology*, Vol. 30, No. 2, February, 1971, pp. 281–286.

[33] Furness, T. A., "The Effects of Whole-Body Vibration on the Perception of the Helmet-Mounted Display," Ph.D. thesis, University of Southampton, 1981.

[34] Banbury, J. R., et al., "Visual Perception of Direct-View and Collimated Displays Under Vibration," *Displays*, January 1987, pp. 3–16.

[35] Laycock, J., "A Preliminary Investigation of the Legibility of Alphanumeric Characters on a Helmet-Mounted Display," Royal Aircraft Establishment, Farnborough, England, Tech. Memo FS 155, January 1978.

[36] Magdaleno, R. E., and H. R. Jex, "A Linearized Model for Vibration Effects on the Eye Control System," System Technology, Inc., Working Paper No. 1037–7, 1978.

[37] Morasso, P., et al., "Control Strategies in the Eye-Head Coordination System," *IEEE Trans. on Systems, Man, and Cybernetics*, Vol. SMC-7, No. 9, September 1977, pp. 639–651.

[38] Van Egmond, A. A. J., J. J. Groen, and L. B. W. Jonkees, "The Mechanics of the Semicircular Canal," *J. of Physiology*, Vol. 110, 1949, pp. 1–17.

[39] Young, L. R., "The Current Status of Vestibular Models," *Automatica*, Vol. 5, 1969, pp. 369–383.

[40] Meiry, J. L., "Vestibular and Propioceptive Stabilization of Eye Movements," *The Control of Eye Movements*, New York: Academic Press, 1971, pp. 483–498.

[41] Wells, M. J., and M. J. Griffin, "Benefits of Helmet-Mounted Display Image Stabilization Under Whole-Body Vibration," *Aviation, Space, and Environmental Medicine*, Vol. 55, January 1984, pp. 13-18.

[42] Wells, M. J., and M. J. Griffin, "A Review and Investigation of Aiming and Tracking Performance With Head-Mounted Sights," *IEEE Trans. on Systems, Man, and Cybernetics*, Vol. SMC-17, No. 2, March–April, 1987.

[43] Widrow, B., et al., "Adaptive Noise Canceling: Principles and Applications," *IEEE Proc.*, Vol. 63, No. 12, 1975, pp. 1692–1716.

[44] Widrow, B., "Adaptive Filters," in R. E. Kalman and N. De Claris, eds., *Aspects of Network and System Theory*, New York: Holt, Reinhart and Winston, 1970.

[45] Velger, M., and S. Merhav, "Reduction of Biodynamic Interference in Helmet Mounted Sights and Displays," *Proc. 22nd Annual Conf. on Manual Control*, AFWAL-TR-86-3093, 1986, pp. 139–164.

[46] Young, P., "Some Observations on Instrumental Variable Methods of Time Series Analysis," *Internat. J. of Control*, Vol., 23, No. 5, 1976, pp. 593–612.

[47] Ljung, L., M. Morf, and D. Falconer, "Fast Calculation of Gain Matrices for Recursive Estimation Schemes," *Internat. J. of Control*, Vol. 27, No. 1, 1978, pp. 1–19.

[48] Riedel, S. A., R. E. Magdaleno, and H. R. Jex, "User's Guide to BIODYN-80: An Interactive Software Package for Modeling Biodynamic Feedthrough to a Pilot's Hands, Head and Eyes," Systems Technology Inc., Tech. Report No. 1146–1, December 1980.

[49] Merhav, S. J., and M. Velger, "Compensating Sampling Errors in the Stabilization of Helmet Mounted Displays Using Auxiliary Acceleration Measurements," *AIAA J. of Guidance, Control, and Dynamics*, Vol. 14, No. 5, September–October 1991, pp. 1067–1069.

9

Helmet Design and Integration With HMD/HMS

The traditional helmet, which prior to the introduction of the HMD/HMS was used mainly for protecting the flyer from impact and noise hazards, is now the link, or interface, between the head and the display and sight elements. Being so, it must meet new requirements arising from the need to maintain accuracy of both the sighting system and the display elements. Because the helmet is now part of the optical system, the alignment among the various components, such as the display source, the relaying optics, the visor, and finally the elements of the head tracker is important for the achievement of HMD/HMS performance. Hence, apart from its protective role, the helmet has to be treated as an optical bench for the HMD elements. The implications of that new role include the need for higher rigidity of the helmet shell, so the alignment among the optical elements is maintained during high-acceleration head movements and high-g forces, better helmet fitting and retention to keep the display in front of the pilot's eye, and—obviously—provisions for quick and safe disconnect of the combined helmet/display during ejection or fast egress of the pilot.

With the incorporation of the HMD, not only the helmet shell changed its role. The visor also is often utilized as a part of the optical system of the HMD. Its shape and optical quality must meet the requirements for projecting a high-quality image.

The visor, however, is not the only existing optical element that affects the optical performance of the display and sight system. The aircraft canopy plays an important role in establishing the system performance. Due to its common bubble-like shape, the canopy deflects and deviates the LOS direction and therefore must be considered when a HMD/HMS system is planned to be used in off-axis, high-angle pointing.

9.1 The Helmet

There are three fundamental approaches to the integration of HMD/HMS with the helmet. In the add-on approach, the display and the sight are designed to fit a standard operational crew helmet, such as the USAF HGU-55/P or HGU-53/P, the Navy HGU-33/P, and the Army SPH-4C. The display is attached to the helmet by a mechanical clip or an interface with only minor adjustments of the helmet. The second approach is the integrated approach, in which a completely new helmet design is used to optimally accommodate the display and the sight. The third approach of modular systems uses either a standard or a custom-designed helmet with several elements that are removable and reattachable according to mission requirements. Each approach, naturally, has its own merits and disadvantages [1].

9.1.1 The Add-On Approach

In an add-on system, the display and the sight are designed around the confinements of an existing helmet. The helmet shape dictates the optical design and placement of the components; eventually, the display weight and size are added to the basic weight and size of the helmet. The main advantage of this approach is that it utilizes the proven design of the original helmet. Figure 9.1 illustrates an example of a concept of a high-performance add-on HMD to a standard helmet. The image is conveyed via a fiber-optic bundle to the helmet to relay optics mounted on top of the helmet, and the image is projected on a holographic eyepiece attached to the visor. The display is attached to and movable with the visor to guarantee correct alignment between the optics and the visor.

9.1.2 The Integrated System

In the integrated approach, the design starts clean, with no constraints other than those dictated by the anthropometry of the human head. The helmet is designed as part of the display and the sight. The basic anthropometric data of the head, such as the head's center of gravity, neck support, eye location, and head length, are used as the basic constraints of the design, and the optical path and elements are located according the head anthropometry. The helmet shape is architected to fit the head and the optical design. Usually, it results in a design that is optimal in terms of optical performance, compactness, and system weight.

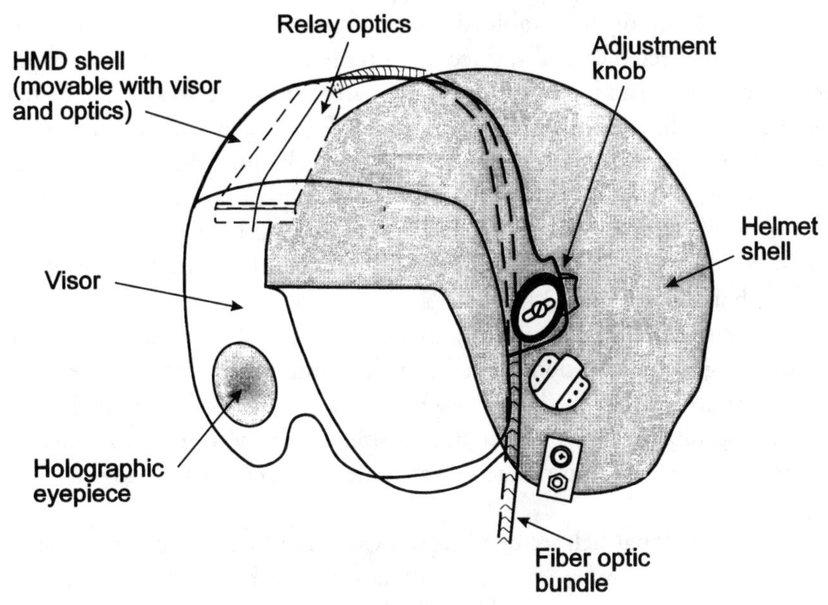

Figure 9.1 HMD integrated with a standard pilot's helmet.

9.1.3 The Modular System

A modular-type system is designed around an existing standard helmet or a new helmet design, and a basic helmet/display configuration is constructed. To the basic design, other configurations are fitted that use different components. The outcome is a basic system with several optional configurations that can be changed easily according to mission requirements. Eventually, there are different configurations for day and for night, for fighter use and for helicopter use. Because not all configurations are optimal, the added size and weight are compensated for by the fact that with each configuration fewer components are used.

9.2 Helmet Designs

Studies have recommended that the total head-supported mass, that is, the helmet and the display, should be less than 1.5 to 1.8 kg, to ensure pilot safety during emergency situations such as ejection or crash landing [2]. The helmet weight must be coupled with proper placement of the center of gravity of the helmet with the display components. The current standard aviator helmet, including oxygen mask, weighs about 1.5 to 1.8 kg. The HGU-33/P helmet, for example, weighs 1.7 kg, and a Kevlar-made HGU-55/P weighs 1.47 kg.

The location of the combined center of gravity of the helmet and the display must be selected carefully, since it is a trade-off among conflicting requirements. For good head mobility while the wearer is looking up and during high-g maneuvers, the center of gravity must be forward of the head's center of gravity; too far forward, however, may increase neck injury during ejection. The best trade-off is to locate the center of gravity as close as possible to that of the head itself. To ensure safe ejection, the recommended helmet center of gravity should be confined to 10 mm forward and upward of the head center of gravity, for a helmet weighing 2 kg [3].

Since the weight of currently used helmets is close to the limit of the allowed weight, the additional weight of the equipment placed on the helmet must be trimmed from the helmet itself. That calls for new helmet designs that are more optimal in terms of weight and size. One way to reduce helmet shell weight is to use composite materials that are stronger yet have lower specific weight [4].

The traditional helmet of fiberglass-cloth reinforced epoxy resin shell is being replaced by newer composite materials such as Kevlar™, by DuPont, or Spectra™, which was developed by Allied-Signal.

Composite materials are used in helmets to reinforce epoxy resin shell. Kevlar is a high-modulus, high-strength organic polymeric synthetic fiber of a class known as aramids with modulus comparable to that of carbon or glass [5]. The fibers are compression molded with the epoxy. Spectra is an ultra-high molecular-weight polyethylene. The fiberglass used in traditional helmets has a specific weight of 2.5 $gr\text{-}cm^{-3}$; Kevlar has a specific weight of 1.44 $gr\text{-}cm^{-3}$; and Spectra a specific weight of 0.97 $gr\text{-}cm^{-3}$ [6].

Apart from the helmet shell itself, lighter yet stronger materials may be considered as a replacement of the metal parts of the helmet. Such materials include aluminum, titanium, and magnesium alloys.

In integrated helmets, the shape of the helmet is partially dictated by the shape of the display, the constraints of the location of the display, and the shape of the visor with which the outline of the helmet must conform so the motion tracks of the visor fit the helmet contour.

The need for precision of the helmet has introduced the need for better definition of the human anthropometry for the helmet design [7]. The helmet has a complex geometry that is hard to design and manufacture. In the past, models were constructed by a trial-and-error process until satisfactory results were obtained. Obviously, that process limited the capability to produce only a few sizes of helmets that were fitted for the individual user by a custom-fitted liner. Evidently, that process was not optimal, and some compromise had to be taken. State-of-the-art three-dimensional measurement techniques, such as sonic and laser scanning methods, are being incorporated to precisely measure the head shapes of the individual users. The measurements are fed to a computer-aided design (CAD) system for completion of the shell design, structural analysis, and even automatic manufacturing of the helmet. Figure 9.2 illustrates the main anthropometric data of the head of an aviator wearing a helmet with a HMD.

The helmet houses several optical and electronic components that must be aligned accurately to ensure required optical performance. The ability to rigidly mount these components and retain accuracy dictates higher stiffness of the helmet shell.

The shell houses a molded urethane energy-absorbing liner that is custom fitted to the individual wearer of the helmet. The liner is covered with a thin layer of protective leather. For extra comfort, the helmet edges also are covered with leather edgeroll. In a similar process, a further mass reduction is achieved by a more economical design of the helmet liner. With a careful design, monitored with comprehensive structural analysis, spaced holes are made in the liner that reduce the material amount and, as a result, also the weight.

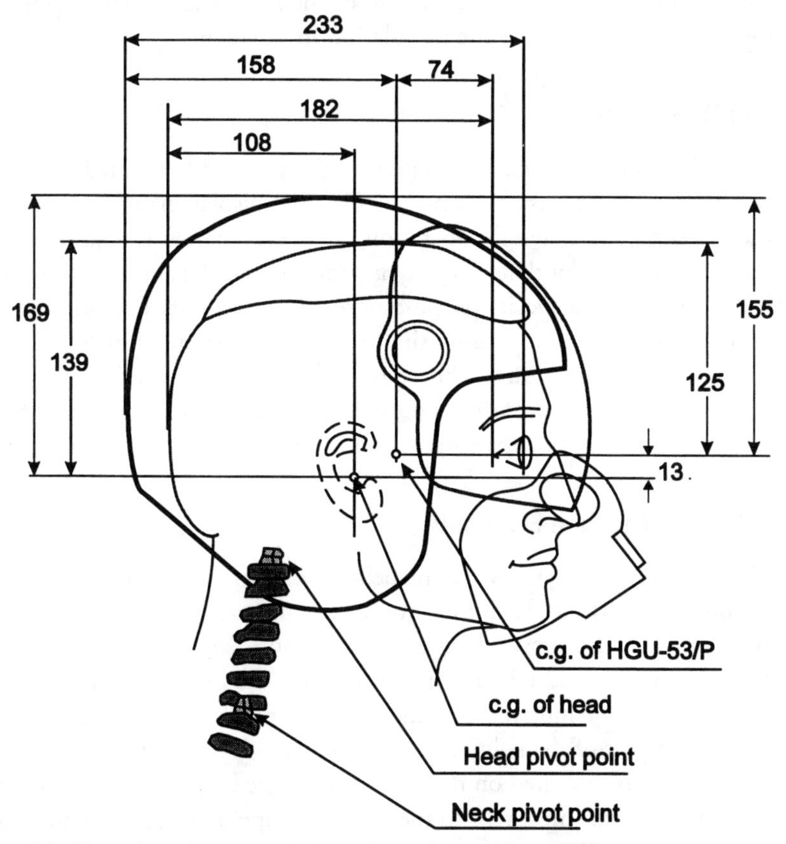

Figure 9.2 The main anthropometric data of an aviator with an HMD.

9.3 HMD/HMS Integration With Helmet

9.3.1 Display and Sight Alignment

The optical elements of the display are aligned as part of the manufacturing and assembly process. Similarly, the projected display also must be aligned to the eye (or eyes), which is carried out as part of the HMD final assembly process. Finally, the display must conform with the outside world; alignment of the display and the real world is the process called boresighting.

A CRT-based display requires less precise factory alignment and calibration because the display image and symbology easily can be offset electrically.

If sights that use an electromagnetic head tracker require only a rough alignment and calibration while being mounted on the helmet, electro-optical based sights rely on accurate position and well-defined geometry and require a precise calibration prior to use. Their accuracy depends also on the stability of this calibration. Therefore, the electro-optical LEDs or detectors usually are mounted on a rigid structure independent of the helmet itself. The metal rigid structure is calibrated prior to the installation in the helmet.

9.3.2 HMS Boresight

The boresight process, which is initiated by the pilot and executed each time the HMD/HMS is used, is performed to align the display or sight reticle with the head tracker. For a boresight operation, the crew member is required to fixate on a reference symbol that generally is projected on the aircraft HUD or, if a HUD is absent, on a special-purpose boresight generator. The boresighting process involves the pilot superimposing the sight reticle on a reticle projected on the HUD or boresight unit. The head tracker computes the LOS direction and its deviation from a prestored boresight. The deviation is used to correct the tracker measurements during operation of the sight.

9.3.3 Helmet Structural Flexibility

From the mechanical point of view, the helmet shape is not optimal in the sense that it is only partially enclosed, thus not firmly supported in its bottom side as it is, for example, in motorcycle helmets. Furthermore, so the wearer will be able to put the helmet on and take it off comfortably, the helmet must be flexible enough to make an opening for the head to pass through. Evidently, that means that the bottom part of the helmet is not rigid enough to ensure no movement of the display if it is mounted on the lower side of the helmet. When using that type of display, the user must rely on the chin and nape straps to maintain secure position of the display. Obviously, that dictates that the image source and all optical components be mounted on a rigid structure other than the helmet.

In the course of the design of the helmet structural analysis, the finite-elements method is employed to determine helmet distortions and displacements of the optical elements due to their own weight, high-g forces, vibrations, and voluntary head accelerations. The finite-elements model includes the helmet shell, the visor, and the display structure. Also, analysis is executed to assess helmet plastic deformations, which may occur in the course of helmet placement and removal from the head.

9.4 Visor Design

The main role of the visor is to protect the pilot's face from windblast during ejection and from flying objects hitting the aircraft and crew. The visor is clear and has high transmissibility, but often two visors are used, one clear and one tinted for high-brightness conditions, in which the display image typically is washed out by the ambient light.

The visor is essentially an optical element, and being so, it almost always creates distortions and LOS deviations.

The visor is located between the outside world and the optical projection system. In some HMDs, the visor is used as a combiner and hence as a part of the optical assembly of the HMD. To fulfill that role successfully, it must have adequate optical qualities in terms of shape, clarity, optical distortions and aberrations, and stability.

9.4.1 Visor Shape

In conventional helmets, the visor geometry is shaped to conform to the geometry of the helmet shell to avoid spoiling the helmet's outline and to protect the visor from breakage. The only optical quality of concern is good clarity and unnoticeable visual distortions. In contrast to the plain visor, in some HMD designs the visor is essentially an optical element, either a simple beam splitter or a complex catadioptric element. In those latter designs, it is shaped as either a parabolic or a spherical visor, as dictated by the complete HMD optical design. By proper design, the shape of the visor can be optimized to a specific helmet contour, so the outer envelop of the helmet-visor combination is minimized [8]. Figure 9.3 depicts a typical visor used with an HMD.

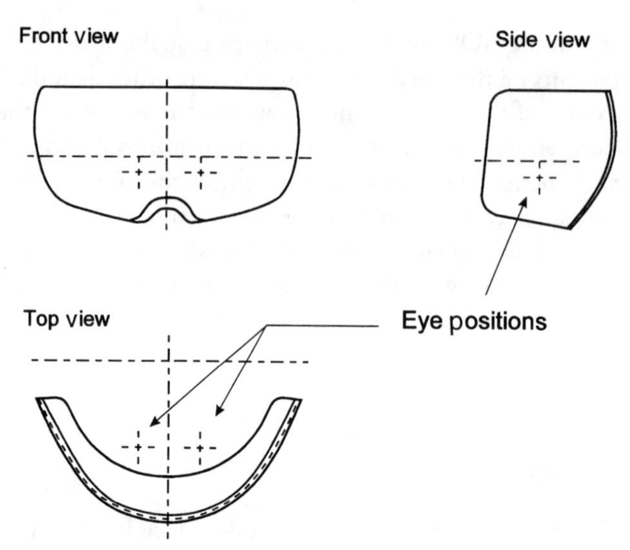

Figure 9.3 Layout of a visor used with an HMD.

9.4.2 Visor Materials

Ordinarily, visors are made of either polycarbonate or methacrylate plastic. Methacrylates, also called acrylics, and polycarbonates are thermoplastics and easily formable to desired shapes. They both have a combination of properties that make them very suitable for visors. They have superb transparency, 89% for polycarbonate and 92% for methacrylate, compared to 91% for clear glass. They have very high impact strength, 50 times stronger than glass, and low specific gravity, 1.19 gr-cm^{-3} for the methacrylate and 1.22 gr-cm^{-3} for the polycarbonate, which is about one-third of glass. They also have excellent dimensional stability and rigidity.

The main distinction—and the reason for the use of polycarbonate as the common material of visors—is heat resistance. Polycarbonate has superior heat resistance; its softening temperature is above 135°C. The methacrylate has a relatively low softening temperature of about 82°C, so it is less adequate for use in HMDs, which rely on precise shape stability of the visor over temperature, especially when exposed to sun radiation for extended periods of time. The polycarbonate also has excellent durability and abrasion resistance compared to methacrylate. Heat resistance is of paramount importance in the production process during visor coating, which normally is done at high temperatures. Visors made of materials with too low heat resistance often deform during coating and lose their precise optical characteristics.

Like metal and glass, polycarbonate can be heated and molded or extruded into various forms or shapes; thus, it readily can be shaped to any required geometry of the visor.

9.4.3 Visor Coating

HMDs that utilize the visor as a projection combiner and thus a critical optical element require consideration of the thin film coating on the visor. The visor surface first has to be finished to final optical figure without flaws or internal stresses that might distort the projected or see-through image. That is followed by a deposition of metal filters and dielectric beam splitters on the inside visor surface and antireflection films on the outside surface. The deposition is performed by thermal evaporation at high temperature, close to or above the softening temperature of the plastic substrate of the visor. The process is rapid, to avoid raising the temperature of the visor substrate and thus to prevent the risk of deformations [9].

A key parameter in the design of the visor is the reduction of ghost images reflected off the outside surface of the curved visor. Those images must have negligible brightness, at least two orders of magnitude smaller, compared to the primary image reflected from the inside surface. To realize that, it is necessary to use a dielectric, multilayer reflective coating on the inside surface and a reflective coating on the outside surface. The coating also is used to protect the polycarbonate visor from scratches.

9.4.4 Variable-Transmission Visor

Ordinarily, pilots have dual-visor helmets, a clear visor for normal and low-brightness conditions and a dark visor for high-ambient-light conditions. The pilot switches between the two visors manually by pulling down one visor while retracting the other.

A better solution is the use of a variable-transmission visor. Such visors have a continuously variable luminous transmittance, which is automatically regulated by a closed-loop control system [10]. The visor transmittance is controlled by liquid electrodipolar crystal light modulators sandwiched between two layers of polycarbonate coated with conductive film acting as electrodes. In nonoperating conditions, the crystal molecules are randomly oriented, thus blocking light passage. When an electric field is applied to the electrodes, the molecules orient in parallel to the electric field, and the previously opaque visor becomes clear. The visor transmittance is determined by the voltage of the electric field. A tiny photodetector behind the visor is used to measure the light intensity and is used by the closed-loop control system to activate the electric field with an optical response time of 50 to 150 ms.

9.5 Helmet-Fitting Techniques

A helmet must be fitted to the individual wearer so the display image is correctly placed in optimum position in front of the eye of the viewer and without compromising the full FOV of the display [11]. The custom fit compensates for variations in head length and eye-to-top-of-the-head variations among the population of potential users. Obviously, a major objective of the fitting also is to obtain a comfortable and stable fit.

Fitting is achieved in several ways. In modern helmets, the comfort liner, made of expanded foam and thermal plastic, is custom-fitted to the user by molding. A wax mold is heated until it softens; it then is pressed against the prospective user's skull to produce an exact impression of the user's head. After the wax mold hardens, it is removed and used to mold a urethane liner. Soft pads are added to increase comfort and eliminate residual spacing. The form fitting usually is done for the top and the back parts of the head. Ear caps are fitted to the ear positions separately, and the helmet eventually is adjusted by the wearer, who fastens the nape and chin straps to keep the helmet firmly in position on the head. If performed properly, the process ensures that the helmet and the display are placed and retained in the correct position during high-g maneuvers without causing discomfort, fatigue, or strain. Sometimes during the process, the pilot wears a boresight rig to align the helmet correctly while the liner is being formed.

In another technique of pseudo-custom fit, which is used, for example, with the IHADSS system, the custom-fit mold is replaced by a web suspension cap with a crown pad drawstring. The wearer fits the cap to the shape of the skull by adjusting the straps. The advantage of using the web suspension is the better ventilation it provides. Furthermore, this type of helmet can fit every casual HMD user, does not require special workshops for molding the liners, and enables the air crew to replace a helmet in cases of damage of a personalized one in a very short time. Figure 9.4 shows the structure of the IHADSS helmet with the various components.

Attempts to form-fit helmets by employing inflatable rubber liners have been found to be uncomfortable, due to excessive pressure on the skull and lack of proper ventilation.

Laser surface three-dimensional scanning technologies that can measure accurately the shape of an arbitrary body are available. Such techniques have been employed so far for better helmet design [7] but can be readily used for helmet fitting by creating more accurate molds from which liners can be produced.

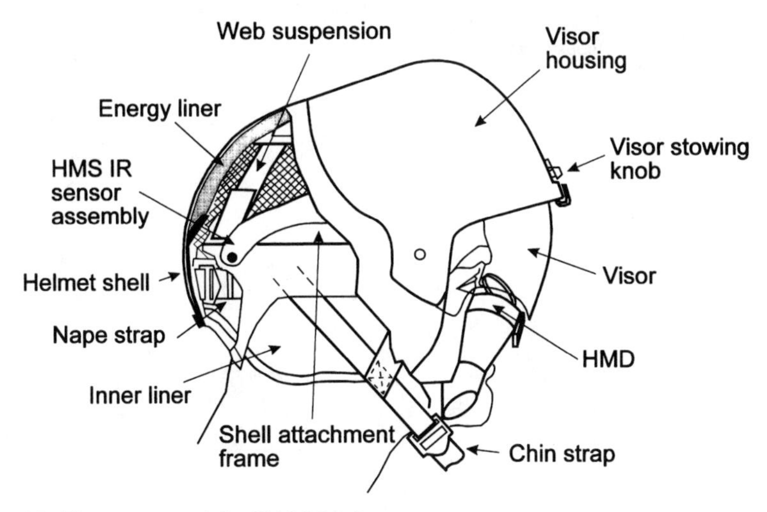

Figure 9.4 The structure of the IHADSS helmet.

9.6 Sound Attenuation and Active Noise Control

Lighter and thinner helmet shell designs usually reduce a helmet's ability to attenuate acoustic sound. That impairment must be compensated for by the use of better isolating materials and design. Passive noise isolation in the form of ear caps and liners is effective mainly in reducing high-frequency noise; it is less effective for low frequency or longer wavelengths. In many vehicles, including aircraft, high levels of low-frequency noise are present. Cockpit noise is both random broadband and periodic harmonic. It is generated by engines, turbine blades, rotor blades, transmission gears, and airframe noises. The noise is particularly high in helicopters. Current combat helicopter noise levels range from 90 to 100 dBA to 105 to 115 dBA [12]. Earcaps attenuate the noise by 24 dBA at 1,000 Hz and by 43 dBA at 4,000 Hz. At lower frequencies, which include rotor blade noise and some transmission gear noise, earcaps are less effective. Further attenuation of acoustic noise can be achieved by use of techniques of active noise control or cancellation.

Active noise control, a technique that is effective in reducing noise only for in the low-frequency range of up to about 500 Hz, reduces noise by producing destructive interference with the main interfering acoustic wave [13].

Active noise control is implemented by using a miniature microphone embedded in the helmet itself. The microphone picks up the noise and, after appropriate processing and filtering, generates a signal with the same frequency but with antiphase injects the noise into the earphones. The active noise control scheme is illustrated in Figure 9.5.

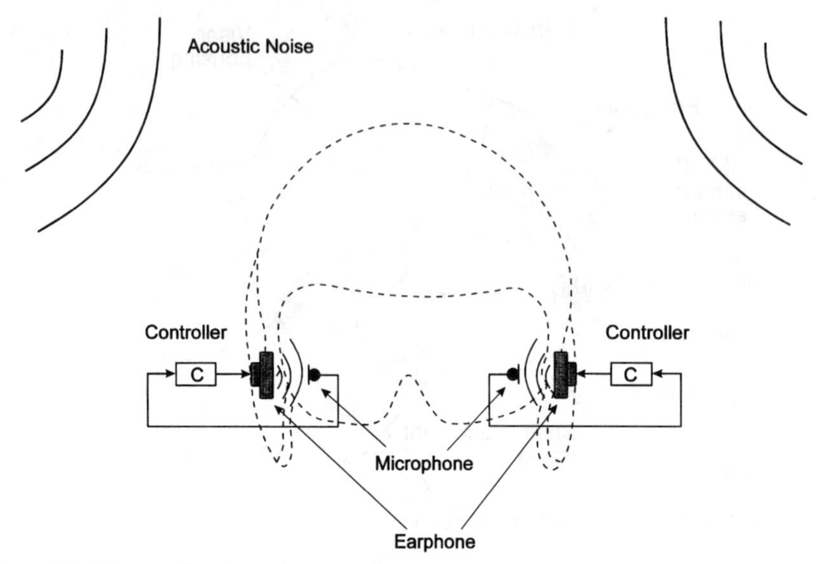

Figure 9.5 Scheme of active noise control.

The active noise control scheme basically is a closed-loop control system, as shown in Figure 9.6. The controlled "plant," $G(s)$, contains the acoustic characteristics of the path between the earphone and microphone and the electroacoustic response of both earphones and the microphone. The feedback, $H(s)$, includes the electrical transfer function from the earphone to the microphone output. $V(s)$ is the voice and communications to the pilot, $d(s)$ the external acoustic noise, and $R(s)$ the sound transmitted to the pilot after the external noise is attenuated.

From the scheme shown in Figure 9.6, the transmitted sound to the pilot is

$$R(s) = \frac{1 - C(s)}{1 - G(s)H(s)C(s)} V(s) + \frac{1}{1 - G(s)H(s)C(s)} d(s) \qquad (9.1)$$

If both the microphone and the earphones have a flat transfer function over the frequency range of interest, that is, $G(s) = 1$ and $H(s) = 1$, then (9.1) reduces to

$$R(s) = V(s) + \frac{1}{1 - C(s)} d(s) \qquad (9.2)$$

and the controller of the scheme basically is a simple gain with sign inversion. The sound in that case can be attenuated arbitrarily simply by increasing the gain. However, because both always have finite bandwidth, with phase lag thus increasing with frequency, a shaping filter that guaranties the stability and

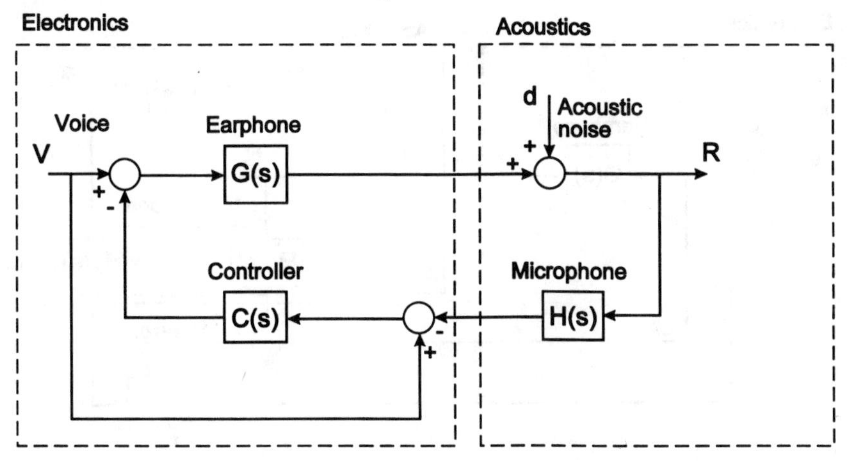

Figure 9.6 Block diagram of the closed-loop active noise control.

response of the loop is used. That filter, or controller, is adjusted to certain dynamic responses of the system. However, often with acoustic noise, the transfer function of the acoustic system changes during operation, mainly due to the speed of the noise-generating source, temperature, and even the relative location of the user from the noise source. In such cases, adaptive filtering techniques are used to guaranty optimal performance of the noise-control scheme over all operating conditions.

In adaptive noise cancellation, a secondary microphone external to the helmet is employed to sense the ambient acoustic noise. The output of the secondary microphone is compared to the measurements of the internal microphone and the difference between the two signals is used to automatically adjust the adaptive filter parameters. The adaptive filter operates so the power of that difference is minimized. The minimum of the power obtained for the adaptive filter output equals the acoustic noise. The output of the adaptive filter again is injected into the pilot's earphones, canceling the external noise. Figure 9.7 describes the adaptive acoustic noise canceling scheme.

The acoustic noise, together with the voice and communications, is sensed by the primary helmet-embedded microphone, whose acoustic and electric transfer function is $H_p(s)$. A secondary microphone, external to the helmet, $H_s(s)$, is used to pick up the ambient noise. The adaptive filter error $\epsilon(s)$ is

$$\epsilon(s) = H_p(s)R(s) - e(s) - y(s) = G(s)H_p(s)e(s) + H_p(s)d(s) - e(s) - y(s) \tag{9.3}$$

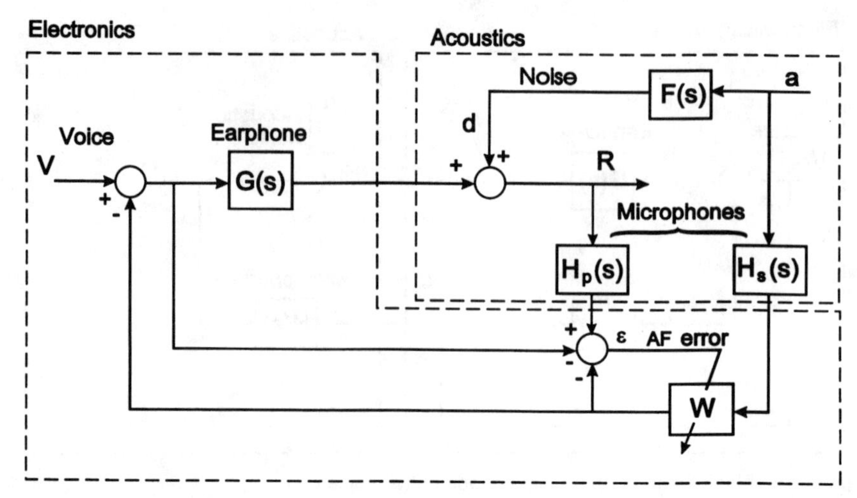

Figure 9.7 Block diagram of the adaptive filtering scheme for active noise control.

Now, $d(s) = F(s)a(s)$ and $y(s) = W(s)H_s(s)a(s)$ and assuming again $G(s) = 1$, $H_p(s) = 1$, and $H_s(s) = 1$, (9.3) becomes

$$\epsilon(s) = e(s) + F(s)a(s) - e(s) - W(s)a(s) = [F(s) - W(s)]a(s) \quad (9.4)$$

From (9.4), it is evident that the adaptive filter error is minimized for $F(s) = W(s)$; then $R(s) = V(s)$, that is, voice communications are transmitted without interfering noise.

The adaptive filtering technique eliminates the need to design a fixed filter or controller, and it adjusts itself to any variations in the path of the noise to the helmet as a result of different head positions and orientations. Typically, the adaptive noise canceling scheme employs the LMS algorithm [14–16] and is implemented using digital signal processors.

In practice, the active sound control system is adjusted to maintain a certain level of external noise to enable the helmet wearer to remain in touch with the external world. It is still useful for the aviator in the aircraft to be able to judge whether everything is running properly.

9.7 Helmet Quick Disconnect

Operating the HMD/HMS in the cockpit of a fixed-wing aircraft requires a quick disconnection of the helmet gear from the electronics unit during emergency ejection, emergency quick egress, and even in routine disconnection. The harness typically transfers both low- and high-voltage signals. The high voltage required for a

CRT-based display usually is 7 to 15 kV, which must be disconnected automatically without pilot intervention and without arcing or producing external sparks or exposing hot embers to the potentially explosive cockpit environment. The quick disconnect connector (QDC) must act as a conventional connector with excellent electrical characteristics under normal operating conditions.

A special concern is the mating and demating forces of the QDC. On one hand, the forces must not be too large, so the connector can be operated easily by the pilot and disconnected during ejection. On the other hand, it must remain mated when subjected to forces associated with the normal movements of the pilot in the cockpit and not subject to accidental disconnection during flight. Furthermore, the shape of the connector must be unobtrusive to the pilot to guarantee freedom of movement and to prevent injury during ejection or rapid ground egress [17]. Figure 9.8 depicts such a quick-disconnect connector

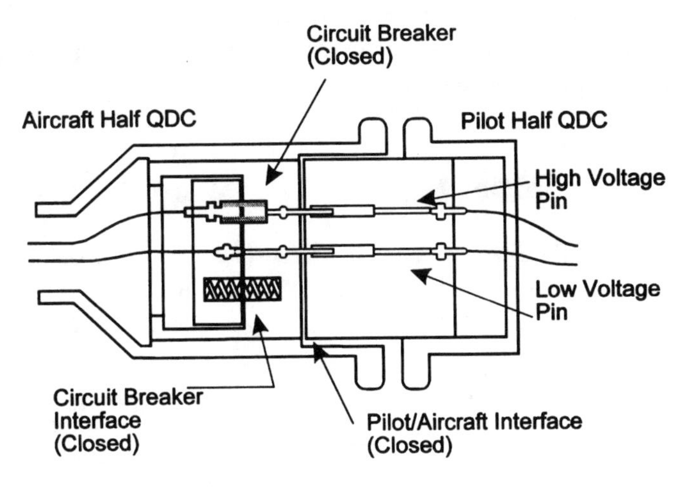

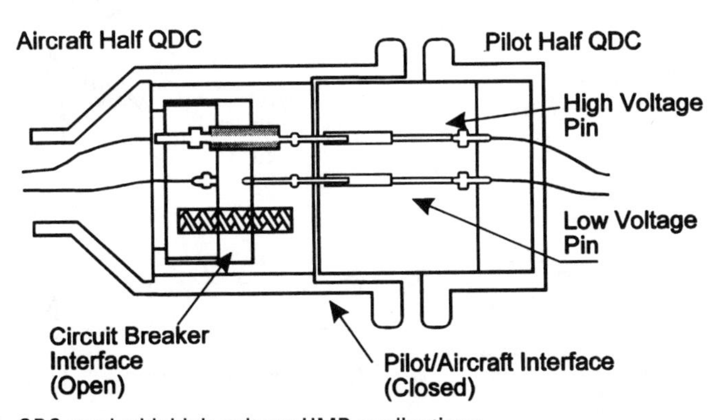

Figure 9.8 QDC used with high-voltage HMD applications.

based on the concept of a two-stage connector, having a circuit breaker that disconnects before the two connector parts are fully demated. The approach ensures that the display-aircraft contacts are never hot when fully demated.

Harness disconnection must be performed at a specific point on the torso to ensure there are no long, loose cable ends during ejection. Loose cables could physically harm the pilot by the windblast during ejection.

The problem is equivalently acute when the HMD image is conveyed to the helmet via a fiber-optic image guide. Due to the high mating accuracy required with such displays, a QDC is not feasible, so the fiber-optic cable must be cut during ejection. The force needed to cut the fiber-optic image guide is in the excess of 300 kg. The separation can be performed by the use of a mechanically or pyrotechnically activated guillotine. Both methods must have extremely high reliability; failure during ejection could result in severe damage to the pilot's head or neck. Figure 9.9 is a drawing of a pyrotechnically activated guillotine. It uses pyrotechnic cartridges, similar to rifle cartridges, which typically are activated by an electrical command.

9.8 Canopy Refraction Mapping

A high-speed and high-performance fighter aircraft canopy is constructed as a large bubble with curved surfaces and low inclination angles, to provide the pilot a large, unobstructed FOV and to retain good aerodynamic performance

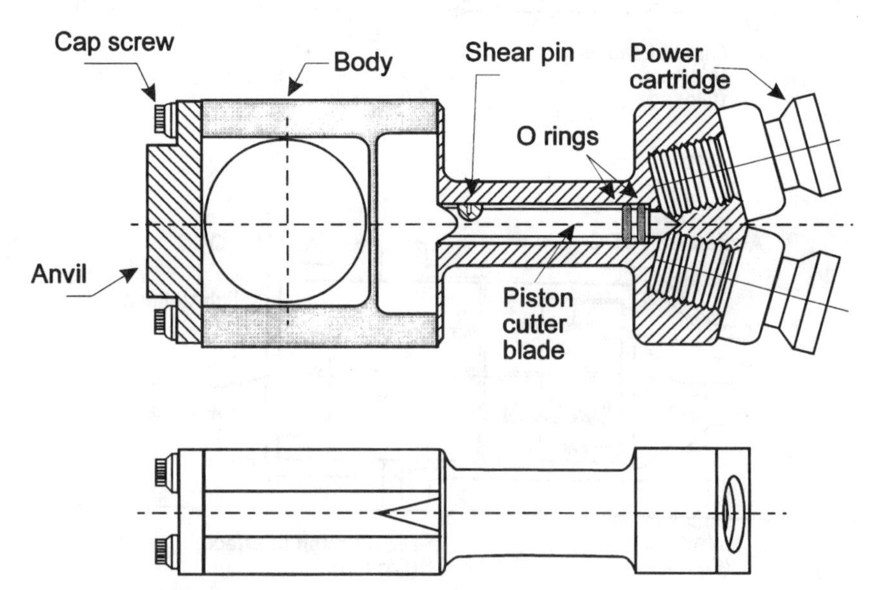

Figure 9.9 Electrically activated pyrotechnic guillotine.

of the aircraft. That shape, however, impairs the optical quality of the canopy compared to that of a flat, high-inclination canopy, as found, for example, on helicopters. The optical effects of the canopy include changes in both the position and the magnification of the object as seen through the canopy. The angular deviation of light rays by the refraction of the canopy cause sighting errors, because the actual position of the target does not correspond with the apparent position seen by the pilot from the cockpit. Lateral displacements cause only parallax effects, which become insignificant beyond a few meters. Figure 9.10 illustrates the lateral displacement and angular deviation of a light ray due to refraction through the canopy.

In the design and manufacture of a canopy, the refraction error in the small forward area of the canopy, where the HUD usually is located, is minimized. Residual errors are corrected by the fire-control computer. When an HMD/HMS is used, however, no such small area exists, and correction for the whole canopy must be performed. Because light rays are deviated differently by each small area of the canopy (Figure 9.11), the mapping has to be performed for each eye position and LOS direction within the cockpit [18].

The mapping of the canopy is performed by the projection of an L-shaped collimated image along the optical path of dual and orthogonal CCD linear arrays. The image is placed outside the canopy while the CCD arrays are placed on the other side of the canopy. A beam splitter directs the image to two channels, one measuring the elevation and the other the azimuth. Each CCD array is offset from the optical path so that each leg of the L-shaped image inter-

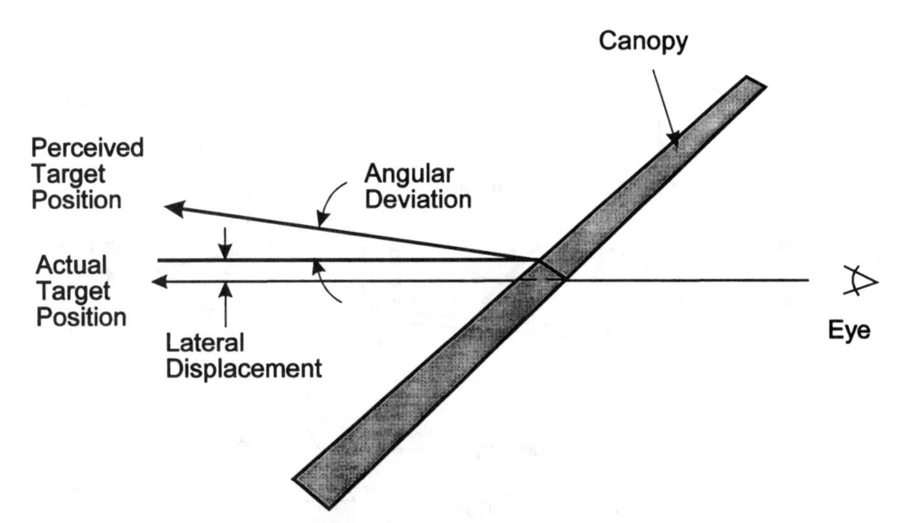

Figure 9.10 Lateral displacement and angular deviation of a light ray through an aircraft canopy.

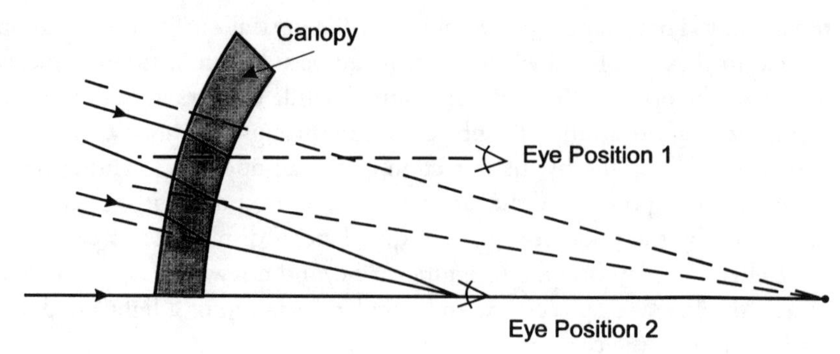

Figure 9.11 Effect of canopy curvature on sighting error at two eye positions.

sects one array (Figure 9.12). The image deviation along the two arrays indicates the canopy refraction angles. The process is repeated for all elevation and azimuth angles of the image relative to the eye design point and covers all eye positions within the head-motion box. The mapping data are used to formulate polynomial functions, which are part of the LOS measurement unit.

While mapping generally is performed for each newly manufactured canopy before installation in the aircraft, it must be repeated periodically. A canopy's optical characteristics can change during operational service from extended exposure to extreme temperatures and pressures and from wear, scratches, and physical damage from birds and other object strikes.

Canopy mapping in the aircraft cockpit can be performed using a laser beam projected onto a screen outside the aircraft [19]. The measurements are

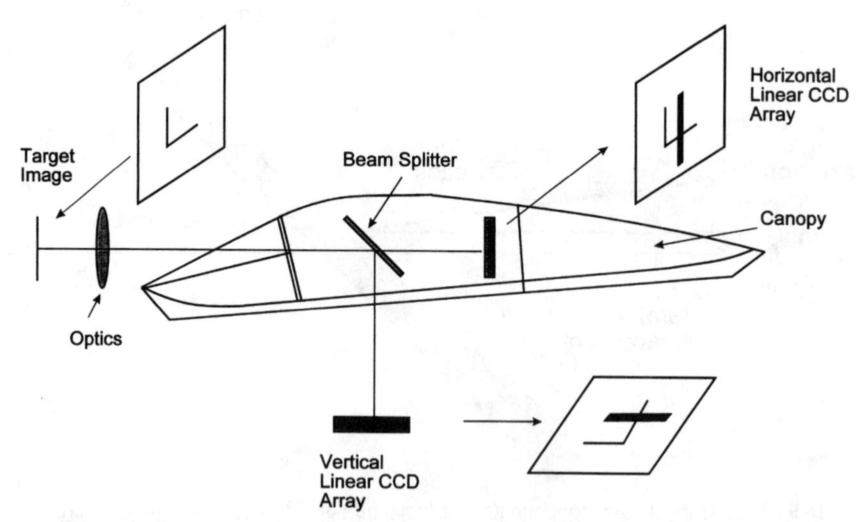

Figure 9.12 The arrangement for mapping canopy distortions.

performed once with the canopy lifted and repeated with the canopy in place. Distortions are deduced from a comparison of the two measurements.

9.9 Summary

An HMD/HMS is not a stand-alone system. It must be integrated with other helmet components to become a complete and effective system [20]. Past designs (and some current designs) that treat the HMD as an add-on to the helmet unit are not optimal in terms of weight, size, and location of the center of gravity.

The concern about the safety of the aviator with the added size and weight led designers of HMDs to consider designing the flying helmet as a sophisticated integrated system in which attention is given to all aspects of the system. It has been found that, by employing more careful design, newer materials, and design methods for the helmet shell, the visor, the inner liner, and the supporting systems (earphones, oxygen mask, etc.), not only can reduce the weight of the helmet but also can make it more comfortable for the pilot, safer, and able to provide better protection.

While weight reduction is crucial for the successful incorporation of an HMD/HMS with the helmet, it has to be achieved without compromising the strength and rigidity of the helmet. On the contrary, the helmet shell must be stiffer, more stable, and more precisely constructed to fulfill its new role as the optical bench for display and sight components. The helmet must be well fitted both on the user's head and also to the display, so the optical path from the display to the eye is correctly defined and placed. It must be integrated with a specially designed visor that functions as the combiner of the HMD image with the outside world and thus must posses high optical qualities and maintain them during all operating conditions.

A consequence of the new lightweight helmet designs is the unavoidable impairment of the isolation of acoustic noise. Therefore, active noise control by either closed-loop control or adaptive filtering must be used to compensate for the loss of noise isolation.

Special emphasis also must be given to safety aspects. The wearer must be able to quickly disconnect the helmet-display combination to allow safe ejection and fast egress from the aircraft in an emergency. For that purpose, special QDCs have been developed which have the capability to demate without producing sparks or endangering the pilot in any other way. QDCs with equivalent qualities must be utilized when a fiber-optic image guide is used to convey the image to the display.

Finally, the aircraft canopy has a major effect on the LOS direction of the pilot. So that aiming accuracy is not compromised, the canopy refraction must be mapped during integration of the HMD/HMS system and be continuously compensated during HMS operation.

References

[1] Jarrett, D. N., and A. Karavis, "Integrated Flying Helmets," *Proc. Inst. Mechanical Engineers*, Vol. 206, 1992, pp. 47–61.

[2] Stiffler, J. A., and L. Wiley, "I-NIGHTS and Beyond," in *Helmet-Mounted Displays III*, SPIE, Vol. 1695, 1992, pp. 13–20.

[3] Leger, A., et al., "Binocular HMD for Fixed-Wing Aircraft: A Trade-Off Approach," in *Display Systems*, SPIE, Vol. 1988, 1993, pp. 160–168.

[4] Ward, I. M., and J. E. McIntyre, "High-Modulus Fibers," in M. B. Bever, ed., *Encyclopedia of Material Science and Engineering*, Vol. 3, Cambridge: MIT Press, 1986, pp. 2138–2140.

[5] Black, W. B., "High-Modulus High-Strength Organic Fibers," in M. B. Bever, ed., *Encyclopedia of Material Science and Engineering*, Vol. 3, Cambridge: MIT Press, 1986, pp. 2140–2145.

[6] Cadogan, D. P., A. E. George, and E. R. Winkler, "Aircrew Helmet Design and Manufacturing Enhancements Through the Use of Advanced Technologies," in *Display Systems*, SPIE, Vol. 1988, 1993, pp. 80–89.

[7] Robinette, K. M., "Anthropometry for HMD Design," in *Helmet-Mounted Displays III*, SPIE, Vol. 1695, 1992, pp. 138–145.

[8] Gilboa, P., "Designing the Right Visor," in *Large-Screen-Projection, Avionic, and Helmet-Mounted Displays*, SPIE, Vol. 1456, 1991, pp. 154–163.

[9] Pratt, P. D., "Advanced Helmet Sight Reticle Assembly (AHRA)," AMRL-TR-73-11, USAF Aerospace Medical Research Laboratory, Wright-Patterson AFB, Ohio, 1976.

[10] Dobbins, J. P., "Variable-Transmittance Visor for Helmet-Mounted Display," AMRL-TR-74-28, Wright-Patterson AFB, Ohio, July, 1976.

[11] Barson, J. V., and R. J. Croft, "The RAF Institute of Aviation Medicine Proposed Helmet Fitting/Retention System," in *Helmet Mounted Displays and Night Vision Goggles*, AGARD-CP-517, 1991, pp. 5.1–5.6.

[12] Hart, S. G., "Helicopter Human Factors," in E. L. Wiener, and D. C. Nagel, eds., *Human Factors in Aviation*, San Diego: Academic Press, 1988, pp. 591–638.

[13] Elliott, S. J., and P. A. Nelson, "Active Noise Control," *IEEE Signal Processing Mag.*, Vol. 10, No. 4, October 1993, pp. 12–35.

[14] Widrow, B., et al., "Adaptive Noise Canceling: Principles and Applications," *Proc. IEEE*, Vol. 63, No. 12, 1975, pp. 1692–1716.

[15] Elliott, S. J., I. M. Stothers, and P. A. Nelson, "A Multiple Error LMS Algorithm and Its Application to the Active Control of Sound and Vibration," *IEEE Trans. Acoustics, Speech and Signal Processing*, Vol. ASSP-35, No. 10, October 1987, pp. 1423–1434.

[16] Eriksson, L. J., M. C. Allie, and R. A. Greiner, "The Selection and Application of an IIR Adaptive Filter for Use in Active Sound Attenuation," *IEEE Trans. Acoustics, Speech and Signal Processing*, Vol. ASSP-35, No. 4, April 1987, pp. 433–437.

[17] Bapu, P. T., et al., "Quick-Disconnect Harness System for Helmet-Mounted Displays," in *Helmet-Mounted Displays III,* SPIE, Vol. 1695, 1992, pp. 91–99.

[18] Clubine, W. R., "Modeling the Helmet-Mounted Sight System," M.Sc. Thesis, Air Force Institute of Technology, Wright-Patterson AFB, Ohio, 1982.

[19] Genco, L. V., and H. L. Task, "Aircraft Transparency Optical Quality: New Methods of Measurement," USAF Aerospace Medical Research Laboratory, Wright-Patterson AFB, Ohio, 1981.

[20] Whitcraft, R. J., "Helmet Integration: An Overview of Critical Issues," in *Helmet-Mounted Displays,* SPIE, Vol. 1116, 1989, pp. 122–125.

10

Applications of HMDs

Despite the great potential for applications of HMDs that are envisioned, only a few operational applications currently exist. Many factors contribute to the scarce use of HMDs: cost, lack of suitability to a particular application, lagging technology, and perhaps insufficient awareness of the potential benefits.

Despite all those deficiencies, the merit of HMDs and HMSs for both fixed-wing aircraft and helicopters is now recognized; undoubtedly, they will become part of the next-generation aircraft, either as a sophisticated all-aspect sight or as wide field of regard HUD. Soon, HMS will be used to exploit the benefits of the new high-offset boresight missiles and the HMD to improve situational awareness of the pilot. HMDs coupled with night vision capabilities will be used for night attacks and for helicopter NOE flights. Among the other applications of HMDs, the virtual cockpit is perhaps the most notable.

Proposals for applications of helmets or HMDs other than for aviation first must answer several questions. What are the benefits of the HMD in terms of information provided to the user? What are the advantages of using such a display over alternatives, such as conventional displays or indicators? Is the current technology mature and sufficient to implement an effective HMD? Finally, does the cost of such a display justify its use?

The potential applications of HMDs in nonaviation uses can be divided into three categories:

1. Applications in which the potential user already uses a helmet and the mounting of a display on that helmet appears natural;
2. Applications in which the addition of an HMD can be tolerated by the user if the benefits of the display justify it;

3. Applications in which forcing the user to wear a helmet may be unrealistic. For these applications, the head-mounted display must be constructed as a very light and compact device, more like eyeglasses or goggles than a helmet display.

10.1 Training and Simulations

Second to aviation, the most widely used applications of HMDs are in training and simulations. The HMD offers a great advantage for training and simulation in the sense that it can simulate the entire environment on a single display and may better replicate the environment than any other alternative. It eliminates the need for highly sophisticated and expensive trainers and simulators and essentially can be used for training in almost any type of task or role [1,2].

Training differs from simulation in the respect that simulation often is used as a tool for development of new designs or to visualize results of complex computations that result with a large amount of, typically, three-dimensional graphics [3]. Training, on the other hand, is used to present repeated sets of graphics scenarios.

The demands on the display for simulation and training are perhaps the toughest of all. Simulation fidelity is the most important factor; thus, the display must show graphics and scenery with the highest resolution and with color correctness and brightness. Also, the dynamic response of the display system in terms of time lags, motion smoothness, and correctness is vital.

A high-performance HMD-based simulator was used by the U.S. Air Force to design the supercockpit [4], a project aimed at the research and study of futuristic concepts for the next generation of aircraft. With the simulator, researchers were able to create a complete high-performance aircraft environment and to test and study new high-technology concepts such as voice commands, three-dimensional audio and graphics information for close-terrain flights, and high-g maneuvering.

In recent years, simulation techniques and requirements have progressed considerably. They have moved from simple flight training and aircraft response behavior testing into war simulations of multiple aircraft or other large-scale forces with multiple participants taking part in the simulation [5]. The participants may be physically located at the same facility or at remote locations and participating via fast computer networks [6].

HMD-based simulators are almost the only way to perform such realistic huge simulations at a reasonable cost. Other alternatives would require extremely expensive simulation facilities for each participant.

In aircraft simulation, mainly in the stage of aircraft design when concepts are explored and visualized, perhaps the greatest benefits of using three-dimensional graphics is the ability of the pilot being able to inspect the aircraft response and, at the same time, the engineers being able to look from the outside at the behavior of both the pilot and the aircraft [7,8]. For example, if the pilot encounters an unexpected situation during simulation, the engineers can "look" at the fluid flow around the aircraft and try to find the cause of the problem.

Although similar capabilities are available with a regular simulator, to obtain the same results, the pilot must repeat the simulated sequence many times for the engineers to view different aspects and views of the aircraft. Simultaneous "experience" can be achieved only with a virtual reality–type, multiperson environment [9].

Figure 10.1 shows a high-performance simulator that employs a high-resolution display. The image of the display is generated by a three-CRT 9-in color display and is conveyed to the HMD via coherent fiber-optic bundles. Previous design of the same fiber-optic HMD (FOHMD) was implemented by using light-valve projectors, in which the color display was realized by the principle of color subtraction [10]. The system is capable of displaying a stereoscopic image with a combined FOV of 100 by 49 degrees and resolution of 2,000 lines and 2,000 pixels per line [11]. The high resolution is given only in a small inset of 3 to 5 degrees, which is the area of interest (AOI). The system uses eye trackers to track the eye movements and place the AOI in the high-acuity zone of the eye.

A unique application of a low-cost trainer is used for training parachutists [12]. It comprises a parachute harness hanging from the top. The harness strings are allowed limited freedom, and their movements are measurable. The

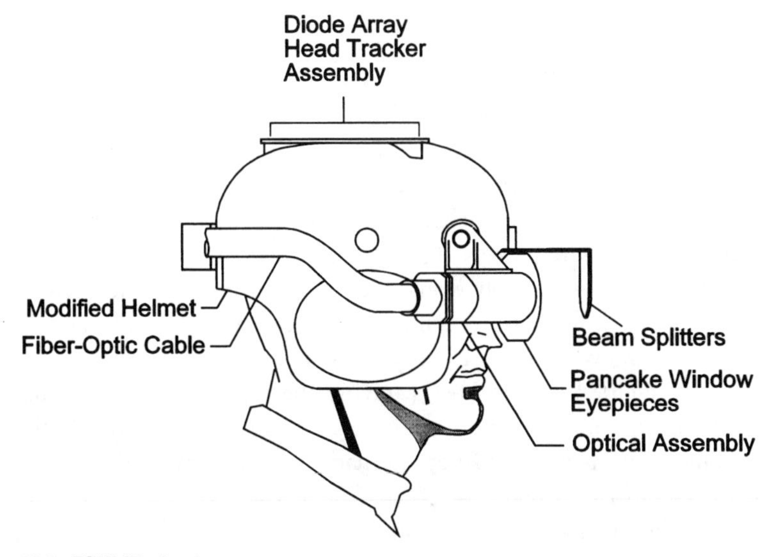

Figure 10.1 FOHMD simulator.

trainee hangs in the harness and by pulling the strings can practice jumps and maneuvers with the parachute. Motion sensation and feedback on the actions of the trainee are obtained by presentation of the surrounding scene on a low-cost commercial HMD. This type of simulator exemplifies how a low-cost HMD can be used to construct an effective simulator. The alternative would have been a complex and expensive projection system surrounding the trainee in the training room.

A full range of HMD-based simulators is used for entertainment. Often referred to as virtual reality simulators or games, they are used in amusement parks and game arcades. These basically interactive three-dimensional games mimic aircraft-type missions, space war games, and adventure type games.

10.2 Vehicular Applications

In general, HUDs are better than HMDs for vehicular applications. Because the driver usually is required to view only the front of the car, it is sufficient to use a small-size HUD located above the steering wheel. Indeed, HUDs are offered in some Japanese and top-of-the-line U.S. cars.

The only exceptions for which HUDs are impractical are for motorcyclists and race car drivers. With the currently available technology, however, HMDs can be readily used in those two applications. Probably the main obstacle that has prevented their use so far is that neither motorcycles nor race cars are prepared for interfacing with electronic displays.

10.2.1 Motorcycle Helmets

Almost all over the world, motorcycle drivers are required by law to wear crash-resistant helmets. In that respect, the addition of an HMD should not contribute to a rider's discomfort. Furthermore, most currently used helmets are relatively heavy, and the extra weight of a compact HMD is not significant or can be easily compensated for by slight modification of the helmet structure.

Motorcycle users probably would settle for a simple alphanumeric display showing numerical information such as speed, gear, hazards, and status indications. In the future, when navigation systems are available, driving instructions also may be presented on the HMD, giving the wearer current location and navigation instructions. For those purposes, a LED display source with sufficient luminance is adequate. Other alternatives are commercial monochrome LCD displays and EL displays.

Due to the small amount of information that is needed, a 5- to 10-degree display FOV is ample. For simplicity and lower cost, the virtual display image does not have to be overlaid on the external scene. A small occluded display with simple magnifying lens–type optics can be placed at the lower section of the helmet, provided it does not obstruct the external view. Because the information would be viewed only when needed, users could shift their gaze at the display to view the image. This type of arrangement enables use of a lower luminance display source and prevents image washout by external high luminance.

10.2.2 Race Cars

Race car drivers operate under very stressful conditions. Like fighter pilots during aerial dogfights, they must process information about their environment constantly. A single lost second can change a winning situation to a losing one.

The driver must be fully aware of the race situation, the road conditions, and the car condition. Because of the high driving speed, the curved path of the road, and the proximity to other cars and edge obstacles, the driver must devote all visual attention to the outside world. In contrast to regular cars, the race car cockpit is small and lacks the space for installing a HUD. Thus, an HMD is a natural alternative to place the display.

Unlike a motorcyclist, a race car driver cannot shift gaze from the road. Therefore, the HMD image must be overlaid on the outside-world scene. Useful information would be mainly hazard indications (e.g., engine temperature, oil pressure, and tire air pressure) and general information such as engine RPM, car speed, time lapsed, etc.

10.3 Nonaviation Military Applications

10.3.1 Armored-Vehicle Crews

Armored-vehicle crew members, especially tank commanders, must be able to move freely in their compartments. While they are moving about, they are disengaged from the display and controls; hence, they lose the ability to supervise the tank's status and environment. Tank commanders must function in three main positions: closed-hatch position, open-hatch position, and popped position [13]. In all three positions, it is necessary to provide the commander with full control of the tank and the ability to monitor the tank systems. An HMD is helpful in retaining the ability to command and control the weapon systems and the tank. With an HMD, a tank commander could ascertain the sight view of the gunner and thus shoot the gun while being outside the turret or confirm that the gunner is laying the gun on the proper target.

Another intriguing HMD application for tank commanders is for situations of sealed-hatch warfare. That type of operation would be necessitated by chemical or biological warfare conditions. A continuously rotating electro-optical sensor installed on the turret would scan the surrounding environment and acquire a panoramic image of the external world. The image would be stored in a large image memory and be updated during the sensor rotation. By head rotation, the tank commander would be able to view each segment of the image. For any head orientation, the proper segment of the panoramic image would be presented on the display. In that way, the tank commander would get the perception of a transparent turret and retain situational awareness even while in a closed environment.

The HMD for tank crew members would be constructed in two basic forms: attached to the helmet in a manner similar to an aviator's HMD or as a part of the standard sand, wind, and dust (SWD) crew goggles [14].

Figure 10.2 is a photograph of a prototype HMD for a tank commander.

10.3.2 HMD for the Infantry Soldier

Most high-technology advancements seem to have skipped the infantry soldier. Today, the infantry soldier lacks almost any source of information [15]. HMDs could improve significantly infantry soldiers' situational awareness and survivability. The new gear would incorporate a lightweight display, coupled with night vision thermal or intensified light sensors, a GPS receiver, and a magnetic compass or perhaps an IMU-based attitude and position reference unit. The system would include a body-hugging small computer, with enough storage for a digital map of the close arena. The map would show on the HMD the soldier's current position as reported by the GPS receiver. The helmet would house a wireless voice and data communication transmitter/receiver. The display also could be linked to the gun sight with a camera actually located on the gun, so it could be sighted and fired at arms-length, for example. The system also could integrate a night vision sensor.

Because the soldier would have to carry all the gear for extended periods of time and in tough environmental conditions, the requirements for such a system would be very high. It should weigh less than 1 kg and consume less than 5W, be powered by a single battery, possibly rechargeable by solar panels, and obviously be rugged enough to endure battle conditions [16].

Figure 10.3 shows a conceptual illustration of an HMD for a futuristic infantry soldier. The helmet is constructed of lightweight materials such as Kevlar or Spectra, which, as a side benefit, would provide superior protection from bullets and small artillery splinters. A hardened transparent plastic visor would protect the soldier from projectile fragments. The special coating on the visor would protect the soldier from laser-beam warfare, thus reducing the risk of blindness [17].

Figure 10.2 Prototype of an HMD for a tank commander (courtesy of ELOP Electro-optics industries).

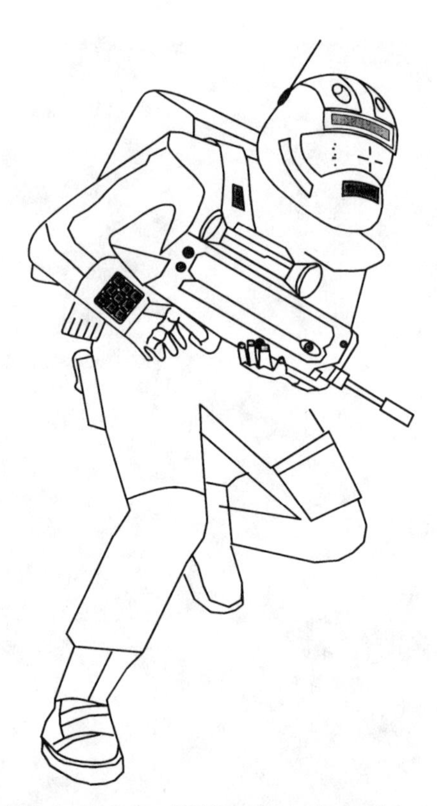

Figure 10.3 Conceptual illustration of an HMD and gear for an infantry soldier.

This concept of HMD is not restricted to the infantry soldier. It also could be useful for other persons, for example, fire fighters, who use helmets for protection from heavy smoke. With an HMD, fire fighters could view drawings, in the form of graphics information, of the burning building. Furthermore, fire fighters could use some sort of external vision-enhancement sensors, such as IR or LLTV cameras located on their heads as well as mask protection from smoke inhalation.

Similarly, deep water divers could use an HMD to project such useful information as water depth, oxygen-tank status, time, even navigational information obtained from an inertial navigation system.

10.3.3 Antiaircraft Aiming and Tracking Systems

A military application for which HMDs/HMSs may prove extremely advantageous is in antiaircraft batteries. One of the greatest threats to military forces is high-speed, low-level ground attack aircraft. These suddenly emerging all-aspect attackers leave very little time to acquire the target and to direct the battery guns at the attackers. For example, an aircraft flying at the speed of 300 m/s at tree-top level, if detected at the range of 6 km, leaves the defense unit about 10 sec to organize before it opens fire.

A human observer can quickly detect the swiftly approaching threat and lay the HMS on the target with speed that practically cannot be matched by any other equipment. Furthermore, it is a low-cost alternative to the more sophisticated target acquisition equipment.

The sight is a simple reticle focused to infinity. The helmet incorporates a magnetometer to measure the observer's head azimuth and two miniature accelerometers to measure the head tilt. The head-measured orientation is transmitted to a computer, which computes the slaving commands to each gun in the battery while taking into account the parallax between each gun and the observer. By tracking the head orientation, the location of the target can be calculated and transmit aiming commands to the guns. Figure 10.4 illustrates the concept of laying guns on an attacking aircraft.

10.4 Low-Cost Consumer Systems

Low-cost consumer HMDs are used mainly for two purposes: as low-weight and low-power portable computer displays and for entertainment, mainly computer and TV games.

With the miniaturization of portable, hand-held, or belt-worn personal computers, the need for suitable displays with a sufficient resolution and low-

Figure 10.4 The concept of aircraft tracking and antiaircraft gun battery aiming

power consumption was identified. Small computers have only a small display, which obviously is unreadable and hence impractical to use. The HMD is the solution, since it creates an image larger than the display screen itself and essentially gives the user the perception of viewing a normal-size display. Such displays, commonly termed personal displays (PDs), are more head attached than HMDs. They are worn by using a crown-like ring that is easily placed on and removed from the head.

Most low-cost HMDs are based on monochrome or color LCD matrix displays with standard VGA 640 by 480 pixels. A low-cost alternative solution for the consumer head-mounted display is the Private-Eye display (see Figure 10.5) [18]. It is based on a resonating spring-mounted mirror that scans a column of LEDs.

Another exciting method for implementing a low-cost HMD is the VRD, which scans low-power light directly onto the viewer's retina [19]. This personal display was conceived at the Human Interface Technology Laboratory of the University of Washington in Seattle and is presently being developed as a commercial product by a private company. In the VRD, shown schematically in Figure 10.6, there is no display or real image. Rather, the image is formed directly on the retina with a coherent source, such as a laser diode, beam scanned by a pair of miniature rotating mirrors similar to the Private-Eye. The greatest advantage of the VRD is the high brightness of the perceived image, because it is directly projected onto the retina; hence, a low-cost see-through display in daylight conditions can be constructed without the need for a high-power image source. The main obstacles to utilization of the VRD are the lack of a good solution for a compact scanning device and the difficulty in achieving a sufficiently large exit pupil. Moreover, efficient blue and green laser diodes still are not common; thus, full color cannot be generated yet.

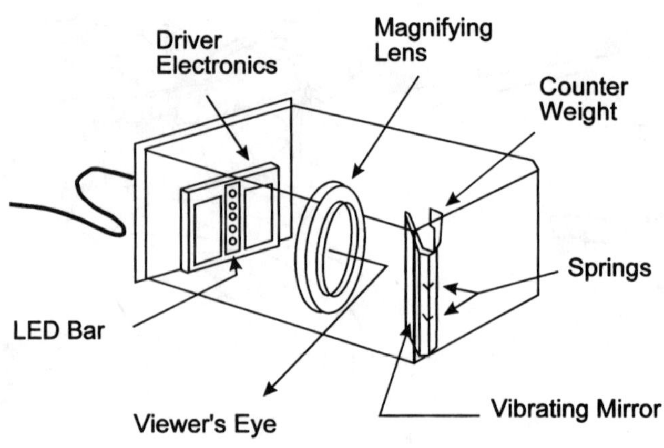

Figure 10.5 The Private-Eye HMD.

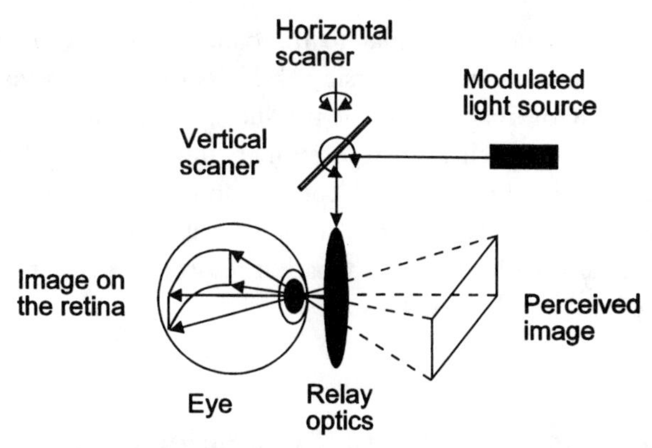

Figure 10.6 The concept of the VRD. A low-power laser beam is scanned by a pair of mirrors and forms the image on the retina.

Many applications have the potential to be transported to the miniature personal computer. Such applications include, among others:

- *Portable pagers and fax machines:* A miniature computer with a built-in modem that is hooked up to a cellular phone can receive and display on the HMD text messages, e-mail, and faxes. Faxes essentially would be unreadable on a small computer display.
- *Hands-free paperless maintenance manuals:* Technicians and maintenance personnel could perform complex repair jobs without carrying paper manuals. That would be very beneficial when the repair task has to be performed in conditions where access to the manual is impractical, for example, in confined spaces or under uncomfortable conditions such as performing maintenance in an aircraft body, in space performing extravehicular activity (EVA) [20], or on an electrical pole.
- *Hand-held personal computers:* Portable and mobile computers that can be used by passengers during air-line flights would replace the bigger and heavier laptop computers.
- *Hand-held electronic instruments:* Instruments such as oscilloscopes and voltmeters are easier to use with HMDs since they do not require the users to shift their gaze in order to view the instrument.
- *Miniature TV screens:* Portable and mobile TV sets for use by car and airline passengers. The display could be used to show TV broadcasting or as screens for TV computer games. Such displays also could be used in fitness clubs to help patrons pass the time during prolonged cycling, treadmill, and other exercise routines.

- *Sales and marketing:* Similar to the concept of scientific visualization and complex engineering design, HMDs can help in visualization of nonexisting or unapproachable products. For example, interior designers could show their ideas to customers more easily by letting the customers "walk" through the decorated house and inspect the proposed furniture, rearrange the furniture, shift walls, and so on [21]. Travelers could take a stroll through a vacation resort and inspect the attractions before booking flight tickets.

10.5 Medical and Therapeutic Applications

10.5.1 Low-Vision Enhancement

Over three million people in the United States are affected by low vision. Low vision is defined as any chronic condition, uncorrectable by lenses, that impairs the ability to perform everyday functions [22]. One type of low vision is degenerative macular disease, which progressively leads to a complete loss of function for the affected patches of the retina. The patches become blind spots, or stocomas. Stocomas in the central field result in reading problems, since the blind spots obscure parts of the text that is imaged onto the stocoma. Persons with such a disability often are unaware that part of the text is missing; instead, they see text that appears meaningless and garbled.

A vision-enhancement device based on an HMD uses a CCD video camera and a display, both mounted on a headset, and an image processing unit. Using mapping of the stocoma on the retina, the device repositions or manipulates the image on the display screen in a way that the text is not positioned on a stocoma (Figure 10.7). For that purpose, an eye tracker is used to measure the eye position so at each moment the stocoma position relative to the display is known. An alternative that avoids the need for an eye tracker is a fixation symbol that is presented on the display; text is scrolled on the screen, so the viewer is not required to execute eye movements to read the text.

10.5.2 Phobia Exposure Therapy

In exposure therapy, a phobic subject is exposed to a virtual environment that contains the feared stimulus. Such therapy has been shown to be an effective treatment approach for various phobias such as fear of heights and fear of flying.

Conventional methods to treat fear of flying include therapy during real flights, obviously an expensive proposition. The virtual reality method is a relatively simple treatment, with similar results at a fraction of the cost.

(a)

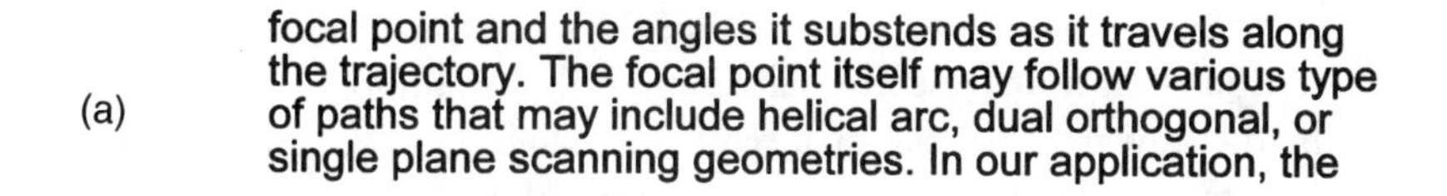

(b)

focal point and the angles it substends as it travels along
the trajectory. The focal point itself may follow various type
of paths that may include h cal arc, dual orthogonal, or
single plane scanning geometries. In our application, the

(c)

focal point and the angles it substends as it travels along
the trajectory. The focal point itself may follow various type
of paths that may include hcal arc, dual orthogonal, or
single plane scanning geometries. In our application, the

(d)

focal point and the angles it substends as it travels along
the trajectory. The focal point itself may follow various type
of paths that may include h elical arc, dual orthogonal, or
single plane scanning geometries. In our application, the

Figure 10.7 The principle of low-vision reading-aid display. When parts of the text (a) are projected on a stocoma (the gray circle) on the retina (b), it is not seen by the viewer. Instead the viewer sees only a part of the text, which looks meaningless (c). The aiding device uses warping for mapping the text of the display away from the blind area (d), so it is fully seen by the viewer.

Basically, the therapy to treat fear of flying involves simulation of an airliner flight, in which the subject is placed in a virtual airplane. The headset includes a sound system that reproduces the environment of a real airplane. The therapy goes through all the stages of flight, emphasizing situations that evoke anxiety, such as turbulence, landing, and storms [23].

The key to the success of the therapy is the realism of the stimulus. Since the visual system plays a major role in the sensation of the stimulus, the quality of the image must be very high.

Most therapy systems use low-cost displays, capable of showing low-quality animations that more resemble cartoons, and have limited effectiveness. To make the therapy valuable, the quality of the image must be improved at least to the quality of video image. For that purpose, the systems have to include a large image bank covering all scenarios, and the subjects must be able to access those images as they act as they would in the real world.

10.5.3 Motion Sickness Treatment

The motion sickness often experienced by sailors and naval vessel passengers is explained by a theory that suggests that it results from conflicting cues. The theory is termed neural mismatch or sensory rearrangement theory [24]. Its principal concept is that motion sickness occurs when sensory information about body movement provided by the eyes, the vestibular apparatus, and other receptors stimulated by forces acting on the body is at variance with the inputs that the central nervous system is expected to receive.

The theory suggests that people in enclosed rooms in vessels sense the ship's motions through their vestibular systems, but those senses are not supported by visual cues, which are of a nonmoving world. Indeed, in lower-deck compartments, people experience motion sickness syndromes more often and more severely than people on the deck, even though those on the deck are exposed to larger body movements.

By the rationale of the conflicting cues theory, if visual stimuli conforming with the vestibular stimuli are provided to the passenger, the motion sickness can be reduced or avoided. A natural way to present the visual stimuli is to use an HMD. On the display, the inertial information can be shown as an artificial horizon or as a symbology related to the ship motions in the present location. The information can be transmitted as a very low power radio frequency by a transmitter located in the various compartments on the ship.

10.5.4 Surgery and Medical Examinations

Augmented reality, which presented graphics on a see-through display, may prove beneficial in certain types of surgeries, for example, in removing cancerous growths. The procedure requires knowledge of the precise location of the growth, especially when the tumor is hidden behind another organ. The display can show graphically the location of the tumor on the obstructing organ, ensuring that the procedure is performed safely and without harming the organ. Obviously, such a system would rely heavily on accurate head trackers.

Another medical application is in radiology and fluoroscopic examinations. Currently, a radiologist observes the patient with real-time x-rays, adjusts the patient on the table to the desired position, and then executes the examine. In the future, the radiologist could position the patient while observing the x-ray image on an HMD. Similarly, dentists can benefit from HMDs, by having an x-ray image of a tooth during treatment or, alternatively, a miniature TV camera or fiberscope pointing at the ill root and showing an image of the treated point.

10.6 Virtual Reality

Virtual reality (VR) is a field that spans a vast number of applications that employ HMDs. Although it is sometimes difficult to distinguish applications that belong to simulations, training, entertainment, or therapy from those that are in the category of VR, there is a clear distinction between the "plain" HMD application and VR application. In a VR system, the user of the HMD interacts with the virtual environment and affects it by employing a virtual interface such as gloves or a pointing device. According to that distinction, a flight simulator in which the subject uses a physical joystick or controller is merely an HMD flight simulator and not an actual VR application. A training simulator in which the subject performs a maintenance task by accessing parts of a machine, disassembles and manipulates it by just executing the proper hand movement without having the physical part in hand is a VR application [25]. Yet in spite of that important distinction, applications of VR overlap with other applications, and the separation line often is very thin.

A photograph of a typical commercial head-mounted display, used mainly for VR applications, is shown in Figure 10.8. It has two miniature, 1.3-in diagonal, active matrix LCDs, each of which can display 742 by 230 pixels. The displays are mounted on a lightweight crown. The display accepts video signal and stereo audio. Its total weight is under 1 kg. Many VR applications are aimed at displays of this type.

A successful trial to use VR systems for training was done at NASA as part of the repair and maintenance of the Hubble Space Telescope. The primary training goal was to familiarize the ground support team, for example, engineers, technicians, and flight controllers, with the location, appearance, operability, repair, and maintenance procedures of the telescope [26]. The team numbered about 100 members, so training them all in the existing simulators and training facilities would have required use of the training facilities for extended period of time. Instead, the trainees were given commercial VR HMDs and control interfaces in the form of joysticks, manipulators, head trackers, and gloves. A special three-dimensional graphics simulator

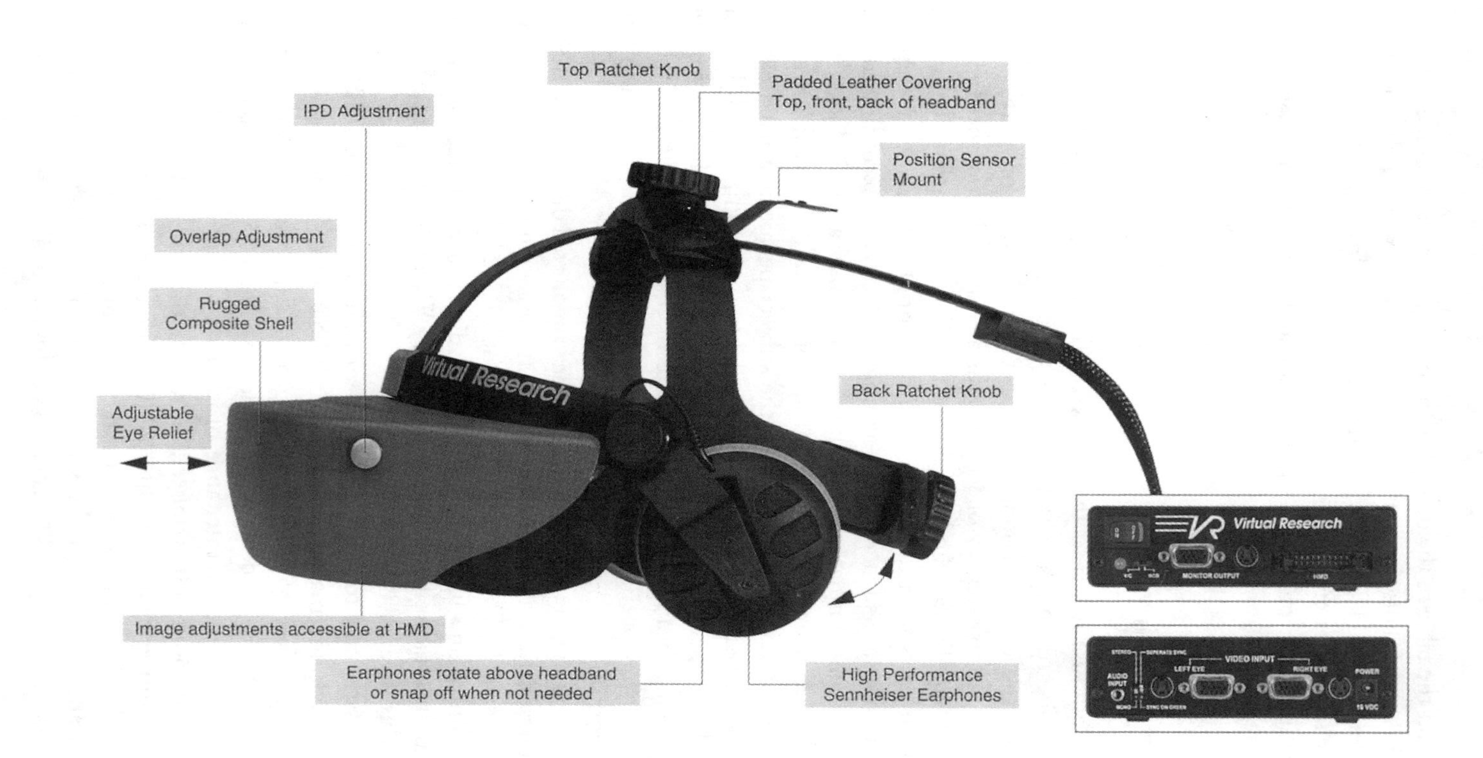

Figure 10.8 A typical commercial head-mounted display, commonly used for VR applications such as training, entertainment, and therapy (photograph courtesy of Virtual Research Systems, Inc., Santa Clara, CA).

was constructed. It included all training scenarios spanning all duties and tasks. With VR, the training time was significantly reduced, and more trainees could be trained without occupying the heavily used training facilities.

VR techniques are explored for use in the design of multipart products. A virtual prototyping of mass-produced products can save a lot of money in building mock-ups and in reducing design errors. That has a large impact on the aerospace and automotive industries. Part fitness can be examined before mass production assembly release. Flaws can be detected at an early stage, long before the parts have been manufactured, the production lines constructed, and assembly instruction constituted. With VR, immediate corrections can be tested and implemented in a much shorter time than with prototypes. The savings in development cost and time have a tremendous impact on manufacturers of mass-produced commercial products.

VR training can be used to train physicians in performing dangerous and uncommon medical treatments without endangering the patient. Critical operations such as brain and neurologic surgery can be rehearsed and taught by having student doctors perform surgery on virtual humans instead of cadavers or real patients. Future VR medical training systems are expected to have gloves to give doctors the tactile sensations associated with surgery.

Although it is a common perception that in VR the subject has to be totally immersed in the virtual world, in some applications of so-called augmented reality, the HMD is a see-through and its image is overlaid on the physical world. One such application was developed by Boeing to assist in the assembling of complex wire harnesses for aircraft [27]. Traditionally, the harnesses were assembled on a board on which a paper template was placed, and the worker had to follow the scheme without confusing the wire paths. In the VR application, the worker uses a blank board and the scheme is projected on the display. With the new system, the worker has to follow only the emerging path on the display without being confused by the rest of the scheme. Furthermore, if the worker uses a glove that measures hand movements, the system, in principle, can validate the correctness of the assembly process and warn the worker of any mistakes.

Another field of VR is teleoperation, or telepresence, in which the teleoperator interacts with the real world via some teleoperated system or vehicle: robot, car, remotely piloted vehicle (RPV), submarine, or camera. The manipulated object has some type of artificial vision device through which the operator can see the remote world and manipulate the object as if the operator were physically there. Obviously, the remote image can be enhanced by overlaid graphics and information symbology.

10.7 Summary

HMDs have the potential to be incorporated in many applications. The large number of envisioned applications that currently are in the stages of study, design, and trials indicate that miniature, head-worn, high-resolution, and lightweight displays are a real necessity and it is only a matter of time before they become common in many products.

The PD is being proposed as the logical solution for readable and usable display for small systems and computers that must present high-resolution text and graphics that easily can be perceived by the users. PDs open the way for a wide variety of portable and mobile products. Applications for PDs include miniature personal computers, mobile fax machines, and video games, electronic maintenance manuals, and low-cost trainers and simulators.

The other area of commercial products spans the systems that can be categorized as VR systems. Those systems have a great potential in training, aiding devices in production lines, design of complex systems, and visualization of complex scientific data.

Much research and development activity are taking place in military applications. These efforts are aimed at the development of HMDs for armored-vehicle crews and for infantry soldiers. Armored forces tend to adopt electronics and electro-optics systems used by aviators, so as HMDs become accepted by the air force communities, they infiltrate the armored forces. In the infantry, the need has been long recognized to equip soldiers with modern information sources such as electronic maps, modern navigation aids, and personal computers. The natural way to combine all those sources of information is to use HMDs.

An area that seems to greatly benefit by the use of HMDs is medical treatments and therapy. A huge number of potential applications are being explored and researched. Applications include training surgeons in performing high-risk operations; remote surgery, in which a distant specialist guides a less experienced surgeon who is performing a complex operation; removing a hidden cancerous growth with the aid of an x-ray image of the tumor overlaid on the physical organ; helping dentists see the inside of a tooth while treating it; and helping radiologists position patients in the proper position more quickly and more accurately before an examination.

Head-mounted displays have been tested successfully as enhancement devices for low-vision people and are showing success in treating people who suffer from phobias or motion sickness.

As more HMD-based applications and products are conceived and become available, there is pressure to accelerate technology advancement, lower the cost of HMDs, and increase acceptance and awareness of the benefits. It is expected that the number of new applications will grow dramatically in the forthcoming years.

References

[1] McCarty, W. D., et al., "A Virtual Cockpit for a Distributed Interactive Simulation," *IEEE Computer Graphics and Applications*, January 1994, pp. 49–54.

[2] Dannenberg, K., and P. Sanders, "New Directions in Modeling and Simulation," *Aerospace America*, January 1997, pp. 34–39.

[3] Casey, C. J., and J. E. Melzer, "Part-Task Training With a Helmet Integrated Display Simulator System," in *Large-Projection, Avionic, and Helmet-Mounted Displays*, SPIE, Vol. 1456, 1991, pp. 175–178.

[4] Lerner, E. J., "Toward the Omnipotent Pilot," *Aerospace America*, October 1986, pp. 18–22.

[5] Karr, C. R., D. Reece, and R. Francheschini, "Synthetic Soldiers," *IEEE Spectrum*, March 1997, pp. 39–45.

[6] Waters, R. C., and J. W. Barrus, "The Rise of Shared Virtual Environments," *IEEE Spectrum*, March 1997, pp. 21–25.

[7] Lapiska, C., L. Ross, and D. Smart, "Flight Simulation: An Overview," *Aerospace America*, August 1993, pp. 14–17.

[8] Schilling, L. J., and D. A. Mackall, "Flight Simulation Takes Off," *Aerospace America*, August 1993, pp. 18–21.

[9] Noor, A. K., et al., "A Virtual Environment of Intelligent Design," *Aerospace America*, April 1997, pp. 28–35.

[10] Mooij, H. A., "Technology Involved in the Simulation of Motion Cues: The Current Trend," in *Motion Cues in Flight Simulation and Simulator Induced Sickness*, AGARD-CP-433, 1987, pp. 2-1-2-14.

[11] Fernie, A., "Helmet-Mounted Display With Dual Resolution," *J. Society of Information Displays*, March-April 1995, pp. 151–153.

[12] "Simulator Trained Bush for Voluntary Jump," *Aviation Week and Space Technology*, April 28, 1997, pp. 62.

[13] Brooks, R. L., "Helmet Mounted Display for Tank Applications," in *Imaging Sensors and Displays*, SPIE, Vol. 765, 1987, pp. 19–21.

[14] Nelson, S. A., "CVC HMD—Next Generation High-Resolution Head-Mounted Display," in *Helmet- and Head-Mounted Displays and Symbology Design Requirements*, SPIE, Vol. 2218, 1994, pp. 7–16.

[15] Urban, E. C., "The Information Warrior," *IEEE Spectrum*, November, 1995, pp. 66–70.

[16] Kennedy, A. J., "Helmet-Mounted Display for Infantry Applications," in *Imaging and Sensor Displays*, SPIE, Vol. 765, 1987, pp. 26–28.

[17] O'malley, T. J., "Soldier 2000," *Armada International*, No. 2, April–May, 1995, pp. 30–42.

[18] Becker, A., "Design Case Study: Private Eye," *Information Display*, Vol. 6, No. 3, March 1990, pp. 8–10.

[19] Tidwell, M., et al., "The Virtual Retinal Display—A Retinal Scanning Imaging System," *Virtual Reality World '95 Conf. Doc.*, IDG Conf., 1995, pp. 325–333.

[20] Tritsch, C. L., "A Helmet Mounted Display Application for the Space Station Freedom Extravehicular Mobility Unit," in *Helmet-Mounted Displays*, SPIE, Vol. 1116, 1989, pp. 76–79.

[21] Adam, J. A., "Virtual Reality Is for Real," *IEEE Spectrum*, October 1993, pp. 22–29.

[22] Dagnelie, G., and R. W. Massof, "Toward an Artificial Eye," *IEEE Spectrum*, May 1996, pp. 21–29.

[23] Hodges, L. F., et al., "Virtually Conquering Fear of Flying," *IEEE Computer Graphics and Applications*, November 1996, pp. 342–349.

[24] Reason, J. T., and J. J. Brand, *Motion Sickness*, London: Academic Press, 1975.

[25] Kalawsky, R. S., "The Realities of Using Visually Coupled Systems for Training Applications," in *Helmet-Mounted Displays III*, SPIE, Vol. 1695, 1992, pp. 72–82, .

[26] Loftin, R. B., and P. J. Kenney, "Training the Hubble Space Telescope Flight Team," *IEEE Computer Graphics and Applications*, September 1995, pp. 31–37.

[27] Noor, A. K., and S. R. Ellis, "Engineering in a Virtual Environment," *Aerospace America*, July 1996, pp. 32–37.

About the Author

Mordekhai Velger is head of the control engineering section of ELOP Electro Optics Industries, Israel, where he has been actively involved in the design and development of an advanced HMD/HMS system since 1987. He also led the development of a helmet tracking system based on Inertial Measurement Unit for tank commanders. Dr. Velger received his B.Sc. in 1977, his M.Sc. in 1980, and his D.Sc. in 1985, all in aerospace engineering from the Technion-Israel Institute of Technology, Haifa, Israel. Between 1985 and 1987 he was a National Research Council (NRC) research associate at NASA Ames Research Center, Moffett Field, California, researching the reduction of biodynamic interference and HMD image stabilization. Dr. Velger has published nearly 30 technical papers, holds two patents, and holds the title of scientist at ELOP. He was adjunct professor at the Technion and is presently at the Coventry University Program in Israel. He is a member of the IEEE and a senior member of the AIAA.

Index

The Artech House Optoelectronics Library

Brian Culshaw and Alan Rogers, *Series Editors*

Chemical and Biochemical Sensing With Optical Fibers and Waveguides, Gilbert Boisdé and Alan Harmer

Coherent and Nonlinear Lightwave Communications, Milorad Cvijetic

Coherent Lightwave Communication Systems, Shiro Ryu

Elliptical Fiber Waveguides, R. B. Dyott

Frequency Stabilization of Semiconductor Laser Diodes, Tetsuhiko Ikegami, Shoichi Sudo, Yoshihisa Sakai

Fundamentals of Multiaccess Optical Fiber Networks, Denis J. G. Mestdagh

Handbook of Distributed Feedback Laser Diodes, Geert Morthier and Patrick Vankwikelberge

Helmet-Mounted Displays and Sights, Mordekhai Velger

Highly Coherent Semiconductor Lasers, Motoichi Ohtsu

Integrated Optics: Design and Modeling, Reinhard März

Introduction to Lightwave Communication Systems, Rajappa Papannareddy

Introduction to Radiometry and Photometry, William Ross McCluney

Introduction to Semiconductor Integrated Optics, Hans P. Zappe

Optical Document Security, Second Edition, Rudolf L. van Renesse, editor

Optical FDM Network Technologies, Kiyoshi Nosu

Optical Fiber Amplifiers: Design and System Applications, Anders Bjarklev

Optical Fiber Amplifiers: Materials, Devices, and Applications, Shoichi Sudo, editor

Optical Fiber Communication Systems, Leonid Kazovsky, Sergio Benedetto, Alan Willner

Optical Fiber Sensors, Volume Four: Applications, Analysis, and Future Trends, John Dakin and Brian Culshaw, editors

Optical Measurement Techniques and Applications, Pramod Rastogi

Optical Network Theory, Yitzhak Weissman

Optoelectronic Techniques for Microwave and Millimeter-Wave Engineering, William M. Robertson

Reliability and Degradation of LEDs and Semiconductor Lasers, Mitsuo Fukuda

Reliability and Degradation of III-V Optical Devices, Osamu Ueda

Semiconductor Raman Laser, Ken Suto and Jun-ichi Nishizawa

Smart Structures and Materials, Brian Culshaw

Wavelength Division Multiple Access Optical Networks, Andrea Borella, Giovanni Cancellieri, and Franco Chiaraluce